Fundamentals
of Electronic
Image Processing

Fundamentals of Electronic Image Processing

Arthur R. Weeks, Jr.
Department of Electrical and Computer Engineering
Center for Research and Education in Optics and Lasers (CREOL)
University of Central Florida

SPIE OPTICAL ENGINEERING PRESS

A Publication of SPIE—The International Society for Optical Engineering
Bellingham, Washington USA

IEEE PRESS

The Institute of Electrical and Electronics Engineers, Inc., New York

Library of Congress Cataloging-in-Publication Data

Weeks, Jr., Arthur R.
    Fundamentals of electronic image processing / Arthur R. Weeks, Jr.
       p.   cm. -- (SPIE/IEEE series on imaging science & engineering)
    Includes bibliographical references and index.
    ISBN 0-8194-2149-9 (hardcover)
    1. Image processing. I. Title. II. Series
  TA1637.W441996
621.36'7—dc20   96-3272
                                              CIP

Copublished by

SPIE—The International Society for Optical Engineering
P.O. Box 10
Bellingham, Washington 98227-0010
Phone: 360/676-3290
Fax: 360/647-1445
Email: spie@spie.org
WWW: http://www.spie.org/
ISBN 0-8194-2149-9

IEEE Press
445 Hoes Lane
Piscataway, NJ 08855-1331
Phone: 1-800/678-IEEE
Fax: 908/562-1746
Email: ieeepress@ieee.org
WWW: http://www.ieee.org/
IEEE Press No. PC5681
ISBN 0-7803-3410-8

Copyright © 1996 The Society of Photo-Optical Instrumentation Engineers

Printed in the United States of America.

Cover illustration: Images of one of Thomas Edison's cars taken at the Edison Museum in Fort Myers, Florida.

To my wife, Dian,
and my children, Anne and Frank

Contents

Preface xi
Acknowledgments xiii

1 **Introduction to Electronic Image Processing** *1*

    1.1   Historical Background *1*
    1.2   Applications of Image Processing *9*
    1.3   Introduction to Visual Perception *13*
    1.4   Image Formation *25*
    1.5   Sampling and Quantization *26*
    1.6   Image Neighbors and Distances *34*
    1.7   Typical Image Processing Systems *37*

2 **Transforms Used in Electronic Image Processing** *40*

    2.1   The Fourier Series *40*
    2.2   The One-Dimensional Fourier Transform *44*
    2.3   The Two-Dimensional Fourier Transform *48*
    2.4   Important Functions Relating to the Fourier Transform *51*
    2.5   The Discrete Fourier Transform *55*
    2.6   Example and Properties of the Discrete Fourier Transform *59*
    2.7   Computation of the Discrete Fourier Transform *69*
    2.8   Other Image Transforms *71*

3 Image Enhancement by Point Operations *90*

3.1 An Overview of Point Processing *90*
3.2 Constant and Nonlinear Operations *93*
3.3 Operations Between Images *102*
3.4 Histogram Techniques *109*

4 Spatial Filtering and Fourier Frequency Methods *121*

4.1 Various Types of Noise That Appear in Images *121*
4.2 Spatial Filtering *129*
4.3 Spatial Frequency Filtering *144*
4.4 Image Restoration *158*

5 Nonlinear Image Processing Techniques *173*

5.1 Nonlinear Spatial Filters Based on Order Statistics *173*
5.2 Nonlinear Mean Filters *197*
5.3 Adaptive Filters *208*
5.4 The Homomorphic Filter *221*

6 Color Image Processing *228*

6.1 Color Fundamentals *229*
6.2 Color Models *237*
6.3 Examples of Color Image Processing *276*
6.4 Pseudocoloring and Color Displays *288*

7 Image Geometry and Morphological Filters *294*

7.1 Spatial Interpolation *294*
7.2 Image Geometry *299*
7.3 Binary Morphology—Dilation and Erosion *316*
7.4 Binary Morphology—Opening, Closing, Edge Detection, and Skeletonization *333*
7.5 Binary Morphology—Hit-Miss, Thinning, Thickening, and Pruning *347*
7.6 Binary Morphology—Granulometries and the Pattern Spectrum *359*
7.7 Graylevel Morphology *367*

8 Image Segmentation and Representation *387*

8.1 Image Thresholding *388*
8.2 Edge, Line, and Point Detection *414*
8.3 Region Based Segmentation *440*
8.4 Image Representation *452*

9 Image Compression *471*

9.1 Compression Fundamentals *471*
9.2 Error-Free Compression Methods *483*
9.3 Lossy Compression Methods *522*

Bibliography 548
Index 557

Preface

During the last thirty years, electronic image processing has grown from a dedicated scientific research field limited to a small set of researchers to a technical area that has found use in many scientific and commercial applications. During this time, there has been enormous growth in this field due to the improvements in computer and imaging sensor technologies. In particular, electronic image processing of color images, which had been limited to a dedicated few because of its hardware requirements, is now readily available on typical desktop computers. The manipulation and inclusion of color images is becoming standard practice in business presentations.

Digital film is slowly replacing chemical film technology, making electronic cameras that acquire, capture, and store images digitally available to the general public. Many of these camera systems provide an easy means of transferring these images to a standard desktop computer for processing and storage.

The goal of this text is to provide the fundamentals of image processing to practicing engineers or scientists who needs to understand these fundamentals to perform their technical tasks. In many technical fields, images must be processed to enhance features that are present within the image. For example, the microbiologist is typically interested in enhancing images of cells, while the astronomer is typically interested in enhancing images of remote galaxies. It is hoped that this book will bridge the gap between the existing image processing texts dedicated to researchers in the field, and the practicing engineer or scientist who needs to understand and use the various types of image processing algorithms.

Throughout this text are included a large variety of example images to give the reader a better understanding of how a particular image processing algorithm works. Whenever possible, the text goes into detail, explaining the advantages, disadvantages, and when to use each algorithm.

Chapter 1 discusses the fundamentals of image processing, including the spatial sampling digitization of images. Also discussed in this chapter are the fundamentals of the human visual system, since typically the final result of image manipulation is for viewing by a human. Hence, many types of image processing algorithms take advantage of the peculiarities of the human visual system.

Chapter 2 presents several types of image transforms that are commonly used. This chapter follows the standard classical approach of presenting linear system theory expanded into two dimensions for electronic image processing. Based on Fourier frequency analysis, the effect of spatial filtering of an image is presented. Chapters 3 and 4 build upon Chapter 2, giving several types of image enhancement and restoration techniques in both the spatial and the frequency domains.

Chapter 5 covers a new area of image processing and includes several types of nonlinear and adaptive filters. Compared to linear filters, these do a better job of removing noise from an image while preserving the sharpness of the filtered image. In particular, adaptive filters offer the capability of changing their characteristics depending on the noise or image features present within the filter window.

Chapter 6, which covers color image processing, is the newest area of image processing. Included in this chapter are the many color models that are used to represent a color image and also several color image enhancement methods that are commonly used to enhance color images. Because this is an emerging field, the author has included material from two recent technical papers on the subject.

Chapter 7 covers the important area of geometrical operations and morphological filtering. Geometrical operations include rotating, scaling, and zooming of an image. Since many of these geometrical algorithms require pixel interpolation, a section on image interpolation techniques has been included. Morphological filters, on the other hand, are used to geometrically change the shape of objects within an image and to extract key geometrical features.

Chapter 8 covers image segmentation and representation, in particular, the segmentation of objects and features within an image from its background and the detection of edges within an image. Once an object has been segmented or its edges found, several image representation methods are presented that code these objects or edges using different types of data representations.

Finally, Chapter 9 discusses the area of image compression, giving the reader the background behind the different image compression techniques that are commonly used. A comparison is given between lossless and lossy compression.

Arthur R. Weeks
May 1996

Acknowledgments

The author would like to thank the many people who have been involved in the generation of the manuscript. In particular, I would like to thank Edward Dougherty and Eric Pepper for their support and enthusiasm in seeing this text completed, Donald O'Shea for the many hours he spent reviewing and editing this text, Dixie Cheek for her excellent copy editing, and my children, who posed for several of the images presented throughout the text.

The Series Editor and author thank the following individuals who reviewed the manuscript: Nikolas Galatsanos, Jay Jordan, Aggelos Katsaggelos, Murat Kunt, Maria Petrou, and Anastasios Venetsanopoulos.

Fundamentals of Electronic Image Processing

CHAPTER 1

Introduction to Electronic Image Processing

This chapter gives a historical background of image storage and retrieval and the development of electronic image processing. The applications and use of modern electronic image processing are also presented, followed by a short discussion of human visual perception and how it relates to electronic image processing. A model for image formation is presented, followed by the spatial sampling and quantization of images. Finally, two types of electronic image acquisition systems are presented.

1.1 Historical Background

Throughout the history of mankind, there has been the desire of the human race to record an instance of time for future generations via the use of pictures (images). Pictures have also been used in various early languages, providing an easy method of communicating information from one human to the next. The earliest documented use of images to communicate an idea is seen in the drawings created by the early cavemen. Using primitive tools, images were recorded into stone describing the details of everyday life. Important events such as the winning or losing of a battle were often recorded with these man-made images. Even though these images were used for communication purposes, today they give a historical record of early human civilization, providing the details of everyday life. The importance of pictures used as a language to represent ideas is also well illustrated by Egyptian hieroglyphics. The prolific preservation of this language engraved in the stone of various Egyptian remains has provided historians with a detailed story of the Egyptian civilization

Another early form of image generation and storage was the use of various color paints/inks to record scenes observed by the human eye. This required the talent and expertise of humans who could create an image using various colors of paints/inks from a scene that was visualized and perceived by them. For many centuries, this was the only means of recording images. It was not uncommon that an artist would accompany soldiers into battle to record the historic event. During the Middle Ages and the Renaissance, the creation of man-made visual images played an important part in conveying important religious concepts to the average person. Many artists such as Leonardo daVinci and Michelangelo were commissioned to produce paintings of religious scenes within various churches. Artists were also commissioned to produce portrait paintings of royalty. During this time, this was the only means by which the image of a person could be passed from generation to generation. These man-made images had their limitations in that all were created after the visualization and interpretation by another human. The resemblance of the final image to that of the actual scene depended on the artist producing the image. Very seldom would several artists commissioned to produce a painting of the same scene produce exactly the same painting. Each would visualize and perceive something different in the scene. In fact, the differences between the paintings were attributed to an artist's style and perception of the scene being painted.

Since the beginning of time, it has been desired by human civilization to record images of scenes as accurately as possible, removing any human interpretation. During the second half of the sixteenth century an Italian philosopher, J. B. Porta, made an important discovery by accident. He discovered that light rays penetrating a small hole in a door enclosing a dark room produced an upside down image of the exterior scene on a white screen placed behind the door in the dark room. His discovery was the forerunner to the modern day pinhole camera. Amazed at his discovery, he shortly modified his experiment, replacing the small hole with a convex lens, and with the additional use of a mirror, Porta was able to produce a right side up image of an exterior scene. Porta realized the importance of his discovery and immediately recommended his device (camera obscura) to artists wishing to record exact images of scenes onto oil paintings. It was an artist, Canaletto, who first used Porta's discovery to produce paintings of Venice.

The discovery of the camera by Porta was the first step toward modern day photography. At about the same time as Porta's camera, it was discovered by a chemist in France that silver chloride changes its characteristics from clear to black when exposed to light. It was not until two centuries later that Jacques Alexandre Charles, the inventor of the hydrogen balloon, produced simple photographs of silhouettes of his students' heads using writing paper immersed in silver chloride. Unfortunately, Jacques Alexandre Charles, like many before him, did not have a way to stop the development process. Once a sheet of silver chloride paper was exposed to a light source to produce an image, it could be viewed only in very low light levels such as candle light. Eventually, the long term exposure to light would turn the whole silver chloride paper to black. The last piece of the puzzle remained to be discovered, that of stopping (fixing) the development action.

It was not until 1835 that Henry Fox Talbot was able stop the development process. His first images were of leaves and flowers, formed by pressing them against silver nitrate immersed paper and then exposing the paper and the objects to sunlight. Once the paper was exposed to light, producing the image of a leaf or flower, Talbot's final process was to "fix" the paper to prevent any further degradation of the image. Later, in the summer of 1835, Talbot used the camera obscura to produce images of his house. On this day, the age of modern photography was born. It was not until 1839, when Talbot perfected his approach, that his work was presented to the Royal Society. It should be pointed out that even though Talbot was able to produce a stored image without it first being interpreted by a human, his images were negatives of the original objects. It was not until years later that the two step process of producing a negative and repeating the exposure and development processing one more time to produce a positive image was used. Even though the earliest photographs were on paper, in the middle part of the nineteenth century glass plates were used as a means of recording an image using chemical photography. The stability of these plates are evident in the number that remain over one hundred years later. Today they are collector items among historians and photographers.

The invention of modern photography brought with it the desire of the human civilization to record images for observation by future generations and to provide the means of freezing the "images" of history. For example, the importance of early photography is evident in the still photographs that remain and clearly document the U.S. Civil War. This war was one of the first major conflicts in history to be completely documented by modern photography. Over one hundred years later, historians are still referring to these photographs to describe key battles. For most of the nineteenth century, photography was limited to a specialized few who were trained to use the various complex cameras and were familiar with the use of the complex chemical process used in the development of photographs. Photography was used only for special events or by the wealthy as a mean of capturing images to record key events or people. It was the invention of the simple roll-camera ("Kodak" camera) in 1884 by George Eastman that made modern photography available to the average individual. What this camera accomplished was the replacement of large photo sensitive glass plates with a flexible roll of film that was easy to load into the camera. George Eastman also provided a network of laboratories that would develop and process these film-rolls making the use of modern photography as easy as possible for the general public. The technical knowledge of the chemical development process was no longer required to produce photographs using modern photography.

As the field of black and white photography was maturing in the last part of the nineteenth century, researchers were pursuing the generation of color images. It was the work of James Clerk Maxwell and James D. Forbes that showed that the infinite number of colors available in the visible spectrum could be reproduced using a three primary color system. Using a spinning top, they placed circles of different color papers on top of the top. When the top was spinning at a high rate, the different color circles blurred together, producing a new observable color. Maxwell and Forbes were able to produce the color yellow from a red and a green circle. In fact, they found that they could produce any color from a combination of three colors, red, blue, and green. Maxwell suggested a method of proving his three color theory, by photographing a color image using three (black and white) photographic plates taken of exactly the same scene but with a red, blue, or green color filter placed in front of each plate. The three plates, which represented the three primary color images of red, green and blue, were then used to produce three images that were overlaid on top of each other on a white screen. In the generation of the images, a red light source was used to illuminate the red filtered plate, a green light was used to illuminate the green, and a blue light was used to illuminate the blue. Essentially, overlaying the three images produced a composite image composed of a red, a blue, and a green image of the original color scene. On May 17, 1861, Maxwell gave a lecture to the Royal Institute, where he stunned the audience with the presentation of a color image of a plaid ribbon generated from three red, green and blue photographic plates. Maxwell's work not only laid the groundwork for modern color photography, but as a result of his research, every color television/computer system today uses the three primary colors of red, green, and blue to generate color images.

It was also during this period that research began on combining still photographic images to produce moving images. The concept of motion pictures is based on the visual phenomenon known as persistence of vision. The images received by the brain from the eye are stored for about 60 milliseconds after viewing by the eye. In this way, objects moving across the field of view of the eye in faster than 60 milliseconds are ignored by the brain. For example, consider a set of cards containing a set of images of a person walking. Each adjacent card contains the image of the person walking as he or she moves just slightly to the left. Placing these cards in front of an observer and fanning them at a rate faster than about 15 cards per second produces the illusion that the person is walking smoothly to the left. This is the fundamental concept of motion pictures.

The goal of Thomas A. Edison and his assistant William Kennedy Laurie Dickson was to produce a set of time elapsed photographs to give the illusion of a moving image. It was the invention of the roll-film by George Eastman (modified for a positive image) that made the invention of motion pictures possible. In 1889, Edison designed a camera (kinescope) that automatically advanced a roll of film past a shutter and a lens. At a periodic rate of 48 images per second, the film was held still and the shutter was opened, imaging a scene onto the roll of film. The film was then advanced and the process was repeated. When finished, the roll of film contained a time sequence of images, presenting the evolution of a scene as a function of time. After successful use of their camera, Edison and Dickson designed a projector that rapidly projected the time sequence of images saved on the roll of film onto a screen. What they observed was a smoothly moving image. Within 60 years of the invention of still photography, the recording of real scenes as a function of time was accomplished. With the advent of combining sound with moving pictures, a complete record of an important historical event and prominent people was now possible. An example is the millions of feet of newsreel film that was generated in the early part of the twentieth century. These films clearly document the important historical events of time for future generations.

Before moving onto the discussion of still images, which is the emphasis of this text, a brief history of television is in order. This invention, and the efforts to transmit images across the Atlantic Ocean began the history of transmitting images via electronic methods. These efforts laid the groundwork for the electronic generation, manipulation, transmission, and storage of still images. One of the earliest accounts of an electronic television system was given in a letter to *Nature* in April, 1880, by John Perry. He proposed an electronic camera that used an array of selenium detectors as a means of converting an image intensity into an array of electrical signals. The interesting thing about this camera was than it did not use any type of scanning mechanism that is typically found in many early electronic cameras. In fact, the camera proposed by Perry's paper would be referred to as a focal-plane camera and is very similar to the modern charge couple device, CCD, camera.

There were also two types of receivers described. The first used an array of magnetic needles, one for each selenium detector, to open and close a set of apertures. The size of the aperture was directly proportional to the light incident on its corresponding selenium detector. The second system used the Kerr effect to rotate the polarization of linearly polarized light passing through a small crystal. As an electric field is placed on a Kerr cell, the angle in which the polarization of light is rotated is varied. In this system, an array of small crystals equal to the number of selenium detectors was proposed. In front of each Kerr cell, a polarizer was placed so that the combination of the linearly polarized light intensity emerging from the Kerr cell and the polarizer produced an intensity variation proportional to the electric field on the Kerr cell. The goal of this receiver was to have the electric field induced on the Kerr cell be proportional to the light intensity incident on its respective selenium detector. It is interesting to note, this is the same concept that is used for flat screen liquid crystal displays, except the angle in which the polarization light is rotated is controlled by the liquid crystal elements. It took approximately one hundred years after the original concept paper by Perry before this type of receiver system was implemented. Unlike modern television systems, Perry's proposed system requires no point-by-point scanning of a scene. The major limitation of his system was that only a limited number of array elements could be used because of the complexity of the wiring and the number of parts required.

The importance of Perry's paper is that it proposed to spatially sample a continuously varying spatial image, using an array of detectors to produce an array of electric signals that contained all the information necessary to regenerate an image of the original scene. Many of the other early proposed systems were based on electro-mechanical systems that used the concept of scanning an image element by element. One of the first systems, designed by Paul Nipkow, used two rotating discs (one located at the transmitter and one located at the receiver) containing 24 holes oriented in helical fashion on each disc. The discs were then synchronized and rotated at a rate faster than the persistence rate of the eye. Located at the transmitter was a selenium detector that produced an electrical signal that varied as the scene was sequentially scanned. This electrical signal then modulated a light source at the receiver to produce a modulated intensity of light that was synchronized to the sampling of the original scene. The net effect was a perceived image of the original scene at the receiver.

Modern day television had to wait for the electronic vacuum tube invented by Lee De Forest in 1906. On December 29, 1923, Vladimir K. Zworykin of the Westinghouse Electric and Manufacturing Company applied for a patent for a complete electric television system containing no moving parts. Key to a complete electric television was the development of the cathode ray tube (picture tube) used to convert electrical signals into varying light intensities, an all electronic scanning method, and an electronic camera. It was not until 1929 that Zworykin produced a satisfactory picture tube (kinescope) that enabled the development and design of affordable television receivers. By 1933, RCA built and operated a complete television system using an improved electronic camera, which was developed by Zworykin who was now at RCA Incorporated. This system was completely electronic except for the use of a mechanical synchronization generator. By 1940, the mechanical synchronization system was replaced by an electronic system, providing the way for the design of a complete electronic television. During the week of June 25 to 28, 1940, RCA demonstrated the use of its television system by broadcasting the Republican National Convention in Philadelphia to the NBC studios and transmitting this signal from the Empire State Building to approximately four thousand television receivers located throughout New York City.

On January 27, 1941, the National Television Standards Committee (NTSC) sent its recommendations to the Federal Communications Commission (FCC) for the approval of a commercial television system. This committee recommended that each channel be 6 MHz in frequency, with the sound carrier placed 4.5 MHz above the video carrier. The recommendation specified that the video information as well as the synchronization signals be transmitted using amplitude modulation, AM, while the sound be transmitted using the newly developed (by Armstrong) frequency modulation, FM. It also recommended that a 4 (horizontal) by 3 (vertical) aspect ratio as well as a vertical resolution of 441 horizontal lines scanned at a rate of 30 frames per second be used in the transmitting of the picture. Finally, the committee recommended that interlaced scanning should be used, generating two fields (one for the even lines and one for the odd lines) at the rate of 60 times per second. On May 2, 1941, the FCC adopted the recommendations of the NTSC and set the date of July 1, 1941, as the official start date for commercial television in the United States. However, the start of World War II delayed the introduction of commercial television into the American household until the end of the 1940s. By the middle of the 1950s, with the invention of color television, a second set of NTSC standards was generated. These required that the transmission of color images be compatible with existing black and white television standards set forth by the original NTSC committee in 1941, which are the standards still in practice today, some 40 years later.

The desire to transmit an image to a distant location was not limited to motion pictures alone. Several of the initial concepts that led to the invention of the television dealt directly with the transmission of still photographs. On March 17, 1891, Noah Steiner Amstutz applied for a patent for the first device that would transmit photographs. The key to Amstutz's system was the special preparation of photographs to be transmitted. Prior to transmission, photographs underwent a chemical process to provide a surface height variation proportional to the silver density on the photograph. Amstutz's system used a needle attached to a variable resistor that produced varying voltage output as the needle was scanned across the surface of the specially prepared photograph. The receiver discussed in the patent used a special material that when developed produced dark and light areas proportional to the height distribution of the material. Amstutz's receiver was simply a tracing tool that varied the height of this special material proportional to the height of the modified photograph. A cutting tool scanned and cut this special material in synchronization with the scanning of the photograph. The final process was to chemically develop this cut material, producing a copy of the original photograph. In May, 1891, Amstutz successfully sent pictures over a 25-mile line using his system.

Prior to Amstutz's system, it took approximately a week for photographs to cross the Atlantic Ocean from Europe to the United States. In the early part of 1920, several leading newspaper companies in London and New York established a system to transmit photographs across the Atlantic Ocean in approximately two to three hours, which became known as the Bartlane system. This system scanned the input image element by element, producing a paper tape record of the gray tones within the original photograph. The first Bartlane system used a standard Baudot 5-bit telegraph with modified typeface keys. Similar to the concepts of dithered printing methods used today, each key had a different impact width, so when impacted against a piece of paper each produced a different sized ink dot. By 1921, this system was abandoned in favor of an off-line photographic method that was much more accurate at reproducing the gray tones of the original photograph after it had been converted to paper tape. The corresponding holes in the paper tape controlled the opening and closing of a set of shutters that in turn controlled the exposure on an unexposed photographic negative. As the density of holes increased on the paper tape, the light illuminating the unexposed negative increased. By scanning the unexposed photograph element by element and modulating the light illuminating the unexposed photograph, accurate reproduction of the gray tones contained within the original photograph was possible.

To convert a photograph into a digital number for storage on paper tape required the use of photoengraving techniques. Several metal plates were coated with a photosensitive insulating material and then exposed to the original photograph using different exposures. For each successive exposure, the light used to illuminate the photographic plate was increased by a factor of two. After chemically developing the exposed photographic plates, the insulating material that remained on the plate exposed using the lowest light level corresponded to the black gray-tones of the original photograph. The plate exposed using the next lowest light level had both black and dark gray tones recorded in the insulating area. Hence, each successive plate had more gray tones stored in its insulating area. Essentially, the binary pattern of insulating material and metal on each plate corresponded to the bit pattern representation of the gray tones of the original photograph. The number of plates represented the number of bits used to encode the different gray tones. This system, as incorporated into the Bartlane system, used five plates (5 bits) to encode the different gray tones. By 1927, the technology existed to electronically record gray tone variations within a photograph using a photoelectric cell. The photograph was scanned element by element and the light reflected off the photograph was recorded, converted to a digital number, and then saved on paper tape.

The requirement for good imagery for military reconnaissance during the second World War generated further advancements in photographic methods. As an example, the new technical field of high speed photography was developed and used extensively for aerial reconnaissance. During this period, the use of photographic manipulation (the predecessor to modern image processing) became important. The first of these methods simply corrected for poor camera exposures due to poor lighting and weather conditions. Individuals specially trained in the art of object recognition and identification were used to locate enemy targets from aerial photographs. A common practice was to enlarge and contrast enhance aerial photographs to highlight key military targets, making the job of the photographic interpreter easier.

It was not until the invention of the digital computer and the requirement of the NASA space program in the early 1960s that digital image processing came into existence. After several earlier unsuccessful unmanned Ranger spacecraft missions, Ranger 7 successfully sent thousands of television images of the lunar surface back to earth. These images were then converted into digital format and digitally processed to remove geometrical and camera response distortions. This initial image processing was done at NASA's Jet Propulsion Laboratory (JPL) in Pasadena, California. The initial digital image processing results of these images were so good that NASA continued its funding, resulting in the development of new image processing methods.

During the mid-1960s, NASA had the requirement of photographing the lunar surface to determine valid landing sites for the manned Apollo program. The Surveyor series of unmanned spacecraft were designed to survey the lunar surface to locate and verify valid landing sites. Surveyor 7 sent several thousand television images of the lunar surface from its landing site back to earth. Many of these images were converted to digital format and image processed in an attempt to determine the composition and structure of the lunar surface.

The Mariner series of unmanned spacecraft returned digital images of Mars, Venus, and Mercury during the late 1960s and early 1970s. Mariners 4, 6, and 7 returned digital images of Mars as they flew by that planet. Mariner 9 was placed into orbit around the planet Mars and sent back to earth digital images of the Martian surface that were used to map its surface. The Mariner 10 mission was designed so that the spacecraft would fly by Venus and Mercury and around the sun. Thousands of images were obtained and enhanced using digital image processing methods during the life of the Mariner program. The Viking unmanned spacecraft series also used digital image processing techniques to enhance images that were sent back to earth from the spacecrafts during the middle and late 1970s. Two orbiting and two landing Viking spacecrafts sent back high resolution images of the Martian surface, while two other Voyager spacecrafts flew by Jupiter and Saturn, sending detailed images of these planets never seen before.

The use of digital image processing techniques was initially limited to NASA's unmanned space program to evaluate the surface terrain of other planets and the moon, but with the large success of electronic image processing of images obtained from these missions, the tools of electronic image processing were extended to other NASA programs. NASA also launched a series of satellites that orbit the earth, which include LANDSAT, SEASAT, TIROS, GEOS, and NIMBUS, providing multispectral images of the earth's surface. These satellites include spacecrafts. These satellites provide detailed images of the earth's surface and weather information on a daily basis. For example, the formerly difficult task of predicting the formation and path of a hurricane has become routine with the use of weather satellite imagery.

1.2 Applications of Image Processing

The use of image processing techniques to enhance gray tone images has found applications beyond the initial use by NASA. Today, image processing is used in the fields of astronomy, medicine, crime and fingerprint analysis, remote sensing, manufacturing, aerospace and defense, movies and entertainment, and multimedia. Image processing has benefited the field of astronomy by partially removing the effect of the blurring of astronomical images by the earth's atmosphere using restoration and phase estimation techniques. An excellent example of the efficacy of digital image processing is in the restoring of images obtained from the Hubble telescope. After initial deployment, it was found that one of its optical telescope mirrors was incorrectly designed and contained aberrations. Images received from the telescope were badly blurred. Until NASA could design a set of corrective optics, which took several years, the use of the Hubble was limited. As a stopgap measure, image restoration methods were applied to the blurred images, removing a majority of the blurring that was present and making the incoming images usable until the corrective lenses were inserted in the optical train of the Hubble telescope.

The medical field has benefited greatly from the use of digital image processing. Image processing techniques have been applied to ultrasonic imagery, improving the evaluation and monitoring of the fetus during prenatal care. Image processing has also made the early diagnosis of breast cancer much easier by enhancing X-ray images of the chest. With the suspect tissue highlighted, a medical technician can concentrate on these areas, improving the accuracy of the diagnosis. Image processing methods have become standard practice in the generation of magnetic resonance images. Digital image processing methods have also improved standard diagnostic procedures that used to require several days before a medical doctor could make a decision. For example, the upper-GI imaging series that is commonly used to evaluate stomach difficulties used to take several days before the doctor received the developed X-rays for evaluation. Now, using digital imaging methods, real time video is available for evaluation during the procedure, providing immediate feedback to the doctor. X-rays are also collected during the procedure as a double check because of their high spatial resolution. Finally, image processing has been used extensively in the various microsurgery techniques. For example, in vitro fertilization techniques use a fiber optic imaging system to provide real time video so that a medical doctor can easily locate the ovaries for egg extraction.

Both crime and fingerprint analysis have benefited from the various image processing techniques that exist. For example, given an old photograph of a young child who has been reported missing, image processing methods can be used to predict the changes in the facial features as the child ages. This allows law enforcement agents to estimate the appearance of a child who has been missing for many years. Image processing methods have also been used to enhance slow rate surveillance systems used in banks and stores. These low resolution systems are key in many instances in providing the evidence necessary to convict the appropriate criminal. During a bank robbery, these surveillance systems provide images of the robbers. Image enhancement is used to highlight their faces, providing the police with an actual image that can be used as evidence. Automatic fingerprint analysis is another area that has received a lot of attention in the field of image processing. Due to the large number of fingerprints that are kept on file, techniques have been developed that help in their automatic classification, reducing the time required to identify an unknown fingerprint.

Both manufacturing and the aerospace and defense industries have benefited from modern image processing techniques. Such methods are routinely used in the automatic recognition and identification of faulty parts during the manufacturing process. The use of "smart" bombs throughout the 1991 Persian Gulf War has shown the decisive advantage of incorporating state-of-the-art image processing methods into weapons design. The pinpoint accuracy that was achieved by these weapons easily surpassed the accuracy of conventional weapons, as seen by daily new releases of the various bombing missions.

In recent years, the motion picture industry has applied image processing methods in the restoration of old films and in the use of creating special effects in new movies. During the initial part of the twentieth century, millions of feet of newsreel film and silent motion pictures were made on a film substrate that is now physically degrading. In the last several years, many of these original works have been completely restored using image processing methods. For example, before the last release of the movie *Snow White* by Walt Disney Productions, this film underwent a complete restoration. When the master photographic plates for the movie were created, they contained dust spots that became observable in the movie. In addition, since the creation of these final print plates, the original colors had yellowed. This was more evident in the white areas of the movie. During the restoration process, the spots were removed and the color was readjusted, correcting for the yellowing process. This was a tedious, difficult process, because over 200,000 individual plates that comprised the movie were restored. Moviemakers are just now learning what they can do with image processing. Recent movies have incorporated special effects that would not be possible without the use of image processing. For example, "morphing" of one object into another is easily accomplished using digital techniques.

With the advancement in the computing power of desktop computers, the interest in incorporating still images as well as video clips in presentations has increased drastically in the last couple of years. Still image digitizer hardware that acquires a single frame of color video from a camera and stores the image within a computer is available today for as little as $200. A few years ago this same hardware cost several thousand dollars. The prices of color desktop scanners have dropped to levels low enough that these devices are becoming standard within an office environment. The ability to display true-color images on a standard desktop computer is now very inexpensive ($150). Just a few years ago, image processing researchers had to buy very expensive dedicated computer hardware and image display hardware. The digitization of real-time video has also become very popular. Hardware that was limited to video production studios a few years ago is now affordable and readily available. It is now possible for an average individual to create and edit a complete video clip using digital techniques within his/her computer.

Recently there has been a large number of manufacturers producing digital still frame cameras. These cameras use a CCD detector as the electronic detector and provide the analog to digital conversion of the acquired image within the camera electronics. Several of these cameras contain some form of magnetic media, allowing for the onboard storage of several images. Software and interface hardware are included that allow downloading of the images to a standard desktop computer and the storage of the images in many of the popular image formats. The new PhotoCD$^©$ system created by Kodak Inc. has generated a lot of interest in digital image capture and storage. This system provides an easy and inexpensive method of digitizing (at high resolution) photographs from 35 mm film. A photographer uses a standard 35 mm camera to shoot photographs and then sends the exposed negatives to a Kodak laboratory for development. Upon completion of the development process, the photographer receives both the negatives and a CD-ROM containing digitized images of the photographs. The ease of use of this system has made it very popular.

After an image has been acquired and converted into a digital format, it can then be modified using electronic image processing techniques. In many instances, the goal is to use this image in the automatic recognition of objects present within the image. One example could be the automatic recognition and removal of a faulty part during a manufacturing process. Figure 1.1 is a block diagram describing the typical steps used in the processing of an image using digital image processing methods. The final output described in the block diagram is to perform a task based on the recognition of objects present within the acquired image. The first step is to acquire an image using an electronic camera. Next, the analog signal(s) from the camera is converted to a digital format using digitizing electronics. This piece of hardware can be a standard NTSC video acquisition system or it can be a dedicated piece of hardware that interfaces to a specific vendor's camera system. Once the image is digitized and stored within the computer, image preprocessing techniques can be used to improve the quality of the image.

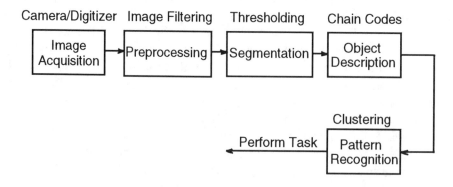

Figure 1.1: Acquisition and processing of an image.

Figure 1.2(b) shows the result of using histogram equalization on the low contrast image shown in Figure 1.2(a). Features not observable in the original image are easily seen in the output image. In particular, the details of the fruit in the basket are now much more evident. Another example of image preprocessing is the use of image restoration techniques to remove blurring from an image. Figure 1.3(a) is a degraded image that has been blurred in the horizontal direction. In this example, the source and type of the blurring degradation is given and is used in the restoration process. Figure 1.3(b) is the restored, unblurred image after application of Wiener filtering. Notice the increased sharpness of this image over the original image given in Figure 1.3(a).

After preprocessing of an image to enhance or restore key features that were degraded during the acquisition process, the next step in building an autonomous recognition system is to segment objects/features from the image background using segmentation techniques. A very common method is to base the segmentation decision on a gray tone threshold value. For example, gray tone values within the image greater than the threshold value are considered object elements. The next step in the recognition system given in Figure 1.1 uses object description techniques to reduce the amount of data that must be recognized. In many situations, only the contours of the object are needed for classification. Keeping only the contours reduces the amount of data that must be classified by the pattern recognition system. This reduces the complexity of the recognition process. For example, a set of fixed contours representing a set of known objects can be stored by the pattern recognition system and compared against an unknown contour in the input image. The contour that most closely matches the unknown contour is then classified as the object represented by the unknown contour. The output of the pattern recognition stage is a decision that is performed by the overall autonomous system.

1.3 Introduction to Visual Perception

For many image processing applications, the purpose of modifying an image is to improve its overall visual quality for viewing by humans. It is the purpose of this section to give the reader the necessary background to understand the concepts of the human visual system and how these concepts are related to electronic image processing.

As input for light stimuli, the eye converts the light energy incident on it into electrical signals that are transmitted and interpreted by the brain. Figure 1.4 gives a simplified representation of the eye. At birth, the average size of a human eye is approximately 16 mm in diameter and grows to approximately 24 mm in diameter in a full adult. At the outermost membrane is the *sclera*, which is contiguously connected to the *cornea* at the front of the eye. The cornea contains no blood vessels and is optically clear, allowing light to enter through

(a)

(b)

Figure 1.2: An example of image enhancement: (a) the original low
contrast image and (b) the enhanced image.

(a)

(b)

Figure 1.3: An example of image restoration: (a) the original blurred image and (b) the restored image.

the front of the eye. At the rear of the eye is the *retina*, which converts a light stimulus into electrical signals that are transmitted to the brain via the *optic nerve*, which contains about 1 million nerve fibers. Adjacent to the retina and connected to the sclera is the *choroid*, which consists of blood vessels that provide nutrition to the eye. At the front of the eye is the *lens*, which is composed of approximately 70% water and absorbs about 10% of the light incident on it. The lens images objects in front of the eye at distances from about 25 cm to infinity onto the retina. Focusing by the lens is accomplished via *cilliary muscles/fibers* that are used to expand or contract the lens, changing its radius of curvature. This effectively changes the focal length of the eye, enabling it to image objects at both far and near distances onto the retina.

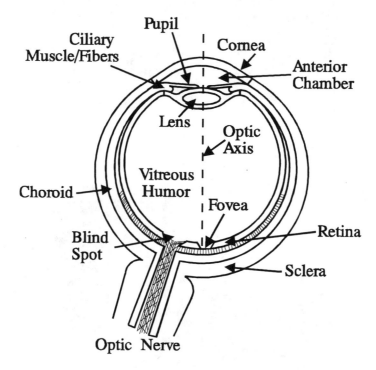

Figure 1.4: A diagram of the human eye.

In front of the lens is the *iris*, which controls the amount of light stimuli entering the eye. The iris varies in size from about 2 mm to 8 mm as the light stimuli entering the eye changes from low light to high light levels. Located just behind the cornea is the *anterior chamber*, which contains *aqueous humor* used to provide nutrients to the cornea. Also in the center of the eye is the *vitreous humor*, a clear fluid providing nutrition to the eye.

Key to visual perception by the eye is the *retina*, occupying an area of approximately 1000 mm^2. Composed of approximately 100 million sensors, each sensor converts only a portion of the light stimuli of the image on the retina to an electrical signal that is processed by the brain. Considering that state-of-the-art cameras typically have about 16 million sensors, the resolution of the eye far exceeds the resolution of today's cameras. The retina is composed of two types of optical sensors, rods and cones. The rod sensors are used to provide a coarse view of an image over a large field-of-view, while the cone sensors provide a very high resolution image within a very narrow field-of-view. Figure 1.5 shows the density of each sensor as a function of spatial position on the retina of the eye. Located at the optical center of the eye (central vision), on the retina, is the *fovea*. Distributed ± 80° field-of-view about the fovea there are approximately 90 million rods, which are very sensitive to very low light intensities (*scotopic vision*) and insensitive to color (the light's wavelength). This is why dark images appear as if they are in black and white. It is also interesting to note that a 1° field-of-view corresponds to approximately 290 μm on the retina surface.

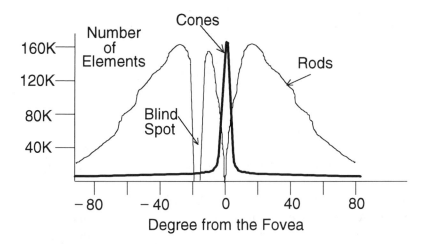

Figure 1.5: Typical rod and cone density as a function of distance from the fovea (adapted from Moses, 1970).

Located within the central 20° about the fovea are approximately 10 million cones that are very sensitive to color but need a moderate amount of light intensity (*photopic vision*). There are three types of cones (red, blue, and green), each sensitive to the wavelength of the light imaged on the retina. At the fovea, each cone is about 0.5 μm in diameter with a cone to cone spacing of 3 μm. Due to the high density of the cones within the fovea, the optical resolution of the eye within this region is close to diffraction limited imaging, and within this region the greatest acuity is achieved. Figure 1.5 shows the lack of cones and rods approximately 20° from the fovea at the location of the optic nerve. The lack of optical sensors at this point produces a blind spot in the field-of-view of the eye. To reduce the effect of this blind spot, the eye performs micro scans (*saccadic motion*) on an image at a rate faster than the persistence rate of the eye, producing an image that is perceived by the brain as containing no blind spot.

Objects placed within the field-of-view of the eye are imaged onto the retina via the cornea and the lens. Figure 1.6 gives an optical layout of the imaging system used by the eye, showing the refractive optical properties of the cornea and the lens. Given the front radius of the lens $-r_2$ and the back radius r_1, the focal length f of the lens in term of its index of refraction of light n is

$$\frac{1}{f} = (n-1)\left(\frac{1}{r_1} - \frac{1}{r_2}\right) . \tag{1.1}$$

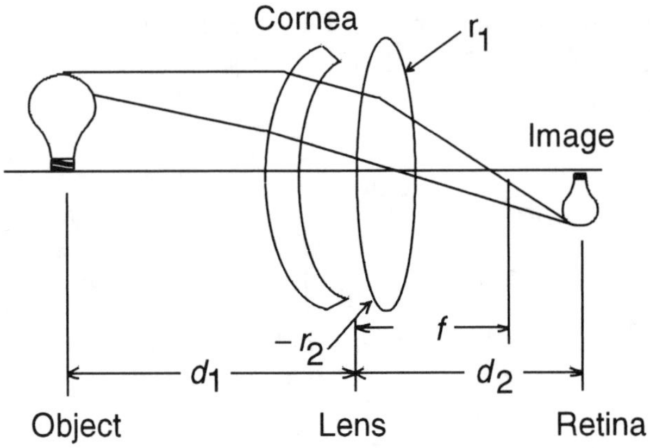

Figure 1.6: Imaging by the eye (images, ©New Vision Technologies).

Controlled by *ciliary muscles*, the radius of curvature of the lens increases automatically when viewing from far distances to close distances. Imaging by the cornea and the lens is predicted using the standard lens equation

$$\frac{1}{f} = \frac{1}{d_1} + \frac{1}{d_2} . \tag{1.2}$$

The refractive power of the cornea and the eye's lens can be combined together as one effective lens with an effective focal length that varies between about 14 mm when the eye is viewing close objects to about 20 mm when the eye is viewing distant objects. Consider an effective focal length for the cornea and the lens of 14 mm and a cornea/lens to retina distance d_2 of 20 mm. Using Equation (1.2) gives that the object is located a distance of 46 mm or approximately 18 inches in front of the eye (near vision). For far vision, the lens is expanded producing a larger radius of curvature (a flatter lens), increasing the effective focal length to approximately 20 mm.

The dynamic range of the eye is approximately 10 orders of magnitude, ranging from scotopic light levels to high light levels that produce glare.

Unfortunately, the eye cannot view this entire range at once but adapts to a smaller range of light levels. The adaptation point in which the eye operates is determined by three processes: the aperture size of the iris, the change in neural activity, and the bleaching and regeneration of receptor pigments. The inhibition of neural activity is accomplished when a portion of the retina is stimulated. At this point, the region of the retina that has been stimulated and adjacent regions are inhibited, resulting in a change in perceived brightness. The result of a steady state light stimulus on the retina results in a chemical change by the process of beaching and regeneration of the cones and rods. This photochemical process takes about a minute and results in a lower sensitivity by the receptors on the retina.

Figure 1.7 gives the perceived brightness as a function of the light intensity incident on the eye. The graph shows the variation in perceived brightness over the range of scotopic to glare light levels. Also shown in Figure 1.7, for a given intensity *I*, is the result of adaptation. After adaptation by the visual system, the level of perceived brightness from black to white no longer follows the total curve of Figure 1.7 but is limited to a small range of intensities, as shown by the short straight line.

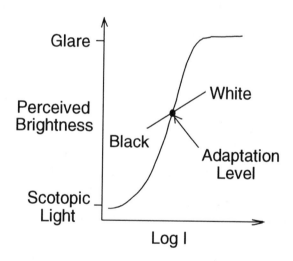

Figure 1.7: Brightness adaptation as a function of average intensity.

Another important characteristic of the eye is the minimum observable contrast difference between an object and its adjacent background. Figure 1.8 shows a simple test image, a square with a rectangle located within the center, used in determining the minimum perceived contrast as a function of illumination intensity. The test begins with the intensity level of the rectangle set to the same level as that of the background, which an observer gazes at for a short period of time. Next, the intensity level of the rectangle is increased just until the rectangle is observable 50% of the time. At this point, the value of ΔI is defined as the minimum observable contrast. Figure 1.9 shows a typical graph of the minimum observable contrast as a function of background intensity. It is interesting to note that as the background intensity increases, so does the minimum observable contrast. Another way of presenting the data given in Figure 1.9 is to normalize the graph by $\Delta I/I$. This ratio is known as the *Weber ratio*. Unlike the curve in Figure 1.9, the Weber ratio increases for both scotopic and glare light levels and is minimum at moderate levels .

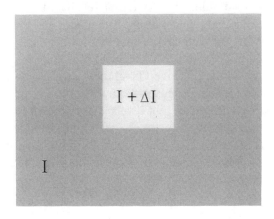

Figure 1.8: A test object used in determining the Weber ratio.

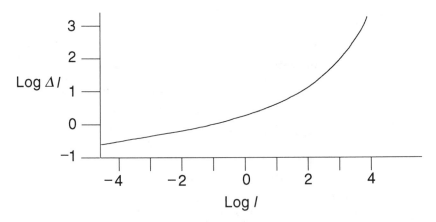

Figure 1.9: Typical plot of ΔI as a function of background intensity I
(adapted from Moses, 1970).

The minimum perceivable contrast depends not only on the background intensity level but also on the size of the object under test due to the limited spatial resolution of the eye. Figure 1.10 shows a typical plot of the minimum observable contrast as a function of object size. The minimum observable contrast increases for both large and small objects. For small objects, the minimum perceivable contrast is limited by the spatial resolution of the eye due to the diameter of the lens, while for large objects, the minimum perceivable contrast is limited by the spatial bandpass filter effect introduced by the visual system. Figure 1.11 gives a model of the eye that includes the bandpass filtering effect of the visual system. At the input to Figure 1.11(a) is the two-dimensional image of spatial light variations on the retina of the eye. This image undergoes a linear bandpass filtering operation that corresponds to the curve of spatial frequencies as shown in one dimension in Figure 1.11(b). Both low spatial frequencies and high spatial frequencies are attenuated by this linear filtering operation. Since small objects contain high spatial frequencies and large objects are limited to low spatial frequencies, attenuation of both the low and high spatial frequencies by this bandpass filter reduces the response of the eye to both of these sized objects, hence, the increase in the minimum observable contrast for both small and large objects.

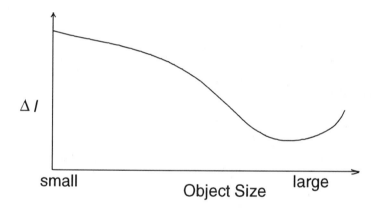

Figure 1.10: A typical plot of ΔI as a function of object size.

The bandpass filter operation also changes the way the human visual system perceives an edge. Figure 1.11(c) gives an intensity profile as a function of spatial position, showing two sharp edges. The perceived image by the human visual system is given in Figure 1.11(d). The limiting of the high spatial frequencies by the bandpass filter has limited the perceived sharpness (slope) of the edges. The limiting of the lower spatial frequencies by the bandpass filter has added both undershoot and overshoot to the edges. The perceived distortion of these edges is called the *Mach band* effect after Ernst Mach, who first described it. The photograph in Figure 1.12 is a linear step of increasing intensity levels moving from left to right. Looking closely at the transitions between the different gray tones, one will see a perceived light band to the right of an edge and a perceived dark band to the left of an edge. These two perceived bands are the undershoot and overshoot of the edge as illustrated in the Figure 1.11(d).

The distortion of an edge is an important feature of the human visual system. The human visual system is poor at determining the actual intensities at an edge.

But, what is most important to the perception of a sharp and focused image is the preservation of the slope of an edge. This visual peculiarity of the human visual system allows for the use of several nonlinear image processing filters that drastically distort the intensities near an edge to preserve its sharpness. Several of these filters such as the harmonic, contra-harmonic, and geometrical mean filters will be discussed in Chapter 5 of this text.

Another phenomenon of the human visual system, called simultaneous contrast, is the dependence of the perceived intensity of an object on its surrounding background. Figure 1.13(a) is an image of a square at intensity I against a dark background. Figure 1.13(b) shows the same square at the same intensity against a light background. Upon comparison of both images, the square in Figure 1.13(b) appears to be darker. Yet both squares were created using the same intensity value I.

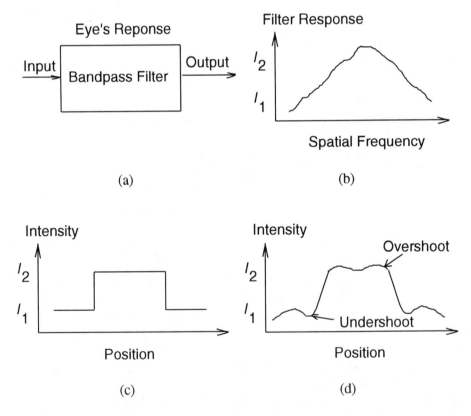

Figure 1.11: The spatial response of the eye showing the Mach band effect: (a) The linear filter model for the visual system, (b) its spatial frequency response, (c) the intensity profile with two edges, and (d) its perceived intensity showing both undershoot and overshoot.

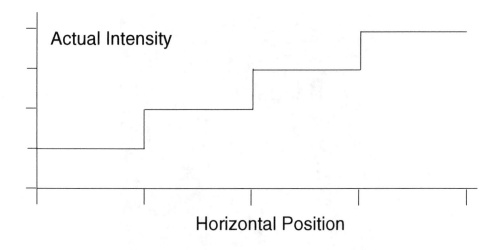

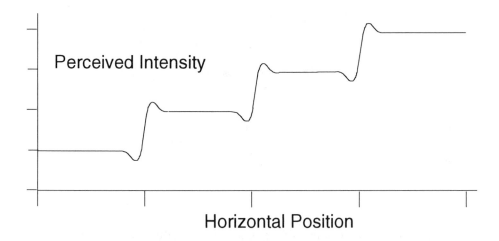

Figure 1.12: An example of the Mach band effect.

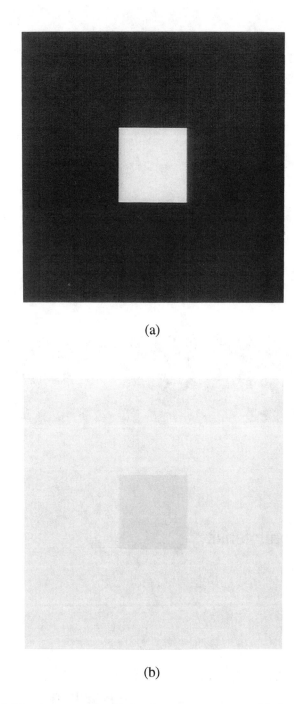

(a)

(b)

Figure 1.13: An example of simultaneous contrast: (a) An example
image of a rectangle with intensity I against a dark
background and (b) the same rectangle with the same
intensity I against a light background.

1.4 Image Formation

An image is a distribution of light energy as a function of spatial position. Figure 1.14 depicts the formation of an image on a video camera. A light source (the sun in Figure 1.14) emits light energy that is incident on an object that will form an image. Light energy can do one of three things when incident on an object: be absorbed by the object, be transmitted through the object, or be reflected from the object. By conservation of energy, the total light energy incident on an object must be conserved. Let R be the percentage of light that is reflected, A be the percentage of light that is absorbed, and T be the percentage of light that is transmitted, then

$$R + T + A = 1 .$$ (1.3)

If an object transmits most of the light energy, it is referred to as a clear object because other objects behind this object can still be seen. Glass is an excellent example of a clear object in that most of the light energy incident on it is transmitted. An opaque object transmits no light energy and simply either reflects or absorbs the incident light. If an object absorbs all light energy, then by the conservation of energy, this object re-radiates this light energy in the form of heat.

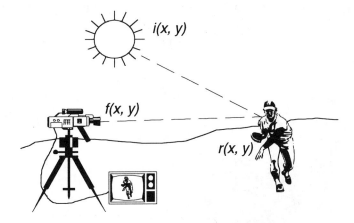

Figure 1.14: An illustration depicting the formation of an image (images, ©New Vision Technologies).

Of the three possible methods in which light energy interacts with an object, light reflected from its surface is the most important in the formation of an image. Figure 1.14 shows that light radiating from the sun is reflected off the surface of the baseball player and then received by the input lens of the camera that forms an image on the camera's electronic detector. The image formed at the camera's lens can be expressed mathematically as

$$f(x, y) = i(x, y) \cdot r(x, y) \; . \qquad (1.4)$$

Equation (1.4) gives the image model at the camera as a function of the light illumination $i(x, y)$ and the surface reflection of an object $r(x, y)$, where $r(x, y)$ varies between 0 and 1. The function $f(x, y)$ describes the light energy of the image at the spatial coordinate x, y. Since an image is the spatial distribution of light energy $f(x, y)$, it can take on only positive, real values. Finally, the image formed at the camera in Figure 1.14 is converted to an electrical signal, which produces the image seen on the television.

1.5 Sampling and Quantization

Before an image can be manipulated using the various image processing techniques, it must be spatially sampled. The process of sampling an image is the process of applying a two-dimensional grid to a spatially continuous image to divide it into a two-dimensional array of elements. Figure 1.15 shows a sampled image containing a total of NM sampled elements using a rectangular grid. Another commonly used sampling grid is the hexagonal grid. Any type of sampling grid can be used, but the rectangular grid is by far the most common simply because of its relationship to two-dimensional arrays. The fundamental unit of a sampled image is a picture element and is typically referred to as a *pel* or a *pixel*. The value of each pixel is equal to the average intensity of the continuous spatial image covered by that pixel.

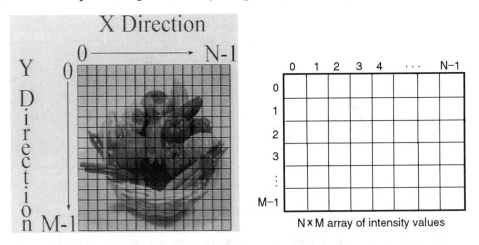

Figure 1.15: A spatially sampled image containing $N \times M$ picture elements.

The result of sampling produces a two-dimensional array of numbers that are directly proportional to the intensity levels of the continuous spatial image. Figure 1.16 shows a two-dimensional array of $N \times M$ elements used to represent a sampled image. $G(x, y)$ defines that intensity value of the sampled image at the pixel coordinate x, y. Figure 1.17 shows a simple pseudo program that accesses each pixel within a sampled image $G(x, y)$ and then multiplies the intensity of each pixel by a constant K. The results of this operation produces a new image $R(x, y)$. The program accesses each pixel in the sampled image $G(x, y)$ using two FOR loops: one in the x direction and one in the y direction.

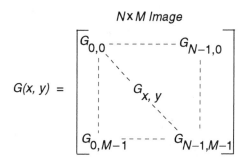

Figure 1.16: A two-dimensional array representing a sampled image.

Today there are a number of image sizes that are standard. Many of these were chosen to be compatible with the spatial size of NTSC video and to meet the storage size requirements of digital memory. In addition, image sizes that are powers of two exist because of the requirements for computing the fast Fourier transform (FFT), to be presented in Chapter 2. Typical spatial sizes are 256×256, 320×240, 512×512, 640×480, 1024×1024, 1024×768, 2048×2048, and 4096×4096. Until recently, because of the high cost of memory, the smaller sizes such as 256×256, 320×240, and 512×512 were standard. For compatibility with black and white NTSC video, an image resolution of 640×480 is used. A black and white NTSC video image of this size is also given the designation of RS170. Figure 1.18 gives an example of the same image presented at several different spatial sizes. Figure 1.18(a) lists the size for each image, varying between 64×64 and 512×512. Each successively larger image was created by doubling each pixel in both directions. For the 512×512 sized image, the individual pixels from the original 64×64 sized image can easily be seen.

```
Integer x, Integer y
For y = 0 to N-1
   For x = 0 to M-1
      R(x, y) = K * G(x, y)
   next x
next y
```

Figure 1.17: An example of a pseudo program that multiplies each pixel in an image by a constant K.

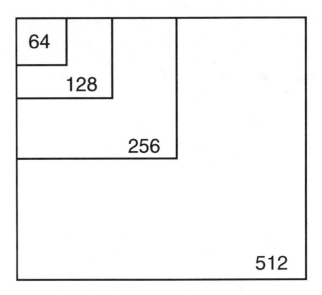

(a)

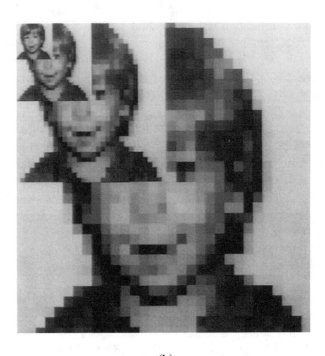

(b)

Figure 1.18: An example of a sampled image for various resolutions: (a) the resolution size of each image and (b) the actual sampled images.

Besides spatial sampling, the intensity level at each pixel must also be digitized into a finite set of numbers. The process of digitization converts an analog intensity value into a set of digital numbers that represent the intensity levels in the image. The quantity of numbers used to represent the intensities in a continuous tone image determines the final quality of the digitization process. We refer to this set of numbers as the *graylevels* or *grayscales* of an image. Since an image is the spatial distribution of light energy, the numbers assigned to graylevels of a digitized image can take on only positive values. Typically though, integer values are used to represent the graylevels present within a digitized image. Figure 1.19(a) gives a 4 × 4 sub-image taken from the lower right side of Figure 1.18(b), which highlights the boundary pixels between the background and the boy's shirt. Figure 1.19(b) gives the corresponding grayscale, with the value of 0 assigned to black and each grayscale value increasing in intensity until the value of 255 is reached, corresponding to white. Because of the low spatial sampling of this image the individual pixels are readily observable.

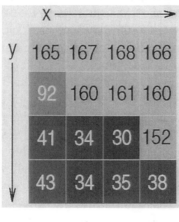

(a)

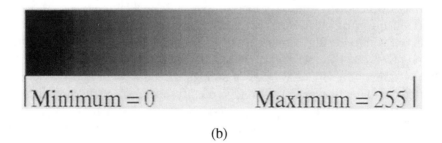

(b)

Figure 1.19: An example of (a) a sampled and digitized 4 × 4 sub-image and (b) its corresponding grayscale.

Consider a continuous tone image that has been digitized into four graylevels: black, dark gray, light gray, and white. Unfortunately, for most images, digitizing a continuous image into four graylevels does not produce an image of high quality. In fact, discrete contours are seen throughout the image at pixel boundaries between the different graylevels. These artificially introduced contours are commonly referred to as *false contours*. Figure 1.20(a) shows an image of a boy's face that has been digitized into 2 graylevels. The features of the face are observable, but this image is of poor image quality. Images that contain only two graylevels are typically called *binary images*. Binary images become important in image segmentation and binary morphological filters. Figure 1.20(b) is the same image of the boy's face but this image contains 4 graylevels. The image in Figure 1.20(b) is perceived as more acceptable than the binary image but is still of poor image quality. Present in this image are false contours, due to the pixel boundaries between the different graylevels. Figures 1.20(c) and (d) are 8 and 16 graylevel images of the boy's face, respectively. Notice the improvement in the image quality of the boy's face as the number of graylevels used to represent the image is increased.

The question comes down to how many graylevels are needed to properly digitize an image. The human eye can typically resolve between 40 to 60 different graylevels. Below this value the eye can detect false contours in an image. Figures 1.20(e) and (f) show the remarkable improvement in the image quality of the boy's face as the number of graylevels within the image is increased from 32 to 64 graylevels, respectively. In Figure 1.20(f), the false contours that were present in the smaller graylevel images has for the most part disappeared. Since most sampled and digitized images are stored as digital numbers, and since digital numbers are stored using bytes of storage, one byte of digital storage is typically used to store one pixel within an image. Since a byte can represent 256 distinct values, one byte per pixel allows up to 256 graylevels to be stored per pixel. This number of graylevels is beyond the range in which the human eye can detect discontinuities in graylevels. A result of using 256 distinct graylevels for each pixel is a digitized image that is perceived as a continuous tone image, which is of high quality.

Many of the video digitizers that are presently available convert the *NTSC-RS170* video format into a two-dimensional array of digital numbers, which are then stored within the digital memory of a computer system using 256 graylevels for each pixel (one byte per pixel). For an RS170 video image, one frame of 640 × 480 video requires 307,200 *bytes* of storage. Even though 256 graylevels for each pixel satisfies the perceived image quality, there are many applications in which 256 graylevels do not provide enough graylevel dynamic range. Under these situations, specialized electronic cameras and image acquisition hardware are available that sample an image using 10 bits (1024 graylevels) or 12 bits (4096 graylevels) per pixel. It has been common practice to use one byte per pixel to store each of the red, blue, and green color components comprising a color image. This results in 3 bytes per pixel or 24 bits per pixel of image storage required for a color image. In fact, digitized and sampled color images are often referred to as 24-bit color images.

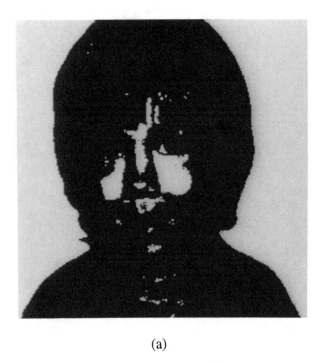

(a)

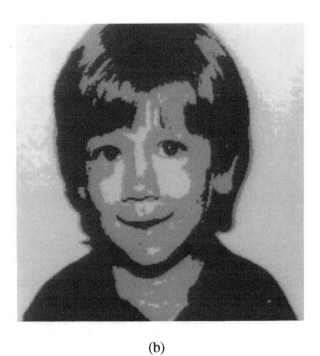

(b)

Figure 1.20: An example of image quantization: (a) 2 graylevels and (b) 4 graylevels.

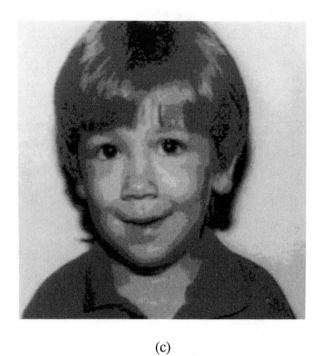

(c)

(d)

Figure 1.20: An example of image quantization: (c) 8 graylevels
and (d) 16 graylevels.

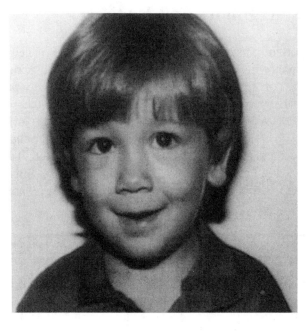

(e)

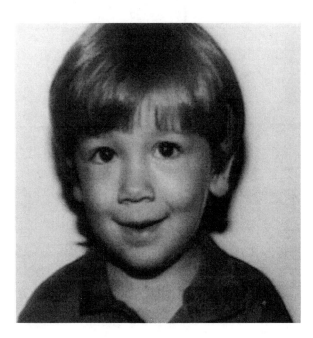

(f)

Figure 1.20: An example of image quantization: (e) 32 graylevels
and (f) 64 graylevels.

1.6 Image Neighbors and Distances

A very important concept in image processing is that of neighboring pixels. The use of neighboring pixels in an image processing algorithm provides a localized window of spatial information.   Consider a pixel $p(x, y)$ within a sampled image at the coordinate x, y.   The collection of the two surrounding vertical and the two surrounding horizontal pixels are defined as the 4-*neighbors* of the pixel $p(x, y)$. Figure 1.21(a) shows the relationship between the four neighboring pixels and $p(x, y)$. These four neighboring pixels define the vertical and horizontal spatial relationship to the pixel $p(x, y)$.   Similarly, the four neighboring diagonal pixels form the set of *diagonal-neighbors* of the pixel $p(x, y)$, shown in Figure 1.21(b).

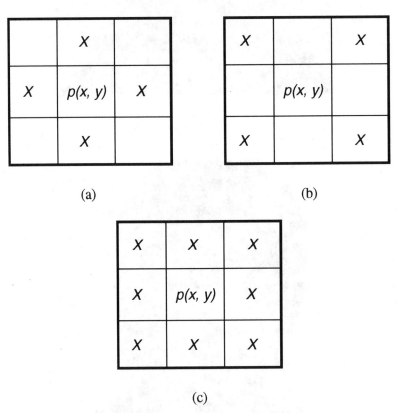

(a)   (b)

(c)

Figure 1.21: Neighbors of pixel $p(x, y)$: (a) 4-neighbors (b) diagonal-neighbors and (c) 8-neighbors.

The union of the set of 4-neighbors with the set of diagonal-neighbors form the set of 8-*neighbors*.   Figure 1.21(c) shows the eight-neighbors relative to $p(x, y)$.   The set of 8-neighbors is comprised of the 8 surrounding pixels of $p(x, y)$.   The concept of neighboring pixels can be expanded to include as many neighboring pixels as desired. For example, it is common to use 24- or 48-neighbors (5 × 5 or 7 × 7) in many image processing techniques involving local neighbors. The use of local neighbors is key in applying spatial filtering methods to an image for image enhancement, as presented in Chapter 4.

Another area important to electronic image processing is that of *connectivity*. Connectivity defines the relationship between neighboring pixels and is used to locate the borders between objects within an image. For example, connectivity defines the similarity between the graylevels of neighboring pixels. Usually, neighboring pixels with large graylevel differences implies pixels that are separated by region boundaries. There are three types of connected pixels that describe how two neighboring pixels are related: 4-connected, 8-connected, and *m*-connected. Consider pixel *A*, which is a neighbor of pixel *B*. Two pixels *A* and *B* are 4-connected provided *A* is a 4-neighbor of *B* and their graylevels meet some predetermined criteria or predicate. One example criteria might be that all pixels must be between ±2 in graylevel to be considered part of the same region within an image. Figure 1.22(a) shows an example of 4-connectivity between pixels *A* and *B* for all pixels with a graylevel of 5. Four-connectivity allows only vertical or horizontal paths to be traced from pixel *A* to *B*.

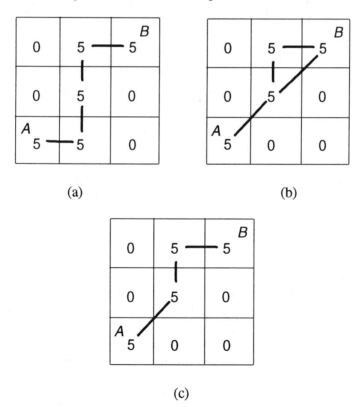

(a)   (b)

(c)

Figure 1.22: Examples of (a) 4-connectivity, (b) 8-connectivity, and (c) *m*-connectivity.

Similar to 4-connectivity is 8-connectivity, which includes the diagonal neighbors in determining a connected path. Two pixels *A* and *B* are 8-connected provided *A* is an 8-neighbor of *B* and their graylevels meet some predetermined criteria. Like 4-connectivity, 8-connectivity requires that a predetermined rule be set about graylevel similarity between pixels. The main difference between 4-connectivity and 8-connectivity is that 8-connectivity allows for diagonal paths between connected pixels. Figure 1.22(b) gives an example of 8-connectivity between pixels *A* and *B* for all pixels with a graylevel of 5. In this example, two paths exist between the two pixels *A* and *B*. In fact, 4-connectivity does not exist between pixels *A* and *B*, since there are no horizontal or vertical paths connecting pixel *A* and the center pixel of the 3×3 sub-image.

The main difficulty with 8-connectivity is that it produces two possible paths, as shown in the upper right hand corner of Figure 1.22(b). However, *m*-connectivity eliminates multiple paths by removing the diagonal path if 4-connectivity already exists between two pixels. This effectively leaves the horizontal and vertical paths and removes the diagonal path. Two pixels *A* and *B* are *m*-connected provided *A* is an 8-neighbor of *B*, their graylevels meet some predetermined criteria, and the 4-neighbor set of *A* with *B* does not intersect the 4-neighbor set of *B* with *A*. Figure 1.22(c) shows the use of *m*-connectivity and the elimination of the double path that was present in the 8-connected example given in Figure 1.22(b).

There are several distance measures that are used in image processing to measure the separation between pixels. The most common is the Euclidean distance. Given two pixels *A* and *B*, at the coordinates (a, b) and (c, d) the Euclidean distance r_e is

$$r_e = \sqrt{(a-c)^2 + (b-d)^2} . \tag{1.5}$$

Similar to 4-connectivity and 4-neighbors, by allowing only vertical and horizontal paths in the calculation of the distance between two pixels, the 4-distance is defined as

$$r_4 = |a-c| + |b-d| \quad . \tag{1.6}$$

The 4-distance is also referred to as the city block distance because this is the distance that an individual would travel ($r_4 = 2$) from one diagonal corner of a city block to the other diagonal corner if the path that was allowed to be traversed was along the sidewalks. Figure 1.23(a) gives the city block or 4-distance from the center pixel.

Another distance measure of interest is the 8-distance, defined as

$$r_8 = \text{maximum}(|a-c|, |b-d|) . \tag{1.7}$$

Figure 1.23(b) gives the 8-distance from the center pixel to neighboring pixels. Note that the contour of constant distances for the 4-distance measures is a diamond [Figure 1.23(a)], while for the 8-distance measures the contour of constant distances follows a square (Figure 1.23(b)).

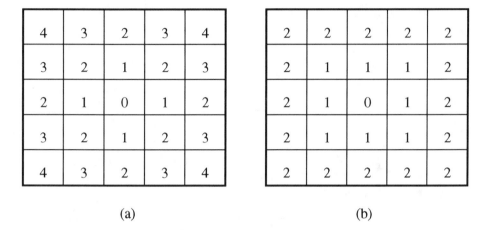

(a)   (b)

Figure 1.23: Examples of (a) 4-distance and (b) 8-distance.

1.7 Typical Image Processing Systems

There are several methods of converting an image into a digital form for processing. The first method is to use a flatbed scanner interfaced into a computer system that scans photographs using a linear array of electronic detectors. The electrical signals from the scanner are then converted into a digital format that is stored within the computer system. The second method of acquiring an image is to use the Kodak PhotoCD© format to store images onto a computer compatible CD disc, as mentioned earlier in this chapter.

Another common method of acquiring images is to convert the standard analog signals from an NTSC RS170 video camera to digital numbers that are stored within a computer system. Figure 1.24(a) shows a block diagram of a very common type of NTSC RS170 video acquisition system. At the heart of the video system is the analog to digital converter that converts the analog NTSC video signal to digital numbers. The "sync generation" section synchronizes all of the video timing with the computer digital hardware. It is the synchronization electronics that separates the good from the average digital video products. Some simple systems use a threshold circuit to generate synchronization pulses from the analog video signal; the more expensive systems use digital phase lock loop circuits with time based correction electronic circuits to regenerate the synchronization pulses. The more expensive method produces better captured pictures in the presence of noise such as those that are typically obtained from videotapes. As discussed earlier, the output of the analog to digital converter is 8 bits, producing a total of 256 graylevels per pixel.

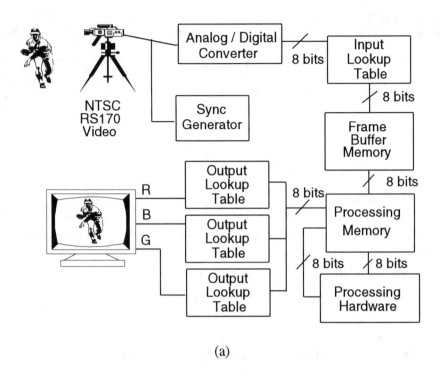

(a)

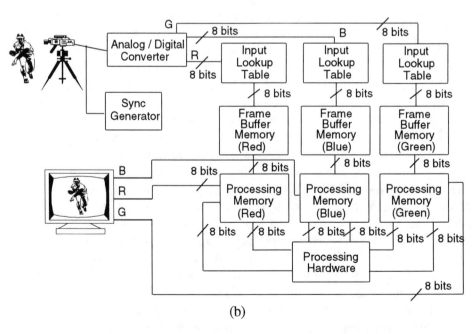

(b)

Figure 1.24: An example of (a) an RS170 video acquisition system
              and (b) a 24-bit color video acquisition system (images,
              ©New Vision Technologies).

The 8 bits of digital data from the analog to digital converter are then used to address one of 256 words of 8 bits of memory used as a lookup table. The input lookup table provides a means of mapping input graylevels to other desired graylevels. This is done by applying the 8 bits from the analog-to-digital converter to the address lines of the lookup table memory. The graylevels from the camera are then mapped to a new graylevel stored within the lookup table. This input lookup table is typically used to remove any nonlinearities that occur from the camera or analog video system. The output of the lookup table is then stored in digital memory, called a frame buffer. At this point, when the frame buffer is filled, one frame of video has been stored in the acquisition system. For digital processing, this data is then copied to a processing memory buffer and the next frame of video data starts to fill the frame buffer memory.

The processing memory is connected to the image processing hardware section, where the various image processing algorithms are applied to the digital image and the results stored back in the processing memory. If the processing unit can manipulate the image and store it back in the processing memory at a rate faster than the video frame rate (30 frames per second) and process the next frame of image data when it appears, this imaging system is referred to as a real-time image processing system. The output of the processing buffer feeds three output lookup tables. The outputs of the three lookup tables are then converted to an analog signal used to drive the red, blue, and green electron guns of the CRT. The three output lookup tables provide a means of adding color to a grayscale image. For example, the lookup tables can be set to produce the color red on the CRT for all pixels in the image containing the graylevel 100. This technique of adding color to a grayscale image is called *pseudocoloring* and will be discussed in more detail in the discussion on color image processing given in Chapter 6.

The block diagram given in Figure 1.24(b) is that of a 24-bit color video acquisition system. This system is basically the same as the NTSC RS170 system given Figure 1.24(a) except that there are now three analog to digital converters converting the red, blue, and green analog video signals to digital numbers. Each of these three digital outputs is then connected to three input lookup tables used typically to remove nonlinearities due to the camera or analog video electronics. The output from each lookup table is then stored in a frame buffer memory using one digital memory for each of the three color images. The processing unit is now expanded from manipulating 8 bits of image data to 24 bits. The processing unit must now process three times the amount of image data as compared to an RS170 image for the same spatial resolution image. If the processing system can process all of the image data within one frame of video data and continuously process incoming frames, this system is then referred to as a real-time color image processing system.

CHAPTER 2

Transforms Used in Electronic Image Processing

This chapter discusses several types of transforms that are commonly used in image processing. The most common transform, the Fourier transform, will be presented first, followed by several closely related transforms such as the Hadamard, Walsh, and discrete cosine transforms, which are used in the area of image compression. The chapter will end with the Hough transform, which is used to find straight lines in a binary image, and the Hotelling transform, which is commonly used to find the orientation of the maximum dimension of an object.

2.1 The Fourier Series

A short review of one-dimensional Fourier transforms will be presented to give the reader the necessary background to understand the properties of the two-dimensional Fourier transform used in electronic image processing. The Fourier transform is the most commonly used transform in image processing because of its relationship to linear system theory. Given an input to a linear system, the output is predicted using the complex operation of convolution or by simple multiplication using the Fourier transform. The Fourier transform/series was originally developed by the French mathematician Baptiste Joseph Fourier (1768-1830) to solve Laplace's differential equation (known as the Dirichlet problem) that described the conduction of heat (distribution of temperature) along an infinite conducting sheet. Fourier used a series of sine and cosine functions as a solution to the differential equation by solving for a set of unknown coefficients.

Since its conception, the Fourier transform has been used to solve a large number of linear problems ranging from the prediction of simple harmonic motion in buildings and bridges by civil engineers to the prediction of periodic upturns and downturns in the economy. It is the foundation of modern electronic communications and the basis of linear filtering by electronic circuits. With the generation of a fast computer implementation known as the fast Fourier transform algorithm by Cooley and Tukey in the middle 1960s, Fourier transform methods have been used to enhance both two-dimensional images as well as one-dimensional signals using digital signal processing approaches.

For a periodic one-dimensional signal $f(x)$, a Fourier series representation of this signal is defined by

$$f(x) = \frac{A_0}{2} + \sum_{n=1}^{\infty} A_n \cos(2\pi n f_o x) + \sum_{n=1}^{\infty} B_n \sin(2\pi n f_o x) . \tag{2.1}$$

The two coefficients A_n and B_n are found by integrating over the fundamental period T of the periodic function $f(x)$

$$A_n = \frac{2}{T} \int_0^T f(x) \cos(2\pi n f_o x) \, dx , \quad \text{for } n \geq 0 \tag{2.2}$$

and

$$B_n = \frac{2}{T} \int_0^T f(x) \sin(2\pi n f_o x) \, dx , \quad \text{for } n \geq 0 , \tag{2.3}$$

with $f_o = 1 / T$. Inspection of Equation (2.1) reveals that the frequencies of each of the sine and cosine functions occur at integer frequencies of the *fundamental frequency f_o* and are commonly referred to as the *harmonic frequencies* of $f(x)$. Equation 2.1 decomposes the function $f(x)$ into a sequence of sine and cosine functions with varying amplitudes depending on the two coefficients A_n and B_n. The use of Equations (2.1) - (2.3) requires satisfying the conditions that

$f(x)$ is a periodic function,
$| f(x) |^2$ is absolutely integrable,

$$\frac{1}{T} \int_0^T | f(x) |^2 \, dx < \infty ,$$

and $f(x)$ has a finite number of discontinuities.

The second condition is easily met for most types of functions in that it states that the energy contained within $f(x)$ must be finite. The third condition is not a major restriction, since most functions of interest will be semi-continuous over some small interval. For the special case of $n = 0$, $B_n = 0$ and

$$A_0 = \frac{2}{T} \int_0^T f(x)\, dx \quad . \tag{2.4}$$

Equation (2.4) is simply twice the average or DC value of the function and occurs at a frequency of zero.

Equation (2.1) can also be written as

$$f(x) = \sum_{n=1}^{\infty} C_n \cos(2\pi n f_o x + \Theta_n) + \frac{C_0}{2} \,, \tag{2.5}$$

where

$$C_n = \sqrt{A_n^2 + B_n^2} \qquad \text{and} \qquad \Theta_n = \tan^{-1}\left(\frac{B_n}{A_n}\right) . \tag{2.6}$$

The coefficients C_n are known as the Fourier magnitude components of $f(x)$ and the coefficients Θ_n as the Fourier phase components of $f(x)$.

Figure 2.1 shows the sine and cosine decomposition of a periodic square wave by the Fourier series. Note, that only the odd harmonics of B_n are nonzero, while all even harmonics of B_n and the coefficients A_n are zero. In particular, A_0 is zero, also yielding a zero average, which can also be determined directly by visual inspection of Figure 2.1. Solving Equations (2.2) and (2.3) yields coefficients of $A_n = 0$ and

$$B_n = \begin{cases} \dfrac{4A}{n\pi} & \text{for } n \text{ odd} \\ 0 & n \text{ even} \end{cases} \quad . \tag{2.7}$$

Only the sine functions from Equation (2.1) contribute to the generation of $f(x)$. Figure 2.2 illustrates the effect that each harmonic component has on the generation of $f(x)$ from its Fourier series. As more harmonics are added to the fundamental sine wave, a better approximation of the original square wave is obtained. The periodic ringing that is present adjacent to the discontinuous edges as shown in Figure 2.2(d) is known as the *Gibbs phenomenon* and is a result of requiring an infinite number of harmonics to properly predict a discontinuous edge.

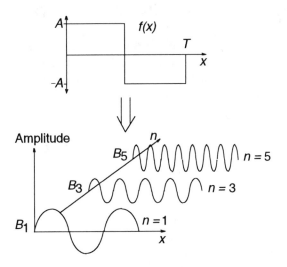

Figure 2.1: The decomposition of a square wave into sine and cosine functions by the Fourier series.

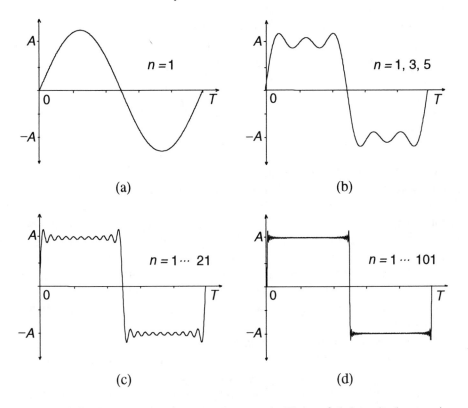

Figure 2.2: Regenerating the square wave in Figure 2.1 from its harmonic components (a) $n = 1$, (b) $n = 1, 3, 5$, (c) $n = 1 \cdots 21$, and (d) $n = 1 \cdots 101$.

2.2 The One-Dimensional Fourier Transform

For a nonperiodic function, the Fourier series as defined by Equations (2.1) through (2.3) is replaced by the Fourier transform. By definition, the continuous one-dimensional Fourier transform $\mathcal{F}\{f(x)\}$ (also called the forward Fourier transform) using complex notation is

$$F(w) = \mathcal{F}\{f(x)\} = \int_{-\infty}^{\infty} f(x)\, e^{-j2\pi wx}\, dx \tag{2.8}$$

and its inverse is

$$f(x) = \int_{-\infty}^{\infty} F(w)\, e^{j2\pi wx}\, dw \quad, \tag{2.9}$$

where $e^{-j2\pi wx}$ is known as the transform kernel and $j = \sqrt{-1}$. Other transforms are easily derived from the Fourier transform definition given in Equation (2.8). For example, if $e^{-sx} = e^{-j2\pi wx}$ is substituted into Equation (2.8) and $f(x)$ is zero for $x < 0$ then Equation (2.8) reduces to the Laplace transform,

$$F(s) = \int_{0}^{\infty} f(x)\, e^{-sx}\, dx \quad . \tag{2.10}$$

The type of kernel used in Equation (2.8) determines the type of transform that is implemented. Even the Fourier transform has two definitions for its kernel. It is not uncommon to have the forward and inverse kernels as defined in Equations (2.8) and (2.9) be swapped, placing the positive exponential kernel with the forward Fourier transform. This is not a major problem, provided care is taken in determining the definition that was used when using Fourier tables from other texts. If the reader desires to use a set of Fourier properties given in another text that uses the opposite Fourier definition given here, one simply replaces the frequency coefficient w with $-w$ in the desired Fourier property of the other text.

The Fourier transform of a nonperiodic function also decomposes a function into a set of sine and cosine functions. Unlike the Fourier series representation of a periodic function, in which the spectral components occur at discrete integer frequencies of nf_0, the components contained within $f(x)$ occur for all frequencies w. Expanding Equation (2.8) using Euler's identity $[e^{jx} = \cos(x) + j\,\sin(x)]$ separates the Fourier components of $F(w)$ into real and imaginary frequency components:

$$F(w) = R(w) + j\,I(w) \quad , \tag{2.11}$$

where $R(w)$ is the real part of $F(w)$ and is defined as

$$R(w) = \int_{-\infty}^{\infty} f(x) \cos(2\pi wx)\,dx \quad , \tag{2.12}$$

and $I(w)$ is the imaginary part of $F(w)$,

$$I(w) = -\int_{-\infty}^{\infty} f(x) \sin(2\pi wx)\,dx \quad . \tag{2.13}$$

Two important properties are easily derived from Equations (2.12) and (2.13). If $f(x)$ is an even function about x, then $R(w)$ is nonzero and $I(w)$ is zero, resulting in only real Fourier components. On the other hand, if $f(x)$ is odd, then $R(w)$ is zero and $I(w)$ is nonzero, producing only imaginary Fourier components. Except for the limits of integration and the division by $1/T$, Equations (2.12) and (2.13) are the same as Equations (2.2) and (2.3), which define the Fourier coefficients A_n and B_n.

The Fourier frequency components can now be written in terms of magnitude,

$$\mid F(w) \mid = \sqrt{R(w)^2 + I(w)^2} \quad , \tag{2.14}$$

and phase,

$$\phi(w) = \tan^{-1}\left(\frac{I(w)}{R(w)}\right) . \tag{2.15}$$

By definition, the magnitude of $F(w)$ is often referred to as the *magnitude spectrum*, while the phase of $F(w)$ is often called the *phase spectrum*. The key to linear filtering is the modification of the magnitude and phase spectra of $F(w)$. For example, a *lowpass filter* attenuates the magnitude components of the high frequencies, while leaving the low frequency magnitude components unchanged. Several types of filters exist that leave the phase spectrum unchanged while modifying the magnitude spectrum. These filters are commonly referred to as *phase preserving filters*.

In the filtering of images for image enhancement using the phase and magnitude spectra, it will be required to leave the phase components unchanged. Hence, phase preserving filters are used. A simple experiment shows the importance of the phase spectrum relative to the magnitude spectrum. Consider an image containing two small circular objects located some small distance from each other. The two-dimensional Fourier transform of this image is computed, from which the magnitude and phase spectra are computed. A filter is applied to the two spectra such that the phase spectrum remains unchanged and the magnitude spectrum is set to a constant value. Inverse Fourier transforming the filtered magnitude and phase spectra yields a distorted image with the two circular objects still observable. On the other hand, if a filter is applied to the original phase and magnitudes spectra such that the phase spectrum is set to a constant and the magnitude spectrum is left unchanged, the resulting inverse Fourier transform yields a degraded image with the circular objects no longer visible.

Consider a pulse of width T, shown in Figure 2.3. The Fourier transform of this pulse is found by evaluating Equation (2.8) with $f(x) = A$:

$$F(w) = \int_{-\frac{T}{2}}^{\frac{T}{2}} A\, e^{-j2\pi wx}\, dx \quad . \tag{2.16}$$

Evaluating Equation (2.16) over the limits of $-T/2$ to $T/2$ and using the identity

$$\sin x = \frac{e^{jx} - e^{-jx}}{2j} \tag{2.17}$$

yields

$$F(w) = AT\, \frac{\sin \pi wT}{\pi wT} \quad , \tag{2.18}$$

which reduces to the well known sinc function $[\text{sinc}(x) = \sin(\pi x)/\pi x]$ $F(w) = AT \cdot \text{sinc}(wT)$.

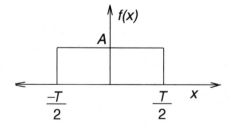

Figure 2.3: An example function for $f(x)$ in which the Fourier components are computed.

The magnitude spectrum for $F(w)$ is given in Figure 2.4. Note the presence of the main lobe and the addition of side lobes that repeat at the rate of one over the width of the pulse ($1/T$). Figure 2.4 illustrates one important property of the Fourier transform: truncation in one domain, which produces an infinite discontinuity, results in the other Fourier domain being infinite in length and containing side lobes similar to those shown in Figure 2.4.

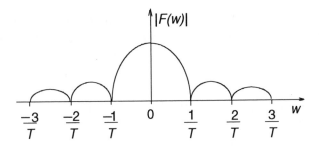

Figure 2.4: The magnitude spectrum of the pulse given in Figure 2.3.

The next example illustrates the effect of translating the pulse of Figure 2.3 by $T/2$ to the position shown in Figure 2.5. The Fourier transform of this pulse is found in the same manner in which the Fourier components of the pulse in Figure 2.3 were found:

$$F(w) = AT \frac{\sin \pi wT}{\pi wT} \cdot e^{-j\pi wT} \quad . \tag{2.19}$$

A comparison of Equation (2.19) with Equation (2.18) shows the multiplication of a complex phase term $e^{-j\pi wT}$. Computing the magnitude spectrum of Equation (2.19) yields the same magnitude spectrum as for the pulse given in Figure 2.3. The translation of the pulse of Figure 2.3 by $T/2$ did not modify the magnitude spectrum but added a phase term. This is an important concept because the location or start of a function does not vary its magnitude spectrum but only its phase spectrum. In terms of electronic image processing, it states that the magnitude spectrum of objects within an image does not depend on their location within the image.

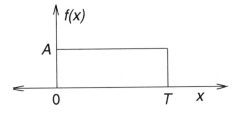

Figure 2.5: The $T/2$ shifted version of pulse given in Figure 2.3.

Figure 2.6 shows the effect that a first order lowpass filter with a magnitude response of

$$|H(w)| = \frac{1}{\sqrt{1 + \left(\dfrac{w}{w_O}\right)^2}} \tag{2.20}$$

has on the rise time of an edge. The effect of lowpass filtering a fast rising edge is to reduce the slope of the edge. For the filter given in Equation (2.20), the edge rises to 63% of its final value within one unit of w_O. In electronic image processing, the reduction of the slope of an edge gives the visual appearance of an image as blurred or out of focus. To maintain a sharp image, the high frequency components of an image must be preserved.

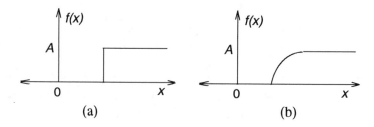

(a)   (b)

Figure 2.6: An example of lowpass filtering an edge: (a) the original edge and (b) the lowpass filtered edge.

2.3 The Two-Dimensional Fourier Transform

Since images are two-dimensional in nature, the concepts and properties of the one-dimensional Fourier transform must be expanded into two dimensions. The Fourier transform $\mathcal{F}\{f(x, y)\}$ is defined in two dimensions as

$$F(w, z) = \mathcal{F}\{f(x, y)\} = \int\limits_{-\infty}^{\infty}\int f(x, y)\, e^{-j2\pi(wx + zy)}\, dx\, dy \tag{2.21}$$

and its inverse as

$$f(x, y) = \int\limits_{-\infty}^{\infty}\int F(w, z)\, e^{j2\pi(wx + zy)}\, dw\, dz\ , \tag{2.22}$$

where the variables w and z are the frequencies in the x and y directions.

Figure 2.7 illustrates the relationship between the various components of the two-dimensional Fourier transform. Although not a requirement of the Fourier transform, $f(x, y)$ typically is a real function. Upon computation of the Fourier transform, $F(w, z)$ can be separated into two components $R(w, z)$ and $I(w, z)$, representing the real and imaginary components of $F(w, z)$. More intuitive to our understanding of Fourier transforms, the phase and magnitude spectra can be computed. Like the one-dimensional Fourier transform, the modification of the magnitude spectrum is the basis for linear filtering of images.

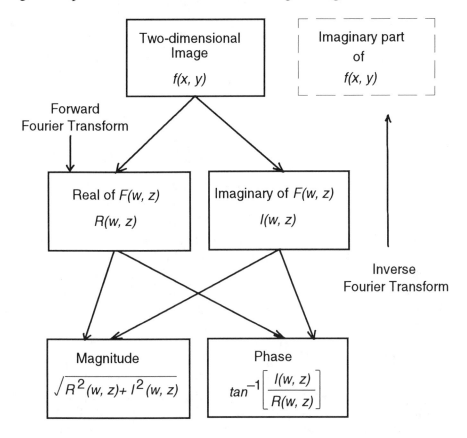

Figure 2.7: A flow chart showing the various components of the two-dimensional Fourier transform.

When using the one-dimensional Fourier transform with time domain signals, the concept of frequency is typically represented as cycles per second or hertz. In two dimensions relating to images, the frequency variables w and z are usually given in cycles per distance, such as cycles/mm or line pairs/mm, and we refer to this as the *spatial frequency*. Spatial frequency gives the periodic rate as a function of spatial distance. Figure 2.8 shows two images of vertical lines with different spacing. Figure 2.8(b) contains more lines per given distance than Figure 2.8(a); therefore this image contains higher spatial frequencies.

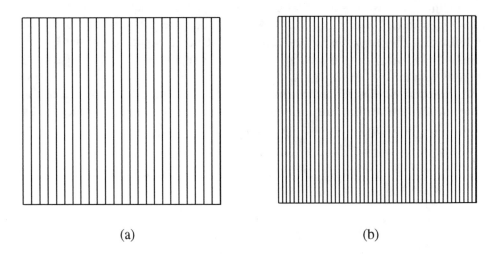

(a)   (b)

Figure 2.8: An example showing the difference between (a) a low spatial
frequency image and (b) a high spatial frequency image.

The use of line pairs/mm as the definition of spatial frequency came about as
a means to measure the maximum spatial frequency of an optical system. A set
of images with white and black line pairs of varied spacing were generated,
similar to the images Figure 2.8. Each was then placed in front of an optical
system under test and observers would indicate if the black and white lines were
observable. If so, another image with still closer line pairs was placed in the
optical system and the observation process repeated. The line pair image in
which the line pairs were just observable determined the maximum spatial
resolution of the optical system in line pairs/mm.

The spatial frequency content of an image provides information about the
size of the objects present. An image containing only low spatial frequencies
implies large objects, while an image containing very small objects will be
comprised of high spatial frequency components. The spatial frequency content
of an image is also related to the correlation between adjacent pixels in a
sampled image. A low spatial frequency image will have neighboring pixels that
are highly correlated (very similar), while neighboring pixels that are highly
decorrelated implies an image that contains high spatial frequencies. In addition,
an image containing very sharp graylevel transitions will also contain high
spatial frequency components.

The accuracy of sampling an image is dependent on the highest spatial
frequency that is present within the image. The well known *Nyquist theorem,* as
it relates to images, states that to properly sample an image, the sampling
frequency must be at least twice the maximum spatial frequency present. If this
condition is not met, aliasing occurs, distorting the objects present in the image
because the size of the pixels determines the sampling frequency. The smaller
the pixels are, the larger the sampling frequency becomes. Figure 2.9(a) shows

an image of a diamond that has been properly sampled. Figure 2.9(b) is an image of the same diamond but the sampling frequency was lower than the Nyquist rate. The distortion present in the improperly sampled image is a result of aliasing.

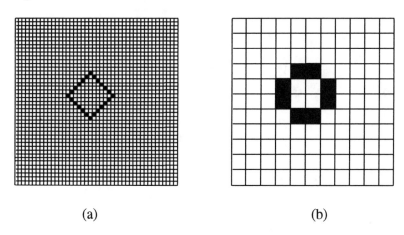

(a)   (b)

Figure 2.9: An example image of a diamond (a) properly sampled,
and (b) improperly sampled below the Nyquist rate.

2.4 Important Functions Relating to the Fourier Transform

There are several functions that are commonly used when dealing with Fourier transforms. Several of these functions will be presented in one dimension and are easily converted to two dimensions using

$$f(x, y) = g(x) \cdot g(y) \quad , \tag{2.23}$$

where $g(x)$ is the one-dimensional function applied in the x direction and $g(y)$ is the same function applied in the y direction. The most commonly used function is the rect function and is defined in one dimension as

$$\text{rect}\left(\frac{x}{a}\right) = \begin{cases} 1 & |x| \le a/2 \\ 0 & \text{elsewhere} \end{cases} \quad . \tag{2.24}$$

Figure 2.10(a) shows the rect function as a pulse of width a centered about the origin. Converting the one-dimensional rect function given by Equation (2.24) to two dimensions is easily accomplished using Equation (2.23):

$$f(x, y) = \text{rect}\left(\frac{x}{a}\right)\text{rect}\left(\frac{x}{b}\right) \quad . \tag{2.25}$$

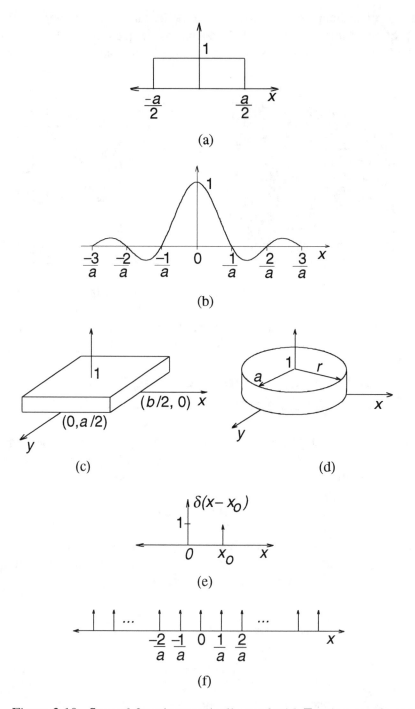

Figure 2.10: Several functions typically used with Fourier transforms:
(a) rect function, (b) sinc function, (c) two-dimensional rect
function, (d) two-dimensional circ function, (e) impulse
function, and (f) comb function.

Figure 2.10(c) shows a three-dimensional plot of this function. A two-dimensional rect function produces a rectangle of width a and height b along the x and y directions, respectively. The two-dimensional rect function is of importance in imaging because it mathematically models a rectangular aperture.

The next most popular function is the sinc function and is defined as

$$\text{sinc } ax = \frac{\sin a\pi x}{a\pi x} \quad . \tag{2.26}$$

Another definition of the sinc function is very similar to the one given in Equation (2.2) except the parameter π is not included. Figure 2.10(b) gives a plot of the sinc function as defined in Equation (2.26). This function has a main lobe around the origin with a value of one for $x = 0$. It also has sidelobes that extend to infinity and that repeat at the rate of $1/a$. Application of Equation (2.23) converts the one-dimensional sinc function to two dimensions. The popularity of this function is due to the fact that the Fourier transform of the rect function yields the sinc function:

$$\mathcal{F}\left\{ \text{rect}\left(\frac{x}{a}\right)\right\} = a \text{ sinc } wa \ . \tag{2.27}$$

The circ function shown in Figure 2.10(d) is typically used in imaging to model circular apertures and is defined by

$$\text{circ}\left(\frac{r}{a}\right) = \begin{cases} 1 & r \leq a \\ 0 & \text{elsewhere} \end{cases} \ , \tag{2.28}$$

where $r^2 = x^2 + y^2$. The two-dimensional Fourier transform of this function produces the well known *Airy* function

$$\mathcal{F}\left\{ \text{circ}\left(\frac{r}{a}\right)\right\} = 2\pi a^2 \frac{J_1(2\pi\rho a)}{2\pi\rho a} \ , \tag{2.29}$$

where $J_1(x)$ is the first order *Bessel* function and $\rho^2 = w^2 + z^2$. This function like the sinc function has a mainlobe centered at the origin and infinite sidelobes. The main difference between the sinc function given in Equation (2.26) and this function is that this one is circularly symmetric.

The *impulse* function is used in image processing to model point light sources. This function is formally defined through a set of limiting functions but for the concepts presented in this book can be defined as

$$\delta(x - x_o) = \begin{cases} \infty & x = x_o \\ 0 & \text{elsewhere} \end{cases} \tag{2.30}$$

and

$$\int_{-\infty}^{\infty} \delta(x - x_O) \, dx = 1 \quad .$$ (2.31)

Figure 2.10(e) shows the representation of an impulse located at $x = x_O$. The value adjacent to the impulse is not its height but the area under the impulse. Like the other one-dimensional functions, the two-dimensional impulse function is found using Equation (2.23):

$$\delta(x - x_O, y - y_O) = \delta(x - x_O) \cdot \delta(y - y_O) = \begin{cases} \infty & x = x_O , y = y_O \\ 0 & \text{elsewhere} \end{cases}$$ (2.32)

and

$$\int_{-\infty}^{\infty} \int \delta(x - x_O, y - y_O) \, dx \, dy = 1 \; .$$ (2.33)

A function closely related to the impulse function is the *comb* function, defined as an infinite series of impulses,

$$\text{comb}(ax) = \sum_{n = -\infty}^{\infty} \delta(ax - n) \quad .$$ (2.34)

Figure 2.10(f) shows the comb function with the impulses repeating at the rate of $1/a$. The comb function is given its name because the plot of this function resembles the teeth of a comb. Both the impulse and comb functions will be used to convert a continuous image to a spatially sampled image. To convert the continuous functions presented in this section to sampled functions, the independent continuous variable is replaced by its sampled version (except for the comb and impulse functions). For example, the continuous sinc function $\sin(\pi x)/(\pi x)$ becomes $\sin(\pi n\Delta x)/(\pi n\Delta x)$, where n is an integer and Δx is the sampling interval. The discrete version of the impulse function as defined in two dimensions takes on the value

$$\delta(n\Delta x - n_O, m\Delta x - m_O) = \begin{cases} 1 & n\Delta x = n_O , m\Delta x = m_O \\ 0 & \text{elsewhere} \end{cases} \quad ,$$ (2.35)

where $n\Delta x$ and $m\Delta y$ are sampled versions of the x and y directions, respectively.

2.5 The Discrete Fourier Transform

The goal of this text is the description and explanation of the processing of sampled images using electronic image processing. To completely understand the application of the two-dimensional Fourier transform to a sampled image, the effects of sampling must first be investigated. What does sampling an image do to its Fourier spectrum? To make the presentation clearer, the results of sampling will first be derived using the one-dimensional case. These results will then be extended to two dimensions, from which the Fourier components of a sampled image will be related to the Fourier components of the continuous image. Consider the continuous function $f(x)$ of infinite extent, shown in Figure 2.11(a), whose frequency spectrum is shown in Figure 2.11(b). The comb function that repeats at the rate of a can be used to sample $f(x)$, producing the sampled function $f_S(x)$:

$$f_S(x) = f(x) \cdot \text{comb}\!\left(\frac{x}{a}\right) . \tag{2.36}$$

Figure 2.11(c) shows the sampled function of $f(x)$ with a sampling rate of a.

To compute the Fourier transform of the sampled function, the well known convolution identity is used. The convolution identity states that multiplication in one domain leads to convolution in the other domain. For example, if $g(x, y) = h(x, y) \cdot f(x, y)$, then by the convolution identity the Fourier components of $g(x, y)$ are related to the Fourier components of $f(x, y)$ and $h(x, y)$ by

$$G(w, z) = H(w, z) * F(w, z) \ , \tag{2.37}$$

where the * operator denotes convolution and is defined as

$$G(w, z) = \int\limits_{-\infty}^{\infty} \int F(w - \alpha, z - \beta)\, H(\alpha, \beta)\, d\alpha\, d\beta \ . \tag{2.38}$$

The Fourier transform of the sampled function is found by taking the Fourier transform of $f(x)$ and the comb(x/a) given in the righthand side of Equation (2.36) separately, and then applying the convolution identity to find the Fourier components of the sampled function $f_S(x)$. The Fourier transform of a comb function that repeats at the rate a is also a comb function with a height of a and repeats at the rate $1/a$. The Fourier transform of $f(x)$ is simply $F(w)$. The Fourier transform of the sampled function is

$$\mathcal{F}\{f_S(x)\} = F(w) = a \sum_{n=-\infty}^{\infty} \delta(aw - n) * F(w) . \tag{2.39}$$

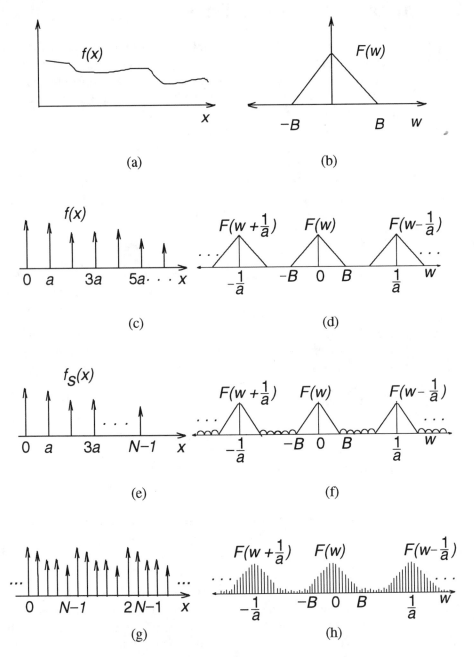

Figure 2.11: Several graphs illustrating the effects of sampling: (a) the
              original continuous function $f(x)$, (b) its spectrum, (c) the
              sampled function, (d) the spectrum of the sampled signal,
              (e) the sampled function truncated to N samples, (f) the
              truncated sampled spectrum, (g) the effect of sampling the
              frequency domain, and (h) the sampled frequency spectrum.

Using the convolution identity and $\delta(ax) = \dfrac{1}{|a|}\delta(x)$, Equation (2.39) becomes

$$\mathcal{F}\{f_S(x)\} = \sum_{n=-\infty}^{\infty} \int_{-\infty}^{\infty} \delta\left(w - p - \frac{n}{a}\right) F(p) \, dp \ . \tag{2.40}$$

The above integral is the convolution of $F(w)$ with $\delta(w - n/a)$. Provided $F(w)$ is a finite function, the integral in Equation (2.40) can be evaluated using a special property of the impulse function known as the *sifting property*:

$$f(x_O) = \int_{-\infty}^{\infty} \delta(x - x_O) f(x) \, dx \ . \tag{2.41}$$

Equation 2.41 states that the integral of an impulse function and the multiplication of a function $f(x)$ is simply equal to the function evaluated at the impulse $f(x_O)$. Substituting Equation (2.41) in Equation (2.40) yields the Fourier components of the sampled function:

$$\mathcal{F}\{f_S(x)\} = \sum_{n=-\infty}^{\infty} F(w - n/a) \tag{2.42}$$

Equation (2.42) is very important because it states that sampling a continuous function $f(x)$ using a sampling interval of a produces a Fourier spectrum in which the spectral components $F(w)$ are now periodic at the rate of $1/a$. In fact, sampling in one domain makes the other domain periodic. For example, sampling of the frequency components $F(w)$ makes $f(x)$ periodic. Figure 2.11(d) shows the spectrum of the sampled function that is periodic at the rate of $1/a$. If a is the sampling rate then $1/a$ is the sampling frequency. From Figure 2.11(d), if the point $1/a - B$ is greater than B, the spectra from each period do not overlap and the original unsampled spectrum can be reconstructed. To guarantee this condition

$$\frac{1}{a} - B \geq B \quad or \quad \frac{1}{a} \geq 2B \ . \tag{2.43}$$

Equation (2.43) states that the sampling frequency must be twice the maximum frequency of the unsampled function, which is the *Nyquist sampling criterion*.

In two dimensions, sampling $f(x, y)$ with sampling frequencies of $1/a$ and $1/b$ in the x and y directions yields

$$\mathcal{F}\{f_s(x, y)\} = \sum_{m = -\infty}^{\infty} \sum_{n = -\infty}^{\infty} F(w - n/a, z - m/b) \quad , \qquad (2.44)$$

which is now periodic in both the w and z frequency directions.

Now that the effects of sampling are understood, the goal is to convert the continuous Fourier transform into the discrete Fourier transform. Sampling an image is only part of the problem; the image must also be truncated to a finite size. Figure 2.10(e) shows the sampled one-dimensional function, truncated to a finite length of N. This is accomplished by multiplying the sampled function $f_s(x)$ by a rect function that has been shifted to the right by $N/2$ and has a width of N. From the convolution identity, multiplication in one domain yields convolution in the other domain. As stated previously, the Fourier transform of the rect function is a sinc function. Convolving the Fourier transform of $f_s(x)$, as shown in Figure 2.11(d), with the sinc function yields the magnitude spectrum shown in Figure 2.11(f). The effect of truncating the sampled function to a finite size distorts the spectral components by adding ringing in the form of sidelobes. The effect of truncation can be partially eliminated by using various types of window functions such as the Hamming, Blackman, and Kaiser windows. These window functions reduce the amplitude of $f_s(x)$ at its edges and hence reduce the discontinuous step added by the multiplication of the rect function. A detailed discussion of these windowing functions are beyond the scope of this text. The interested reader is referred to one of the many linear systems and communication books that are given in the bibliography.

Since the Fourier transform will be computed via a digital computer, both the spatial and frequency domains must be sampled. So far an image has been sampled and truncated to a finite size. The only process that is left is to sample the Fourier domain at the rate of one over the number of sampled points $(1/N)$. Figure 2.11(h) gives the one-dimensional example of sampling the Fourier domain. Using the identity that sampling in one domain makes the other domain periodic, the sampled function $f_s(x)$ is now also periodic at the rate N. Figure 2.11(g) illustrates the periodic nature of $f_s(x)$ due to sampling the Fourier components at the interval of $1/N$. Hence, sampling the spectral components of an image produces a sampled image that is also periodic.

Now that both the spatial and Fourier domains have been sampled, the discrete Fourier transform (DFT) can be defined for an $N \times M$ finite sized image with a sampling interval of Δx and Δy in the x and y directions as

$$F(n, m) = \frac{1}{NM} \sum_{y = 0}^{M - 1} \sum_{x = 0}^{N - 1} f(x, y) \cdot e^{-j2\pi(nx/N + my/M)} \qquad (2.45)$$

and its inverse as

$$f(x, y) = \sum_{m=0}^{M-1} \sum_{n=0}^{N-1} F(n, m) \cdot e^{j2\pi(nx/N + my/M)} . \qquad (2.46)$$

In Equations (2.45) and (2.46) all variables take on integer values. Frequency sampling in the n and m frequency directions are related to the spatial sampling rates by

$$\Delta n = \frac{1}{N\Delta x} \qquad\qquad \Delta m = \frac{1}{M\Delta y} . \qquad (2.47)$$

For a square image, Equations (2.45) and (2.46) reduce to

$$F(n, m) = \frac{1}{N^2} \sum_{y=0}^{N-1} \sum_{x=0}^{N-1} f(x, y) \cdot e^{-j2\pi(nx + my)/N} \qquad (2.48)$$

and

$$f(x, y) = \sum_{m=0}^{N-1} \sum_{n=0}^{N-1} F(n, m) \cdot e^{j2\pi(nx + my)/N} . \qquad (2.49)$$

2.6 Example and Properties of the Discrete Fourier Transform

This section gives several important properties of the DFT that are commonly used in the evaluation and processing of images. Many of the these properties will be listed for completeness, with several of the most important properties illustrated with example images at the end of this section. For ease of notation, both the properties and examples given in this section will be presented for a two-dimensional square image of size N. The first two properties are the direct consequence of the linearity property of the DFT

$$\mathcal{F}\{a\, f(x, y)\} = a\, \mathcal{F}\{f(x, y)\} \qquad (2.50)$$

and

$$\mathcal{F}\{f(x, y) + g(x, y)\} = \mathcal{F}\{f(x, y)\} + \mathcal{F}\{g(x, y)\} . \qquad (2.51)$$

Equation (2.50) states that it does not matter whether a function is multiplied by a constant and then the DFT is taken or the DFT is taken first and the result multiplied by a constant. The second property, Equation (2.51), says that the

DFT of the sum of two functions is equivalent to the sum of the DFT of each function separately.

There are two functions that retain their functional form under the DFT. The first is the comb function as described in the previous section and the second is the *Gaussian* function:

$$e^{-\pi(x^2 + y^2) / a^2} \Leftrightarrow e^{-\pi a^2(n^2 + m^2) / N} \quad . \tag{2.52}$$

The DFT of the discrete *impulse* function yields a Fourier spectrum that is constant in value,

$$\delta(x, y) \Leftrightarrow \frac{1}{N^2} \quad , \tag{2.53}$$

and likewise the DFT of a constant image is

$$A \Leftrightarrow A \cdot \delta(n, m) \quad . \tag{2.54}$$

A spectrum that is constant in value contains all frequencies and is often referred to as a white spectrum. Two important properties known as the shifting properties of the DFT are

$$f(x - a, y - b) \Leftrightarrow e^{-(j2\pi an + j2\pi bm) / N} F(n, m) \tag{2.55}$$

and

$$e^{(j2\pi ax + j2\pi by) / N} f(x, y) \Leftrightarrow F(n - a, m - b) \quad . \tag{2.56}$$

Equation (2.55) states that the shift of $f(x, y)$ by a and b adds only an additional phase to the Fourier spectral components. Equation (2.56) is the result of the reciprocity property of the DFT.

Setting $f(x, y)$ to one in Equation (2.56) and using the identity that $\cos(x) = (e^{jx} + e^{-jx})/2$ yields the DFT of a two-dimensional cosine function as

$$\cos\left(\frac{2\pi ax}{N}\right)\cos\left(\frac{2\pi bx}{N}\right) \Leftrightarrow \frac{1}{4}\, \delta(n - a, m - b) + \frac{1}{4}\, \delta(n - a, m + b) +$$
$$\frac{1}{4}\, \delta(n + a, m - b) + \frac{1}{4}\, \delta(n + a, m + b) \quad . \tag{2.57}$$

The next two properties state the periodicity of the DFT, which has a period equal to the size of the sampling window *N,*

$$F(n, m) = F(n + N, m) = F(n, m + N) = F(n + N, m + N) \quad , \tag{2.58}$$

and that taking the DFT of a function twice yields the flipped, inverted, and scaled version of the original image,

$$\mathcal{F}\mathcal{F}\{f(x, y)\} = \frac{1}{N^2} f(-x, -y) \ . \tag{2.59}$$

Taking the complex conjugate of the Fourier spectral components yields the flipped and inverted version of the original spectral components

$$F^C(n, m) = F(-n, -m) \ . \tag{2.60}$$

The average (mean) or *DC* value is found by evaluating the DFT at $n = 0$ and $m = 0$ in Equation (2.48)

$$F(0, 0) = \frac{1}{N^2} \sum_{y=0}^{N-1} \sum_{x=0}^{N-1} f(x, y) \ . \tag{2.61}$$

The scaling property of the DFT is

$$\mathcal{F}\{f(ax, by)\} = \frac{1}{|ab|} F(\frac{n}{a}, \frac{m}{b}) \ . \tag{2.62}$$

This property illustrates the concept that bigger objects contain fewer high spatial frequencies than smaller objects in an image. Consider an object within an image with spectral components $F(n, m)$ that has undergone the scaling operation of Equation (2.62). If the parameters a and b are greater than one, the scaling operation increases the size of the object, which effectively decreases the number of high spatial frequency components that are present within the image as a result of the object.

The rotational property states that rotating an image by a fixed angle also rotates its Fourier spectrum by the same angle. Consider an image that is circularly symmetric about its center coordinate $(N / 2, N / 2)$. Converting this image to polar coordinates with $x = r \cos \theta$ and $y = r \sin \theta$ yields $f(r, \theta)$. Likewise, its Fourier domain can also be converted to polar coordinates with $n = \rho \cos \phi$, and $m = \rho \sin \phi$ to yield $F(\rho, \phi)$. The rotational property of the DFT states that if an image is rotated by θ_0, then

$$f(r, \theta - \theta_o) \Leftrightarrow F(\rho, \phi - \theta_0) \ . \tag{2.63}$$

The computation of the DFT stores the real and imaginary spectral components in arrays, starting with positive frequency values followed by negative frequency values. Figure 2.12 is an example of the magnitude spectrum from a one-dimensional DFT, showing that the negative frequency components follow the positive frequency components. Normally when plotting these spectral components using the Cartesian coordinate system, negative frequency components are plotted first followed by the positive frequency components. A computer program can be written to swap the first half and last half of the array of Fourier components or the centering property of the DFT can be used:

$$f(x, y) \Rightarrow f(x, y) \, (-1)^x \, (-1)^y \ . \tag{2.64}$$

Equation (2.64) will center the DFT about $n = N/2$ and $m = N/2$, which is the center of the Fourier spectral domain. Figure 2.12(b) shows the output from the DFT after application of the centering property.

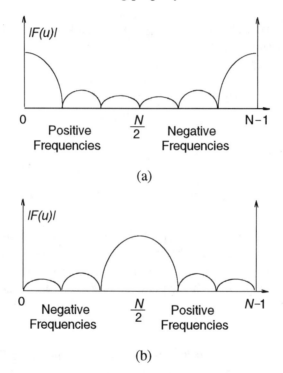

Figure 2.12: An example of the centering property of the DFT:
(a) the uncentered magnitude spectrum and (b) the centered magnitude spectrum.

The last property is the most important and the reason that Fourier transforms are so popular. Convolution is the basis of linear system theory, providing the mathematical foundation that predicts the result of filtering the Fourier components of a function. Its interest in electronic image processing is that it predicts the output of a linear two-dimensional filter applied to an image. The use of the convolution identity was shown in the derivation of the sampling theorem in Section 2.5. Given the Fourier spectral response of a linear filter as $H(n, m)$ and spectral components of an image as $F(n, m)$, the filtered image's spectrum is

$$G(n, m) = H(n, m) \cdot F(n, m) \ . \tag{2.65}$$

The filtered image $g(x, y)$ is then found by taking the DFT of Equation (2.65). By the convolution identity, and by taking the inverse DFT of $H(n, m)$ to produce the filter's impulse response $h(x, y)$, the convolution integral gives the filtered image $g(x, y)$ as

$$g(x, y) = \sum_{m=0}^{y} \sum_{n=0}^{x} f(n, m) \, h(x - n, y - m) \ . \tag{2.66}$$

Equations (2.65) and (2.66) describe two different methods of applying the same linear filter to an image. The advantages of using the DFT over convolution is that it reduces the difficult and time consuming operation of convolution to simple multiplication. The concept of convolution will be discussed further in presenting the topic of spatial filtering in Chapter 4.

Figure 2.13 shows the centered Fourier magnitude spectra for several different types of images. Figure 2.13(a) is an image of a 32×32 pixel white square with graylevel 255 against a black background of graylevel 0. Figure 2.13(a) is simply an image of a two-dimensional rect function with $a = b = 32$ as defined by Equation (2.25). Figure 2.13(b) is its corresponding centered magnitude spectrum displayed as an image by scaling the minimum and maximum values of the magnitude spectrum between 0 and 255. Since Figure 2.13(b) is the centered magnitude spectrum, the average or DC value $F(0, 0)$ is located in the center of the image. Figure 2.13(b) is just a graylevel image of the two-dimensional sinc function, with the mainlobe located at the center of the image and the sidelobes extending from the mainlobe. Figure 2.14(a) shows a four times expanded three-dimensional plot of the magnitude spectrum of this square image. From this plot, the mainlobe and the sidelobes are more easily observable.

Figure 2.13(c) is an image of a 255 graylevel circle with a diameter of 51 pixels. Figure 2.13(d) is its corresponding magnitude image showing one circular mainlobe with several circular sidelobes. This is an image of the Airy function as given by Equation (2.29). Figure 2.14(b) shows the four times expanded three-dimensional plot of the magnitude spectrum for this circular object. Observing the two magnitude spectra of the square and circular objects, the reader who is familiar with Fourier optics might consider these images as familiar. This is because the far-field intensity pattern of a square aperture illuminated by a plane wave aperture is simply the magnitude square of the Fourier transform of this square aperture. Hence, in the far-field, the observed intensity pattern is the square of the two-dimensional sinc function. Likewise, the far-field intensity pattern of a circular aperture is simply the square of the magnitude spectrum of the Fourier transform of this circular aperture. This is the square of the Airy function as described by Equation (2.29), producing the well known *Airy disc* intensity pattern or *Sombrero function.*

(a)

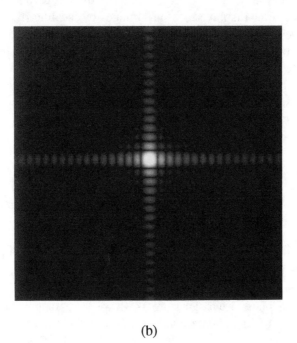

(b)

Figure 2.13: Examples of several images and their corresponding Fourier
              magnitude spectra: (a) an image of a white square and (b)
              its centered magnitude spectrum.

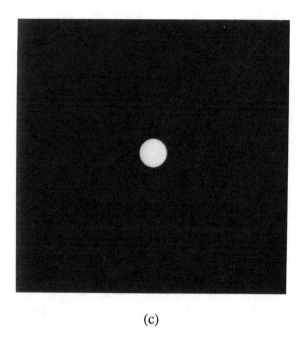

(c)

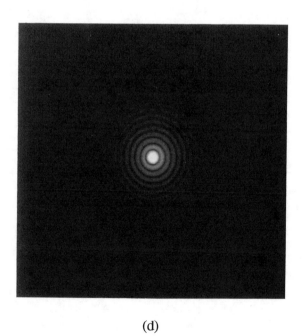

(d)

Figure 2.13: Examples of several images and their corresponding Fourier magnitude spectra: (c) an image of a white circle and (d) its centered magnitude spectrum.

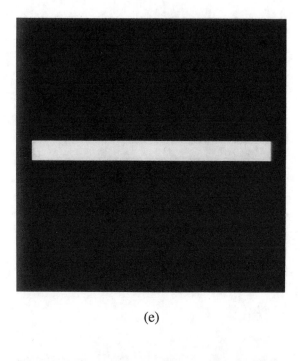

(e)

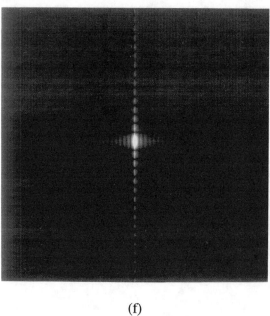

(f)

Figure 2.13: Examples of several images and their corresponding Fourier
              magnitude spectra: (e) an image of a white horizontal line
              and (f) its centered magnitude spectrum.

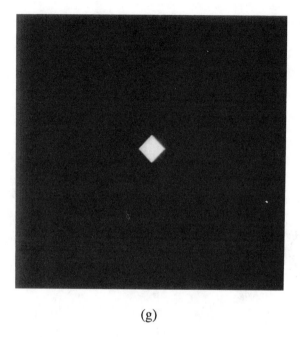

(g)

(h)

Figure 2.13: Examples of several images and their corresponding Fourier magnitude spectra: (g) the square image of Figure 2.13(a) rotated by 45° and (h) its centered magnitude spectrum.

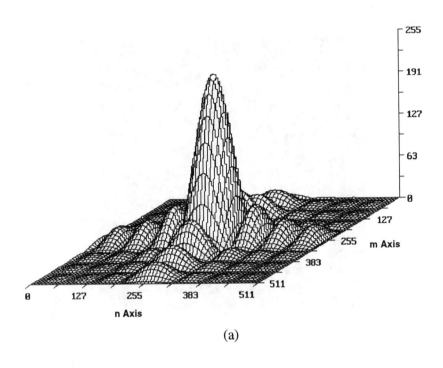

(a)

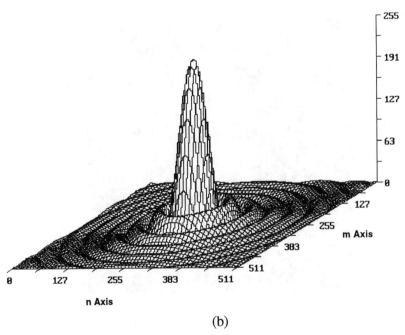

(b)

Figure 2.14: 4 times expanded three-dimensional plots of the Fourier
spectra for (a) the square image given in Figure
2.13(a) and (b) the circular image given in Figure 2.13(c).

Figure 2.13(e) is an image of a horizontal line of length 448 pixels and width of 32 pixels. Figure 2.13(f) is its corresponding magnitude spectrum. Note the predominant sidelobes in the vertical direction due to the narrow width of this line. The spacing between the vertical sidelobes corresponds directly to one over the width of the line. Likewise, the small spacing between the sidelobes in the horizontal direction is due to the long length of this line. If the line were shortened, the spacing between these sidelobes would increase. Figure 2.13(g) is the last example showing the image of the square object rotated clockwise by 45°. Figure 2.13(h) is its magnitude spectrum, showing that it has also been rotated clockwise by 45°. Figures 2.13(g) and (h) show that the rotational property of the DFT holds.

2.7 Computation of the Discrete Fourier Transform

The direct computation of the two-dimensional DFT and its inverse from Equations (2.45) and (2.46) is very inefficient. A better method is to use the fact that the DFT can be separated into a set of one-dimensional DFTs. Because the two-dimensional DFT can be separated into a set of one-dimensional transforms, this transform is known as a *separable transform*. Using a set of one-dimensional DFTs enables the two-dimensional DFT to be computed using the efficient fast Fourier transform algorithm. Consider the two-dimensional DFT as defined by Equation (2.45) for a square $N \times N$ image:

$$F(n, m) = \frac{1}{N^2} \sum_{y=0}^{N-1} \sum_{x=0}^{N-1} f(x, y) \cdot K(x, n, y, m) , \qquad (2.67)$$

where the exponential kernel is defined as

$$K(x, n, y, m) = e^{-j2\pi(xn + ym)/N} . \qquad (2.68)$$

Equation (2.67) is the general definition of a two-dimensional $N \times N$ discrete transform, with the type of transform depending on the kernel $K(x, n, y, m)$.

By definition, a transform is separable if its kernel can be separated into two kernels, one in each direction:

$$K(x, n, y, m) = K_1(x, n) \cdot K_2(y, m) . \qquad (2.69)$$

Also, if $K_1(x, n) = K_2(y, m)$, the discrete transform is also called a symmetric transform. For the discrete Fourier transform

$$K_1(x, n) = e^{-j2\pi(xn)/N} \qquad (2.70)$$

and

$$K_2(y, m) = e^{-j2\pi(ym)/N} \qquad . \qquad (2.71)$$

The DFT is a separable and symmetric transform. Equation (2.67) can now be rewritten using Equations (2.70) and (2.71):

$$F(n, m) = \frac{1}{N} \sum_{y=0}^{N-1} \left\{ \frac{1}{N} \sum_{x=0}^{N-1} f(x, y) K_1(x, n) \right\} K_2(y, m) \quad . \qquad (2.72)$$

Equation (2.72) is simply two sets of N one-dimensional transforms, first in the x direction and then in the y direction.

Another way of representing Equation (2.72) is to expand this equation into two separate one-dimensional transforms:

$$Q(n, y) = \frac{1}{N} \sum_{x=0}^{N-1} f(x, y) K_1(x, n) \qquad (2.73)$$

and

$$F(n, m) = \frac{1}{N} \sum_{y=0}^{N-1} Q(n, y) K_2(y, m) \quad . \qquad (2.74)$$

Figure 2.15 shows graphically the computation of the two-dimensional DFT. First, N one-dimensional transforms are taken along the x direction starting with $y = 0$ and continuing until $y = N - 1$. Then the process is repeated, processing each of the columns of the previous DFT calculation $Q(n, y)$ for $n = 0$ to $n = N - 1$. In this way, $2N$ one-dimensional DFTs of length N are computed.

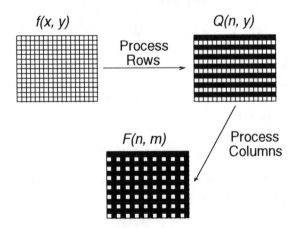

Figure 2.15: Graphical sketch showing the computation of the two-dimensional DFT.

The fast Fourier transform (FFT) algorithm was developed by J. W. Cooley and J. W Tukey in 1965 as a fast means of implementing the discrete Fourier transform. As with the DFT, the FFT algorithm also requires the storage and manipulation of complex numbers. Inspection of Equation (2.73) reveals that there are on the order of N^2 complex additions and multiplications required to compute an N point one-dimensional DFT. The FFT algorithm reduces the number of calculations required to $N \log_2 N$. For a typical 512×512 image, this results in a reduction of computation time by a factor of 57. For a 2048×2048 image, it can mean as much as a 186 to 1 improvement in the time required to compute the DFT.

The improvement in the number of calculations required to compute the DFT using the FFT algorithm comes about from taking advantage of the periodicity of the sine and cosine function used by the DFT kernel. The approach taken by Cooley and Tukey was to divide an N point DFT into two sets of $N/2$ point DFTs. The output from this step is then further divided into four $N/4$ DFTs. The process continues until there are $N/2$ two point DFTs. Since this approach uses a successive division by two to compute the DFT using the FFT algorithm, the total sample size of the input function must be a power of 2. Hence, FFT sizes are limited in size to 2^b, which yield sizes such as 2, 4, 16, 32, 64, 128, 256, 512, 1024, and 2048. Common image sizes are 256×256, 512×512, 1024×1024, 2048×2048, and 4096×4096.

Its not uncommon in electronic image processing to have one of the image dimensions not be a power of 2. In this case, the image size is expanded to the next power of 2 by the process of padding. Padding fills in the new pixels with a set constant graylevel value. In many cases, these pixels are filled with a graylevel of 0. This should be avoided in that it can add an artificial step to the image due to the discontinuity of the original image boundary with the zero padded pixels. Adding an artificial step to the image produces ringing/sideline artifacts in the Fourier spectral components that are not present in the original image's Fourier spectral components. A better approach is to fill the new pixels with the graylevel of the boundary pixels of the original image to reduce the possibility of adding an artificial step. There are many reference books available that provide programming code that implements the one-dimensional FFT. Several of these are given in the bibliography section for this chapter. The use of the separable property of the DFT as presented in this section to separate an $N \times N$ two-dimensional DFT into $2N$ one-dimensional DFTs along with the chosen one-dimensional FFT algorithm is all that is needed to compute the Fourier frequency components of an image.

2.8 Other Image Transforms

As stated earlier in this chapter the Fourier transform decomposes an image into a set of sine and cosine functions. The kernel that is used in Equation (2.67) determines the type of transform that is implemented. In many instances of electronic imaging, it is not required to decompose an image into a set of sine and cosine functions. In the area of image compression it is desired to transform an image into another domain that reduces the number of elements required to represent an image. Consider an image of a constant graylevel K that must be compressed. Taking the DFT of this image produces Fourier coefficients that are zero, except at $n = 0$ and $m = 0$. A simple image compression algorithm can be implemented that stores the locations of only the nonzero Fourier components and their values. In the constant image example, this would require the storage of three values $n = 0$, $m = 0$, and F(0, 0). A more practical example, using the fact that the eye has a limited spatial resolution, is the elimination of high frequency components during the compression process. Though this leads to some degradation in the compressed image, this is the concept behind lossy compression. The difficulty in using the DFT for compression is the computational complexity required by the DFT due to the calculations of the sine and cosine functions. A better transform is one that does not require the use of transcendental functions like the sine and cosine functions.

The *Walsh transform* uses a kernel that produces a sequence of plus and minus one square waves. This transform is much more computationally efficient than the DFT since it does not require any floating point math or transcendental functions. In addition, the Walsh transform can be computed using the FFT algorithm by modifying the sine and cosine kernel used by the FFT to that of the Walsh transform. The Walsh transform in two dimensions is defined for a square image of size $N \times N$ as

$$F(n, m) = \frac{1}{N} \sum_{y=0}^{N-1} \sum_{x=0}^{N-1} f(x, y)$$

$$\cdot \prod_{i=0}^{q-1} (-1)^{b_i(x)\, b_{q-1-i}(n)\, +\, b_i(y)\, b_{q-1-i}(m)} , \qquad (2.75)$$

and its inverse

$$f(x, y) = \frac{1}{N} \sum_{m=0}^{N-1} \sum_{n=0}^{N-1} F(n, m)$$

$$\cdot \prod_{i=0}^{q-1} (-1)^{b_i(x)\, b_{q-1-i}(n)\, +\, b_i(y)\, b_{q-1-i}(m)} , \qquad (2.76)$$

where $b_i(x)$ is a binary bit representation. For example, if the total number of bits (q) equals 4 (hence, $N = 16$) and x equals 7 (0111) then $b_0(7) = 1$, $b_1(7) = 1$, $b_2(7) = 1$, and $b_3(7) = 0$.

A close inspection of Equations (2.75) and (2.76) reveals that the forward and inverse kernels are identical and that the Walsh transform is a symmetric and separable transform. Instead of obtaining spatial frequency components as with the two-dimensional DFT, the Walsh transform components are referred to as sequency components. Figure 2.16 shows the two-dimensional Walsh kernel for $q = 3$ and $N = M = 8$. Given in the horizontal direction are the eight x components and in the vertical direction are the eight y components. Each square represents one unique combination of x and y values. Within each of the 64 squares, the n and m parameters are varied from 0 to 7, with n given in the horizontal direction and m given in the vertical direction. The minus one values of the Walsh kernel have been scaled to black, while the plus one values are given by white. Figure 2.16 shows how the square waves of the Walsh kernel vary as a function of the parameters x, y, n, and m. Note how the rate in which these square waves varies increases as the x and y values increase.

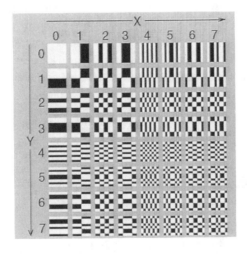

Figure 2.16: The Walsh kernel for $q = 3$ and $N = M = 8$.

A transform similar to the Walsh transform is the *Hadamard transform.* This transform also decomposes an image into plus and minus one square waves. In fact, because of its similarity to the Walsh transform, both are commonly referred to in the literature as the Hadamard-Walsh transform. As defined by Gonzalez and Woods, the Hadamard transform defined in two dimensions for an $N \times N$ image as

$$F(n, m) = \frac{1}{N} \sum_{y=0}^{N-1} \sum_{x=0}^{N-1} f(x, y) \cdot (-1)^{\sum_{i=0}^{q-1} b_i(x) \, a_i(n) + b_i(y) \, a_i(m)} \qquad (2.77)$$

and its inverse

$$f(x, y) = \frac{1}{N} \sum_{m=0}^{N-1} \sum_{n=0}^{N-1} F(n, m) \cdot (-1)^{\sum_{i=0}^{q-1} b_i(x) \, a_i(n) + b_i(y) \, a_i(m)} , \qquad (2.78)$$

where $b_i(x)$ is the binary bit representation and the term $a_i(n)$ is given by

$$\begin{aligned} a_0(n) &= b_{q-1}(n), \\ a_1(n) &= b_{q-1}(n) + b_{q-2}(n) \\ &\vdots \\ a_{q-1}(n) &= b_1(n) + b_0(n) . \end{aligned} \qquad (2.79)$$

Another method of computing the Hadamard transform is to recognize that the transform can be implemented using two-dimensional matrix multiplication, with a transformation matrix defined through a recursion relationship. Given a two-dimensional image $f(x, y)$ of size $N \times N$, it can be written in matrix form as $\mathbf{f}$. Next, writing the Hadamard transformation matrix (kernel) as $\mathbf{H}$, the Hadamard transform of the image $\mathbf{f}$ is defined as $\mathbf{F} = \mathbf{HfH}$. The Hadamard transformation matrix of size N is found starting with the lowest order,

$$\mathbf{G}_2 = \begin{bmatrix} 1 & 1 \\ 1 & -1 \end{bmatrix} , \qquad (2.80)$$

and then using the recursion

$$\mathbf{G}_{2N} = \begin{bmatrix} \mathbf{G}_N & \mathbf{G}_N \\ \mathbf{G}_N & -\mathbf{G}_N \end{bmatrix} , \qquad (2.81)$$

with $\mathbf{H}_N = \mathbf{G}_N / \sqrt{N}$. For example, for $N = 4$ and 8, the Hadamard kernels are

$$\mathbf{H}_4 = \frac{1}{\sqrt{4}} \begin{bmatrix} 1 & 1 & 1 & 1 \\ 1 & -1 & 1 & -1 \\ 1 & 1 & -1 & -1 \\ 1 & -1 & -1 & 1 \end{bmatrix} \qquad \mathbf{H}_8 = \frac{1}{\sqrt{8}} \begin{bmatrix} 1 & 1 & 1 & 1 & 1 & 1 & 1 & 1 \\ 1 & -1 & 1 & -1 & 1 & -1 & 1 & -1 \\ 1 & 1 & -1 & -1 & 1 & 1 & -1 & -1 \\ 1 & -1 & -1 & 1 & 1 & -1 & -1 & 1 \\ 1 & 1 & 1 & 1 & -1 & -1 & -1 & -1 \\ 1 & -1 & 1 & -1 & -1 & 1 & -1 & 1 \\ 1 & 1 & -1 & -1 & -1 & -1 & 1 & 1 \\ 1 & -1 & -1 & 1 & -1 & 1 & 1 & -1 \end{bmatrix} . \qquad (2.82)$$

Figure 2.17 shows the Hadamard kernel for $q = 3$ and $N = M = 8$. This figure was generated in the same manner as the Walsh kernel of Figure 2.16, with the square waves varying as a function of the parameters x, y, n, and m. Note the slight differences in which the square waves vary between the two kernels. This illustrates the difference between the Walsh and Hadamard kernels.

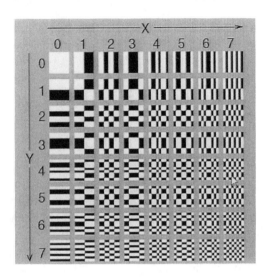

Figure 2.17: The Hadamard kernel for $q = 3$ and $N = M = 8$.

One of the best transforms for use in compressing an image is the *Karhunen-Loève transform* (KLT). Unfortunately, this transform depends on the spatial content in an image and requires calculation of the eigenvalues/eigenvectors of the autocorrelation matrix of an image, which is very computationally intensive. Another transform that has similar compression characteristics is the *discrete cosine transform* (DCT). Like the Walsh transform, it has been easily adapted to the FFT algorithm, resulting in a fast computer implementation. In addition, the DCT does not have the same windowing problems that are associated with the DFT of a finite sized image. Because of these reasons, the DCT is the most popular transform used in the area of image compression. It is the coding method chosen by the *JPEG* (Joint Photographic Experts Group) as its image compression standard. Unlike the DFT, the DCT uses only the cosine function for its kernel. For an $N \times M$ image, the DCT is defined in two dimensions as

$$C(n, m) = k_1(n)\, k_2(m) \sum_{y = 0}^{M - 1} \sum_{x = 0}^{N - 1} f(x, y) \cdot \cos\left(\pi n \frac{x + 1/2}{N}\right) \cdot \cos\left(\pi m \frac{y + 1/2}{M}\right),$$

$$(2.83)$$

where n is defined from 0 to $N - 1$, m is defined from 0 to $M - 1$, and

$$k_1(n) = \begin{cases} \sqrt{\dfrac{1}{N}} & \text{for } n = 0 \\[2mm] \sqrt{\dfrac{2}{N}} & \text{otherwise} \end{cases} \qquad k_2(m) = \begin{cases} \sqrt{\dfrac{1}{M}} & \text{for } m = 0 \\[2mm] \sqrt{\dfrac{2}{M}} & \text{otherwise} \end{cases} \qquad . \qquad (2.84)$$

The inverse discrete cosine transform, which uses the same kernel as the forward DCT, is defined as

$$f(x, y) = \sum_{m=0}^{M-1} \sum_{n=0}^{N-1} C(n, m)\, k_1(n) \cdot k_2(m) \cdot \cos\left(\pi n\, \frac{x + 1/2}{N}\right) \cdot \cos\left(\pi m\, \frac{y + 1/2}{M}\right)$$

$$(2.85)$$

The DCT has the added advantage of mirror symmetry, making it suitable for block encoding of an image (the processing of an image via small non-overlapping sub-images). Because of the symmetry property of the DCT, it produces less degradation at each of the sub-image boundaries than the DFT. Consider an $N \times N$ image divided into a set of smaller $m \times m$ sub-images. The DCT is computed for each of the sub-images. Within each sub-image, only the DCT components that are non-negligible are retained, providing a means of image compression. All other DCT components are assumed to be zero. The remaining nonnegligible DCT components are then used to reconstruct the image. Figure 2.18 gives the kernel for the DCT for $N = M = 8$. Black represents the minimum value of the kernel, while white gives the maximum value. Unlike the Walsh and Hadamard kernels shown in Figures 2.16 and 2.17, the DCT kernel is a continuously varying function. This is observed in Figure 2.18 as the continuous change in graylevel within each of the 64 squares.

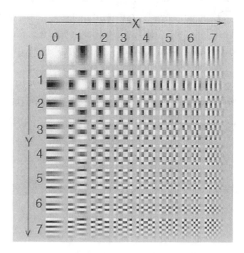

Figure 2.18: The DCT kernel for $N = M = 8$.

The final two transforms that will be discussed in this section operate on objects within an image. Both of these transforms require that the graylevels of an image be segmented into two regions, one describing the background pixels and the other describing the object or objects present in the image. A simple approach is to threshold the graylevel image such that graylevels below a set threshold value are considered background pixels and graylevel pixels above the threshold value are considered object pixels. The concepts of thresholding to perform image segmentation will be discussed further in Chapter 8.

The *Hough transform*, in its simplest definition, finds straight contours within an image. The concept behind the Hough transform is to transform all pixels on a straight line to a single point in the Hough space. A straight line can be described using polar coordinates by the parametric equation

$$\rho = x \cos \theta + y \sin \theta \quad . \tag{2.86}$$

Figure 2.19 shows the relationship of ρ and θ to the description of a straight line. The parameter ρ is the magnitude of the normal vector from the origin to the line and the parameter θ is the angle between the normal vector and the x axis. For example, for $\rho = K$ and $\theta = 90°$, Equation (2.86) reduces to $y = K$, which is a horizontal line. The advantage of using Equation (2.86) over the standard equation for a straight line $y = mx + b$ is that both ρ and θ are finite in value, unlike the Cartesian coordinate description, which requires infinite values for the slope m to describe vertical lines.

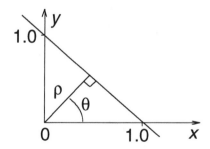

Figure 2.19: A description of a straight line use polar coordinates

Careful inspection of Equation (2.86) shows that all points on the same line have the same ρ and θ value. The Hough transform is the mapping from the Cartesian coordinate space of x and y to the ρ and θ space. All pixels (points) on a straight line map to the same ρ and θ value as shown in Figure 2.20. To implement the Hough transform, a binary or two graylevel image is scanned pixel by pixel, locating the x and y coordinate locations of each contour pixel, each of which is then substituted into Equation (2.86) to produce an equation that relates ρ to θ. There will be a location within the ρ and θ space at which several of these curves will intersect, corresponding to the ρ and θ value of the line as described by Equation (2.86), and the number of lines that intersect

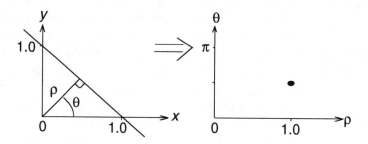

Figure 2.20: An example of the Hough transform.

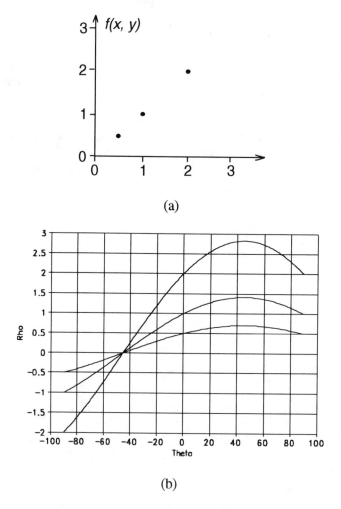

(a)

(b)

Figure 2.21: An example of three collinear points and their mapping in the
Hough space: (a) the original three points defining the line $\rho = 0$
and $\theta = -45°$ and (b) the corresponding Hough transform.

corresponds to the number of points on the line. Figure 2.21(a) shows a plot of three collinear points at the locations (0.5, 0.5), (1.0, 1.0), and (2.0, 2.0). This corresponds to the straight line of $\rho = 0$ and $\theta = -45°$. The three equations describing the Hough transform curves are

$$\rho = 0.5 \cos \theta + 0.5 \sin \theta$$
$$\rho = 1.0 \cos \theta + 1.0 \sin \theta$$
$$\rho = 2.0 \cos \theta + 2.0 \sin \theta \quad . \tag{2.87}$$

Figure 2.21(b) shows these three curves plotted in the Hough space. All three curves intersect at $\rho = 0$ and $\theta = -45°$, which are the ρ and θ values that describe the line given in Figure 2.21(a). Since the combination of any two points can form a straight line (in other words any two pixels within an image define a straight line), to effectively use the Hough transform to find the location of straight lines in a binary image, the minimum length of a line is three pixels. The first step in computing the Hough transform of a sampled image is to generate a two-dimensional array that is initialized to zero, with the indices of the array equal to ρ and θ. For every contour pixel within the image, Equation (2.86) is computed for a selected range of ρ and θ values. Each cell in the array is incremented by one for each occurrence of ρ and θ. The output of the Hough transform is a two-dimensional array, with the values stored at each location giving the number of pixels contained on a straight line and the cell location within the array giving the ρ and θ values describing the line.

Figure 2.22(a) is a graylevel 512×512 image of the University of Central Florida water tower taken from one of the school's parking lots. The goal is to extract only the two vertical lines describing the contour of the water tower using the Hough transform. The first step is to edge detect the original graylevel image using one of the many edge detectors that are available. Because of its simplicity, the *Sobel edge detector* was used here. Edge detection methods will be further discussed in Chapter 8 as a means of segmenting objects within an image. Figure 2.22(b) shows the output from the edge detection operation. This image is a graylevel image, with the graylevel values proportional to the slopes of the edges present in Figure 2.22(a). Since the Hough transform is applicable only to binary images, this Sobel edge image must be thresholded such that edges with graylevels above a threshold value K are preserved and the other edges are removed. The binary edge image of Figure 2.22(c) was generated using a threshold value of 128. This threshold value preserved most of the edges that were present in Figure 2.22(b).

The edge image contains many vertical and horizontal edges due to the cars and the building present in the bottom portion of the image. Application of the Hough transform produces the Hough space image of Figure 2.22(d). This image was generated by scaling the minimum and maximum values stored in the ρ and θ array between the graylevels of black and white. Longer lines obtain a lighter graylevel as compared to short lines, which appear dark. The horizontal

(a)

(b)

Figure 2.22: An example of finding straight contours in an image:
(a) the original graylevel image and (b) the Sobel edge
detected image.

(c)

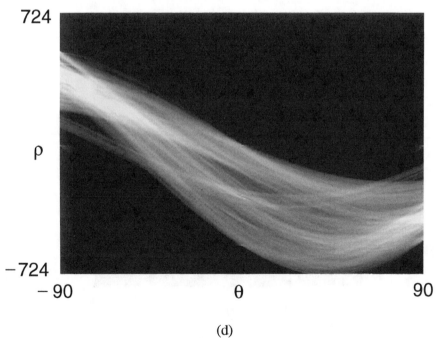

(d)

Figure 2.22: An example of finding straight contours in an image:
(c) the thresholded edge image and (d) the Hough
transform image.

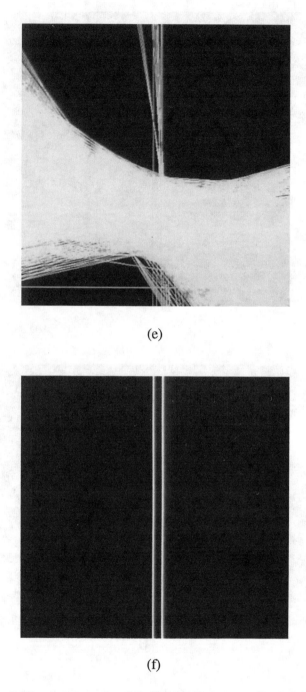

(e)

(f)

Figure 2.22: An example of finding straight contours in an image:
              (e) the inverse of the Hough transform only generating
              lines with at least 128 pixels, and (f) the inverse of the
              Hough transform with the constraint of (e) and $\theta = 0°$.

(g)

Figure 2.22: An example of finding straight contours in an image:
(g) the original graylevel image overlaying the vertical
lines of the water tower found via the Hough transform.

axis gives the change in θ ranging in value from $-90°$ to $+90°$ left to right. The vertical axis gives the change in ρ values ranging from 724 to -724 top to bottom. Since the original image is a 512×512 image, the maximum value that ρ can take on is $\pm\sqrt{2} \cdot 512$ or ±724. Any location within this Hough space image describes a particular set of ρ and θ values. Note the presence of many straight lines near $-90°$ and $+90°$, shown in Figure 2.22(d) as the white graylevel areas. These straight lines are due to the many horizontal lines that were found after edge detection of the parking lot portion of the image. At $\theta = 0$ (located along the horizontal center of the image), several white regions are observable, which represent long vertical lines within the image. The Hough transform, Equation (2.86), can now be used to reconstruct the straight lines. A prior knowledge of the original graylevel image provides the information that the two vertical lines of the water tower are some of the longest lines present in the image. To eliminate short lines of no interest, only lines with lengths greater than 128 will be reconstructed using Equation (2.86).

Figure 2.22(e) gives the reconstruction of all straight lines with lengths greater than 128 pixels. Note the predominance of horizontal lines near the bottom of the image due to the clutter from the parking lot. Also observable in this image are the lines due to the water tower. Since the goal was to find the edges of the water tower, limiting the reconstruction of straight lines from the Hough space to those with $\theta = 0°$ further eliminates undesired lines. Figure 2.22(f) gives the output reconstructed line image with a length threshold value of 128 pixels and $\theta = 0°$. All lines have been eliminated except the two vertical lines describing the edges of the water tower. To make these two lines in Figure 2.22(f) more observable, they were generated four pixels wide. Figure 2.22(g) shows the two vertical lines of the water tower overlaid onto the original graylevel image. The concepts presented here for the Hough transform can be easily expanded to handle any type of geometrical object. For example, instead of the equation for a line, an equation for a circle could be used, generating a three-dimensional Hough transform containing the parameters of the radius of the circle and the x and y center coordinates of the circle.

The final image transform presented is the *Hotelling* or *Karhunen-Loève transform*. There are many uses of this transform, ranging from image compression to pattern recognition. The application that is presented here is the automatic rotation of an object within an image based upon locating the object's major axis. This transform is based upon the eigenvalue decomposition of the covariance matrix of an image, providing a method of determining the orientation of the major axis of an object within an image. Once the orientation of the major axis is found, the object can then be automatically rotated in a predefined direction. Figure 2.23(a) shows the major axis of an object rotated an angle θ about the x axis. Figure 2.23(b) shows how this angle can be used to rotate an object's major axis in the direction of the x axis. Typical applications of the Hotelling transform include airplane silhouette and handwritten character recognition. Automatic rotation of an object prior to object recognition can make the recognition problem easier.

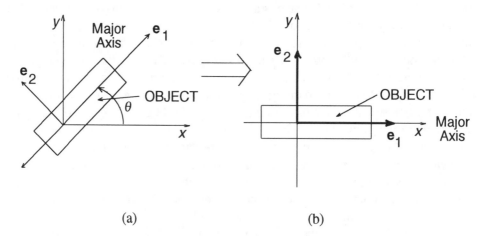

(a)   (b)

Figure 2.23: An example illustrating the rotation capability of the
             Hotelling transform: (a) the original object orientation
             and (b) after implementation of the Hotelling transform.

Let X represent a column matrix consisting of all of the object pixels' x and y coordinates within a binary image $f(x, y)$:

$$\mathbf{X} = \begin{bmatrix} \mathbf{x}_1 \\ \mathbf{x}_2 \\ \dots \\ \mathbf{x}_q \end{bmatrix} , \tag{2.88}$$

where q is the total number of pixels within the object and $\mathbf{x}_i = (x_i, y_i)$ is the coordinate position of the *ith* object pixel. The Hotelling transform requires the computation of both the centroid and covariance of the object. The centroid, or geometrical mean, M is defined as

$$\mathbf{M} = \begin{bmatrix} M_x \\ M_y \end{bmatrix} , \tag{2.89}$$

where

$$M_x = \frac{1}{q} \sum_{i=1}^{q} x_i \tag{2.90}$$

and

$$M_y = \frac{1}{q} \sum_{i=1}^{q} y_i . \tag{2.91}$$

Equations (2.90) and (2.91) are the average of all the object pixels' x and y coordinates. The covariance matrix of the image is found using

$$\mathbf{C} = \begin{bmatrix} C_1 & C_2 \\ C_3 & C_4 \end{bmatrix} , \tag{2.92}$$

where

$$C_1 = \frac{1}{q} \sum_{i=1}^{q} x_i^2 - M_x^2 \qquad C_2 = \frac{1}{q} \sum_{i=1}^{q} x_i y_i - M_x M_y$$

$$C_3 = \frac{1}{q} \sum_{i=1}^{q} x_i y_i - M_x M_y \qquad C_4 = \frac{1}{q} \sum_{i=1}^{q} y_i^2 - M_y^2 . \tag{2.93}$$

The parameter C_1 gives the variance or geometrical size of the object in the x direction, C_4 gives the variance of the object in the y direction, and C_2 and C_3 give the cross correlation of the object's orientation relative to the x and y directions. The goal of the Hotelling transform is to rotate the object's orientation so that the new correlation matrix becomes a diagonal matrix:

$$\mathbf{C} = \begin{bmatrix} A_1 & 0 \\ 0 & A_4 \end{bmatrix} \, , \qquad (2.94)$$

where A_1 and A_4 are the new object's sizes in the x and y directions after the object has been rotated. The Hotelling transform requires the computation of the eigenvalues of the covariance matrix of the unrotated object as defined by Equation (2.92). The eigenvector corresponding to the largest eigenvalue gives the major axis (its vector) of the object. The largest eigenvalue is given by

$$\lambda_{\max} = \frac{(C_1 + C_4) + \sqrt{(C_1 + C_4)^2 - 4(C_1 C_4 - C_2 C_3)}}{2} \, , \qquad (2.95)$$

and its corresponding eigenvector is

$$\mathbf{e}_1 = \begin{bmatrix} 1 \\ \dfrac{C_1 - \lambda_{\max}}{C_2} \end{bmatrix} . \qquad (2.96)$$

The angle the major axis makes relative to the x axis is defined as

$$\theta = \mathrm{atan} \left[\frac{C_1 - \lambda_{\max}}{C_2} \right] . \qquad (2.97)$$

Given θ, the final task of the Hotelling transform is to rotate the object by the inverse of this angle, $-\theta$. Equation (2.98) gives the translation and rotation transformations required to rotate an object's major axis in line with the x axis:

$$\begin{bmatrix} x_{new} \\ y_{new} \end{bmatrix} = \begin{bmatrix} \cos\theta & \sin\theta \\ -\sin\theta & \cos\theta \end{bmatrix} \left\{ \begin{bmatrix} x_{old} \\ y_{old} \end{bmatrix} - \begin{bmatrix} M_x \\ M_y \end{bmatrix} \right\} + \begin{bmatrix} x_o \\ y_o \end{bmatrix} \, , \qquad (2.98)$$

where M_x and M_y are given by Equations (2.90) and (2.91), θ is given by Equation (2.97), and x_o, y_o is the desired x, y coordinate position of the center of the object after the Hotelling transform.

The difficulty of using Equation (2.98) directly on a sampled image is that it misses connect pixels within the object. This is best illustrated by the following example. Consider the mapping $x_{new} = 2 \cdot x_{old}$ and $y_{new} = y_{old}$ that doubles the size of the object in the *x* direction. Going from the old coordinate system to the new coordinate system skips pixels within the new image. Let the original object contain two adjacent pixels at the coordinates 1,*A* and 2,*A*. Using the above mapping produces the two new pixel locations 2,*A* and 4,*A*, skipping the pixel located at 3,*A*. A better approach is to scan all pixels in the new image, mapping them to the old coordinate system and setting the pixels in the new image to object pixels if they map to object pixels in the old coordinate system. In the above example, the pixel at 3,*A* maps to 1.5,*A*. Since only integer values are allowed for a pixel's position, this mapping reduces to 1,*A*, which is an object pixel. Hence, this pixel is set to an object pixel. For more elaborate interpolation methods, such as cubic-spline techniques, the reader is referred to Section 7.1 of this text. The interpolation technique presented here works quite well with binary images and is very simple to implement.

To properly map the translation and rotation transformations given in Equation (2.98) requires solving this equation in terms of x_{old} and y_{old}:

$$x_{old} = [x_{new} \cos\theta - y_{new} \sin\theta - x_o \cos\theta + y_o \sin\theta] + M_x$$
$$y_{old} = [x_{new} \sin\theta + y_{new} \cos\theta - x_o \sin\theta - y_o \cos\theta] + M_y. \qquad (2.99)$$

The procedure just mentioned (scanning the new image pixel by pixel and determining whether the new pixels map to the object pixels in the original image) is then used to determine which pixels within the new image are object pixels.

Figure 2.24 is a binary image of the name Santa Claus handwritten on a white piece of paper, which was scanned using a desktop scanner. The negative of this binary image was then obtained for illustration purposes, producing white text against a black background. Unfortunately, the paper was placed in the scanner incorrectly, producing an incorrectly oriented image. The Hotelling transform was then used to rotate the handwritten name, aligning it to the horizontal axis. As a final step, this name was then located in the center of the image for viewing. The following parameters were computed during the implementation of the Hotelling transform:

The total number of object pixels that were found was 6355, the location of the centroid was calculated as

$$\mathbf{M} = \begin{bmatrix} 199.26 \\ 114.75 \end{bmatrix}. \qquad (2.100)$$

Using Equation (2.93), the four elements of the covariance matrix were

$$\mathbf{C} = \begin{bmatrix} 9472 & 5304 \\ 5304 & 3337 \end{bmatrix}. \qquad (2.101)$$

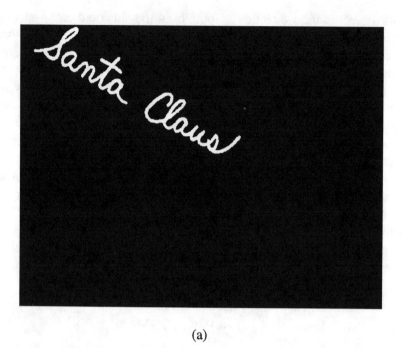

(a)

(b)

Figure 2.24: An example of using the Hotelling transform: (a) the
original rotated signature and (b) the output from the
Hotelling transform.

Solving for the maximum eigenvalue of the covariance matrix using Equation (2.95) yields $\lambda_{max} = 12531.7$ and its corresponding eigenvector,

$$\mathbf{e}_1 = \begin{bmatrix} 1 \\ -0.5769 \end{bmatrix} . \tag{2.102}$$

Using Equation (2.96) to compute the angle of the major axis yields

$$\theta = \tan^{-1}[-0.5769] = -29.97° . \tag{2.103}$$

Figure 2.24(b) shows the output image after the Hotelling transformation, with the name now correctly rotated and centered. The example shown in Figure 2.24 illustrates the effectiveness of the Hotelling transform in aligning objects within an image to a desired orientation. This transform plays a key role in the automatic recognition of handwritten characters. Prior to classification, the Hotelling transform is used to orient the handwritten characters in the predefined orientation. This simplifies the classification process by removing the effect of rotation. Another area in which the Hotelling transform is used is in the rotation of airplane silhouettes to a fixed orientation prior to classification.

In this chapter, several important image transforms were given. The use and explanation of the discrete Fourier transform was given as well as the effect of sampling on an image's Fourier spectral components. The use of the DFT in image enhancement and restoration will be presented in Chapter 4 in the discussion of Fourier filtering techniques. Many other image transforms are available that are beyond the scope of this text. The interested reader may want to refer to several of the imaging texts given in the bibliography. In particular, *Digital Image Processing* by Gonzalez and Woods does an excellent job of discussing several other image transforms that are used within electronic image processing.

CHAPTER 3

Image Enhancement by Point Operations

The goal of image enhancement is to improve the overall quality of an image for use in an autonomous pattern recognition system or for viewing by the human visual system. The first part of this chapter discusses the enhancement of an image by the direct manipulation of its graylevels using a constant or a set of graylevels from a second image. The latter part provides additional image processing methods based upon the histogram of an image to automatically improve the image's brightness and contrast.

3.1 An Overview of Point Processing

Image enhancement can be accomplished in either the spatial domain or in the frequency domain using the discrete Fourier transform, as discussed in Chapter 2. Image enhancement in the spatial domain provides the benefit of being less extensive computationally since the DFT does not have to be computed. Because of the computational simplicity of processing an image in the spatial domain, many image processing algorithms are easily implemented in specialized digital hardware, providing real-time electronic image processing. Pointwise operations can be separated into the modification of a pixel's graylevel by a constant, by algebraic or Boolean methods, and by nonlinear or histogram techniques. Each of these operations will be discussed in detail in this chapter. Pointwise operations provide a means of making an image lighter or darker, enhancing features that would not be otherwise observable. For example, consider an image that was taken in poor lighting, producing an overall dark image. This image can be easily lightened by using one of the constant pointwise operations. Many electronic imaging systems require a one-to-one linear mapping between the input intensity image on an electronic camera and the graylevels of the digital image. Unfortunately, many camera systems produce an electronic voltage that is nonlinearly related to the input image intensities. Nonlinear pointwise operations can easily remove the effects of camera nonlinearity.

The manipulation of an image's graylevels modifies the overall perception of the image, changing its *brightness* and *contrast*. The brightness level of an image is its perceived overall darkness or lightness, while the contrast of an image defines the range of graylevel variations within an image. Typically, zero or low graylevel values are assigned to dark intensities, while high graylevel values are assigned to bright intensity levels. As an example, for a 256 graylevel image, in which 0 is assigned to black and 255 to white, an average graylevel of 200 implies a bright image, while an average graylevel of 40 implies a dark image. The perceived contrast within an image is the amount of change in graylevels throughout the image. An image containing a minimum graylevel of 190 and a maximum of 200 is a bright image with a low contrast. Likewise, an image with a minimum graylevel of 40 and a maximum of 220 has a medium brightness and a high contrast.

The brightness of an image is defined as the average of all the pixels within the image and for an $N \times M$ image $f(x, y)$ is

$$\text{brightness} = B = \frac{1}{NM} \sum_{y=0}^{M-1} \sum_{x=0}^{N-1} f(x, y) \ . \tag{3.1}$$

Figures 3.1(a) and (b) show two versions of an image of the lunar module located at the Kennedy Space Center. The low brightness image of Figure 3.1(a) shows the difficulty of observing features in a dark image. Details of the lunar module are barely visible. Figure 3.1(b) is a high brightness version of the image containing the same contrast as the low brightness version. Note how the detail features of the lunar module are now clearly visible. In particular, the antennas on the top of the lunar module and the landing pads on the bottom are now observable.

The average variation in graylevels within an image is a measure of an image's perceived contrast. For an $N \times M$ image, one method of defining the contrast is

$$\text{contrast} = C = \sqrt{\frac{1}{NM} \sum_{y=0}^{M-1} \sum_{x=0}^{N-1} [f(x, y) - B]^2} \ . \tag{3.2}$$

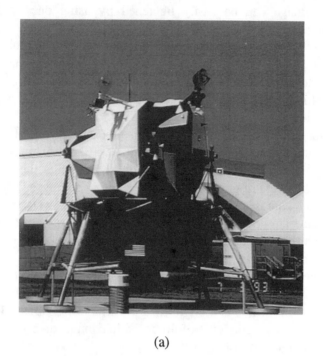

(a)

(b)

Figure 3.1: Examples of (a) low and (b) high brightness images.

Equation (3.2) gives the average graylevel variations within an image and, in fact, is simply the standard deviation of the graylevel variations within an image. Figures 3.2(a) and (b) show two different contrasts of an image of a rocket engine designed for the Apollo space program. Both images contain the same average brightness of 125. The details in the top of the rocket engine are barely visible in the low contrast image but are clearly observable with the high contrast. Application of Equation (3.2) yields a contrast value of 38 for the low contrast image. On the average, there is a difference of only 38 graylevels from the minimum to the maximum graylevels found in this image. The high contrast image shown in Figure 3.2(b), on the other hand, has a contrast value of 82. On the average, this image has a higher graylevel variation than the low contrast image of Figure 3.2(a).

Another common definition used for contrast is

$$C = \frac{\text{maximum}[f(x, y)] - \text{mininum}[f(x, y)]}{\text{maximum}[f(x, y)] + \text{mininum}[f(x, y)]} \quad \text{for } 0 \leq x < N, \; 0 \leq y < M \quad (3.3)$$

The denominator of Equation (3.3) normalizes the contrast value between zero and one. A zero contrast image corresponds to the constant graylevel image and a contrast value of one corresponds to a high contrast image with the graylevel variations taking on the maximum and minimum graylevels possible. The only difficulty in using Equation (3.3) for the definition of contrast is that it does not always match the perceived contrast within an image. For example, consider an $N \times M$ 256 graylevel image with one pixel at graylevel 0, another at graylevel 255 and all other pixels at graylevel 128. Computation of Equation (3.3) would yield a contrast value of one. Yet this image would appear as a zero contrast image with two noise pixels. For most images, in which there are many pixels within the image containing graylevels near the maximum and minimum graylevels of the image, the contrast measure given in Equation (3.3) matches the perceived contrast of the image as given by Equation (3.2).

3.2 Constant and Nonlinear Operations

The ability to manipulate the brightness and contrast of an image is one of the many image enhancement techniques that are available. The importance of contrast and brightness in the observation of features within an image was illustrated in Figures 3.1 and 3.2. By manipulating the graylevels within an image, both the brightness and contrast can be enhanced or changed to that desired by a viewer. What may be acceptable to one person may by completely objectionable to another. Individuals have different opinions of what is perceived as an acceptable brightness and contrast. This is one of the reasons that most television sets include brightness and contrast controls. The

(a)

(b)

Figure 3.2: Examples of (a) low and (b) high contrast images.

manipulation of an image's graylevels is accomplished using a one-to-one mapping of the original graylevels to a new set of graylevels. Given an input image $f(x, y)$ with graylevels q_k, Figure 3.3 shows the mapping of these graylevels to a new set of graylevels p_k, producing a new output image $g(x, y)$.

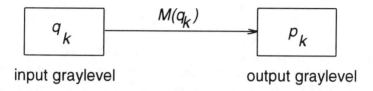

Figure 3.3: A block diagram illustrating the mapping of a one set of graylevels to a new set of graylevels.

The mapping from one set of graylevels to a new set of graylevels is defined by the mapping function $M(q_k)$. This mapping function can be either linear or nonlinear:

$$p_k = M(q_k) \ . \tag{3.4}$$

Figure 3.4 gives a graphical representation of the mapping of an input set of graylevels to a new set of graylevels. The horizontal axis is the set of input graylevels, while the vertical axis is the corresponding set of output graylevels. The curve in Figure 3.4 representing the mapping function $M(q_k)$ shows that the graylevel mapping can be nonlinear. Typically, $M(q_k)$ is a monotonically increasing or decreasing function so that each input graylevel maps to a unique graylevel in the output image. There are four linear mapping functions: addition, subtraction, multiplication and division involving a constant k,

$$p_k = q_k + k$$
$$p_k = q_k - k$$
$$p_k = k \cdot q_k$$
$$p_k = q_k / k \ . \tag{3.5}$$

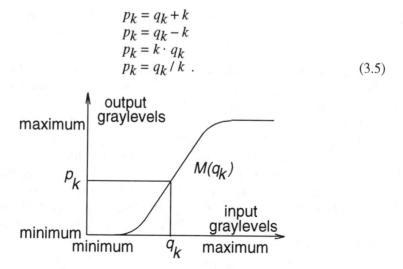

Figure 3.4: The mapping of input graylevels q_k to output graylevels p_k.

Of the four constant linear mappings given in Equation (3.5), only addition and multiplication will be considered, since subtraction is the inverse of addition and division is the inverse of multiplication. In general, the addition and multiplication mappings can be combined together to form

$$p_k = m \cdot q_k + d \ . \tag{3.6}$$

Equation (3.6) describes a straight line with m defined as the slope and d as the offset. It is implemented by scanning the input image pixel by pixel and applying the mapping function given in Equation (3.6) to each pixel in the image. Figure 3.5 shows the mapping of Equation (3.6) and the relationship between m and d. The value of d determines the brightness of the image, with positive values increasing the brightness. The value of m changes both the brightness and contrast of the output image; slopes greater than one result in an increase in both the brightness and contrast.

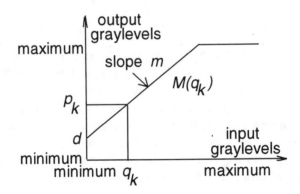

Figure 3.5: The mapping of input graylevels q_k to output graylevels p_k using the linear mapping function given in Equation (3.6).

Equation (3.6) is valid for q_k and p_k only in the valid range of graylevels. For the case when the output graylevel as predicted by Equation (3.6) is less than the allowed minimum graylevel, Equation (3.6) is modified so that the output is set to the minimum graylevel. Likewise, when the output graylevel as given by Equation (3.6) exceeds the maximum allowable graylevel, the output is set to the maximum graylevel. For example, in a 256 graylevel image, negative graylevels computed from Equation (3.6) are set to zero and graylevels above 255 are set to 255. This effect is shown in Figure 3.5 as the saturation of the straight line at the point where the maximum allowable graylevel is reached. When the parameter m is varied, it changes both the contrast and the brightness. Equation (3.6) can be rewritten as

$$p_k = m \cdot (q_k - B) + (B + d) \ , \tag{3.7}$$

where B is the average brightness of the image as defined by Equation (3.1). Equation (3.7) decouples the adjustment of the brightness and contrast. Changing the parameter m in Equation (3.7) changes only the contrast, leaving the brightness of the image unchanged. The process of enhancing the contrast within an image is often referred to as *contrast stretching*. Equation (3.7) was used to create the two different brightness and contrast images given in Figures 3.1 and 3.2.

Autoscaling is another linear mapping function that is quite often used to enhance the brightness and contrast of an image. This function maps the graylevel range of an image to a desired range of graylevels. The mapping function that autoscales the input graylevels of an image to a predetermined range of graylevels is

$$p_k = \frac{(\text{max} - \text{min})}{(f_{\text{max}} - f_{\text{min}})} (q_k - f_{\text{min}}) + \text{min} \quad , \tag{3.8}$$

where f_{max} and f_{min} are the maximum and minimum graylevels found in the input image, and max and min are the desired maximum and minimum graylevels in the output image. Figure 3.6 shows the effect autoscaling has on the rocket engine image given in Figure 3.2(a). During the autoscaling process, the minimum graylevel in the input image was 51 and the maximum was 204. The desired minimum and maximum graylevels in the output image of 0 and 255 produced a scaling factor of 1.67. Since the scaling factor is greater than one, this results in an increase in contrast in the autoscaled image. A comparison of Figure 3.2(a) with Figure 3.6 shows that the contrast has been increased by the autoscaling operation.

Figure 3.6: An example of autoscaling the rocket engine image
of Figure 3.2(a).

The mapping function that produces a negative of an original graylevel image is

$$p_k = G_{max} - q_k \; , \tag{3.9}$$

where G_{max} is the maximum allowed graylevel for the input image. This maps dark graylevels in the input image to light graylevels, while light graylevels are mapped to dark graylevels as shown by the mapping function given in Figure 3.7. Given a 256 graylevel image, Equation 3.9 reduces to

$$p_k = 255 - q_k \; . \tag{3.10}$$

Figure 3.8 shows the negative image of Figure 3.2(b) created using Equation (3.10). Note that the white regions of Figure 3.2(b) are now black.

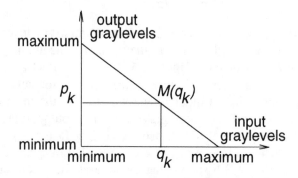

Figure 3.7: A mapping function that produces a negative image.

Figure 3.8: The negative image of Figure 3.2(b).

In addition to the four constant linear mapping functions, there are four Boolean operations that can be used to manipulate each of the bits representing a graylevel:

$$p_k = q_k + k \quad \text{(OR)}$$
$$p_k = q_k * k \quad \text{(AND)}$$
$$p_k = k \oplus q_k \quad \text{(EX-OR)}$$
$$p_k = \overline{q_k} \quad \text{(NOT)} \quad . \tag{3.11}$$

For example, both the EX-OR operation with a constant of 255 and the NOT operation produce a negative image. Consider an image in which the lowest two bits defining the graylevel have been corrupted by noise, making these bits unusable. Performing the AND operation on this image with a constant of 252 sets these two bits of each graylevel to zero. Hence, removing the graylevel noise from the image.

There are several types of nonlinear mapping functions available. The most common are the exponential and the logarithmic mapping functions. These functions are typically used to enhance either dark or light regions of an image. The logarithmic mapping function for a 256 graylevel image is defined as

$$p_k = 45.98 \cdot \log_e(q_k + 1) , \tag{3.12}$$

where the constant in front of the logarithmic function is used to scale an input graylevel of 255 to an output graylevel of 255. In addition, a value of one is added to the input graylevels before computing the logarithmic function so that an input graylevel of 0 maps to an output graylevel of 0. A logarithmic mapping function is used to increase the brightness and contrast of dark regions while reducing the contrast of light regions. This can be seen in Figure 3.9 from the increase in the slope of the mapping function for low graylevels and the decrease for large graylevels.

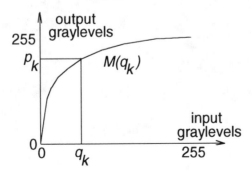

Figure 3.9: The logarithmic mapping function for a 256 graylevel image.

Figure 3.10(a) shows a dark image of a young girl. Lightening this image using a linear mapping function as defined in Equation (3.6) would lighten the

(a)

(b)

Figure 3.10: An example of using the logarithmic mapping function
to brighten dark regions of an image: (a) the original
image and (b) the enhanced image.

image but it would also lighten the bright areas. Eventually, the bright areas would saturate at the maximum graylevel and become washed out. Using a logarithmic mapping function will increase the brightness of the dark regions of the image while only slightly increasing the brightness of the light regions. Figure 3.10(b) gives the enhanced output image, showing the increase in brightness and contrast in the dark regions. The shadows behind the girl are now observable and the details of her hair are clearly visible.

The exponential mapping function is the inverse of the logarithmic mapping function darkening and increasing the contrast of light regions of an image. For a 256 graylevel image, the exponential mapping function is defined as

$$p_k = (e^{q_k/45.98} - 1), \tag{3.13}$$

where the constant of 45.98 is required to map the input graylevel of 255 to an output graylevel of 255. Figure 3.11 gives the exponential mapping function for a 256 graylevel image, showing the increase in slope for light graylevels. Exponential mapping functions on the average will darken an image, while logarithmic mapping functions will lighten an image.

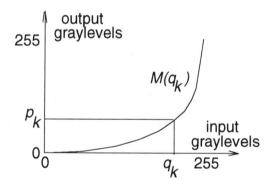

Figure 3.11: The exponential mapping function for a 256 graylevel image.

There are cases when it is desired to have two different graylevels within an image mapped to the same graylevel in the output image. For example, consider an image containing two identical parts. The first part has graylevels ranging from 40 to 70, while the second part contains graylevels ranging from 170 to 200. A mapping function of the form of Figure 3.12 will provide the same perceived contrast and brightness for both parts in the output image. Hence, the mapping function of Figure 3.12 provides a method of eliminating the differences in brightness and contrast between two images so that an unbiased comparison can be made.

Figure 3.13(a) shows two images of a tea cup taken with two different brightness settings and then merged together to form one image. The left image has graylevels ranging from 0 to 121, while the right image has graylevels ranging from 130 to 251. Even though the two images are of the same tea cup, it is hard to compare the two images because of their brightness differences. Figure 3.13(b) shows the result of using the mapping function in Figure 3.12. The brightness levels of the two images have been brought closer together, allowing for an unbiased comparison.

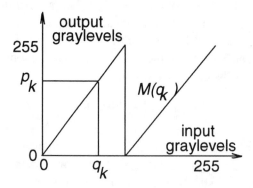

Figure 3.12: A mapping function $M(q_k)$ that makes two objects with different graylevels appear to have the same perceived brightness and contrast.

3.3 Operations Between Images

Similar to the constant operations defined in the previous section, there are four linear operations between images. Given two $N \times M$ images $f(x, y)$ and $g(x, y)$, the four linear operations of addition, subtraction, multiplication, and division are

$$o(x, y) = f(x, y) + g(x, y) \qquad \text{for } 0 \le x < N, 0 \le y < M$$
$$o(x, y) = f(x, y) - g(x, y) \qquad \text{for } 0 \le x < N, 0 \le y < M$$
$$o(x, y) = f(x, y) \cdot g(x, y) \qquad \text{for } 0 \le x < N, 0 \le y < M$$
$$o(x, y) = f(x, y) \div g(x, y) \qquad \text{for } 0 \le x < N, 0 \le y < M . \quad (3.14)$$

Since division and multiplication are inverse operators of each other, only the multiplication operation needs to be considered. Even though subtraction and addition are also inverse operators, they will be considered separately because each has an important application within electronic image processing.

Consider the acquisition of an image $f(x, y)$ with a noisy electronic camera. This camera degrades the noise free image $f(x, y)$ with additive zero mean noise $r(x, y)$, which is uncorrelated with $f(x, y)$:

$$g(x, y) = f(x, y) + r(x, y) . \qquad (3.15)$$

(a)

(b)

Figure 3.13: An example of using the mapping function given in Figure 3.12: (a) the original image and (b) the enhanced image.

If several images of $f(x, y)$ can be acquired with this camera, then image addition can be used to reduce the noise present in the noise degraded image $g(x, y)$ using a method known as *image averaging.*

Consider adding K of these noise degraded images together to produce an average image of the form

$$p(x, y) = \frac{1}{K} \sum_{i=1}^{K} g_i(x, y) = \frac{1}{K} \sum_{i=1}^{K} [f(x, y) + r_i(x, y)] \ , \qquad (3.16)$$

where $g_i(x, y)$ is the *ith* image acquired from the noisy camera. Distributing the summation throughout each term of Equation (3.16) yields

$$p(x, y) = \frac{1}{K} \sum_{i=1}^{K} f(x, y) + \frac{1}{K} \sum_{i=1}^{K} r_i(x, y) = f(x, y) + \frac{1}{K} \sum_{i=1}^{K} r_i(x, y) \ . \qquad (3.17)$$

Equation (3.17) reduces to the sum of the noise free image and the average of the K samples of the noise. The effect of averaging K samples of zero mean uncorrelated noise is to reduce the variance of this noise by K, which on an average reduces the peak-to-peak fluctuations of this noise by $\sqrt{K}$.

Figure 3.14(a) is an image of two coins that has been degraded by zero mean Gaussian noise with a standard deviation of 35.4. Sixteen separate images of Figure 3.14(a) were obtained and then averaged together using Equation (3.16). Figure 3.14(b) is the enhanced image, showing the reduction of noise as a result of image averaging. The grainy appearance due to the noise in Figure 3.14(a) has been almost completely removed in the enhanced image. Some dirt spots that are barely visible in the background region of the noise degraded image are clearly observable after image averaging.

Image subtraction plays an important role in finding changes or differences between images. Image subtraction is typically used in an automatic security system. For example, consider a camera that produces images of the door of a bank's vault at the rate of 30 images per second. Upon closing of the vault door and the bank itself, subtraction of successive images $f_i(x, y)$ and $f_{i-1}(x, y)$ should produce a difference image containing zero graylevels:

$$o(x, y) = |f_i(x, y) - f_{i-1}(x, y)| \quad \text{for } x \in M \text{ and } y \in M \ . \qquad (3.18)$$

If the difference image is nonzero, then some pixels have changed graylevels between successive images. This could be caused by an insect moving on the vault door, or it could be a burglar. A simple way to determine which is to sum the total number of pixels in the difference image that are nonzero. The insect will produce fewer pixels that are nonzero in comparison with the burglar.

(a)

(b)

Figure 3.14: An example of using image averaging to remove noise
from an image: (a) the noise degraded image and (b) the
enhanced image after image addition.

Once the number of pixels that are nonzero reaches a predetermined threshold value, an alarm indicating that a bank robbery is in progress will go off.

Figure 3.15(a) is an image of a basket of miniature ceramic vegetables containing a celery stalk, a squash, a carrot, a tomato and others. Figure 3.15(b) is the same image with one of the vegetables missing. A close inspection of the two images will eventually show that the garlic clove on the left side of the basket is missing. An easier method to automatically find the difference between the two images is to use Equation (3.18). Figure 3.15(c) is the difference image generated using Equation (3.18) and then brightened and contrast enhanced. Careful inspection of Figure 3.15(c) shows that the garlic clove and its shadow were removed, and with them the part of the basket that was in the shadow of the garlic clove. Removing the garlic clove from the basket changed the lighting conditions adjacent to it and hence removed its shadow from the image, as also shown in Figure 3.15(b).

The last operation between images is image multiplication. This operation is typically used to remove a nonuniform camera response. In a typical camera, each sensor element varies in gain sensitivity $g(x, y)$ and in offset voltage $o(x, y)$

$$c(x, y) = g(x, y) \cdot f(x, y) + o(x, y), \quad \text{for } x \in M \text{ and } y \in M. \quad (3.19)$$

Consider an image $c(x, y)$ acquired using the camera described by Equation (3.19). To remove the effects of voltage offset $o(x, y)$, a test image is obtained with the camera shutter closed [$f(x, y) = 0$ for all $x \in M$ and $y \in M$]. This test image is then subtracted from any image $c(x, y)$ acquired with this camera, which subtracts out the offset error associated with each sensor.

A second test image is obtained with the shutter open and the camera illuminated with a known constant intensity. The offset test image $o(x, y)$ obtained previously is then subtracted from this image to remove the offset error, generating a new offset free test image. Any graylevel variations that are present in this test image are due solely to variations in the gain sensitivity between sensors. Finally, this test image is then normalized by the known intensity, producing a new image containing the variations in gain sensitivity $g(x, y)$. Computing the inverse of this image $1/g(x, y)$, and then multiplying this inverse image with any offset corrected image acquired by this camera will remove the effects of gain sensitivity variations.

Figure 3.16(a) is an image that has been degraded by a gain coefficient that decreases radially from the center of the image. This gain coefficient starts at one and decreases to 0.1 at the edge of the image. The result of this degradation is to add a radial shading component to the image. A correction image was obtained using the inverse of the shading degradation. Figure 3.16(b) is the corrected image obtained by multiplying every pixel in the correction image with every pixel in the shaded image. Note the improvement in the overall illumination of the enhanced image.

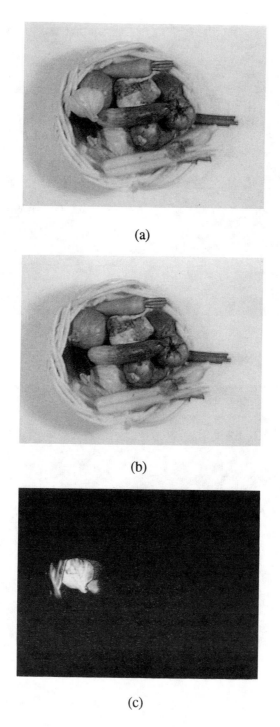

(a)

(b)

(c)

Figure 3.15: An example of using image subtraction to find the difference between two images: (a) the first image, (b) the second image with the garlic clove removed, and (c) the difference image.

(a)

(b)

Figure 3.16: An example of using image multiplication to remove a
              known shading from an image: (a) the original shaded
              image and (b) the enhanced image.

3.4 Histogram Techniques

Histogram techniques play an important role in the enhancement of the perceived brightness and contrast of an image. These techniques can take a dark, low contrast image and automatically increase the image's brightness and contrast, making features observable that were not visible in the original image. The histogram of an image tells a lot about the distribution of graylevels within the image. The perceived contrast and brightness of an image can be determined directly from an image's histogram. The histogram of an $N \times M$ image is defined as the percentage of pixels within the image at a given graylevel:

$$h_i = \frac{n_i}{NM} , \qquad \text{for } 0 \leq i \leq G_{\max} , \tag{3.20}$$

where n_i is the number of pixels at graylevel i, NM is the total number of pixels within the image and $G_{\max}$ is the maximum graylevel value of the image. For a 256 graylevel image, $G_{\max} = 255$. An important property of a histogram is that the sum of each histogram value over the range of graylevels present within an image equals one:

$$\sum_{i=0}^{G_{\max}} h_i = 1 . \tag{3.21}$$

Figure 3.17 gives the pseudo code that computes the histogram of an $N \times M$ image. Upon starting the program, it is assumed that the array *hist*[] is of size $G_{\max}$ and is initially set to zero. The image is scanned pixel by pixel and the graylevel of each pixel is then used to index the array *hist*[] during the increment operation. In this way, every occurrence of a graylevel within the image causes the corresponding cell within the array *hist*[] to be incremented. Upon complete scanning of the image, the values stored in *hist*[] are the number of pixels at each graylevel. For example, cell location 0 has the number of pixels at graylevel 0, while cell location $G_{\max}$ has the number pixels within the image at graylevel $G_{\max}$. The last function performed by this pseudo program is to divide by the total number of pixels to produce the histogram values h_i.

Both the average graylevel and the standard deviation of the graylevels can be computed from an image's histogram. The average graylevel of an image in terms of its histogram is

$$avg = \sum_{i=0}^{G_{\max}} i \cdot h_i , \tag{3.22}$$

and its standard deviation is

$$std = \sqrt{\sum_{i=0}^{G_{max}} i^2 \cdot h_i - avg^2} \; . \qquad (3.23)$$

Equations (3.22) and (3.23) are analogous to Equations (3.1) and (3.2). The average given in Equation (3.22) relates directly to the perceived brightness of an image. A low average value implies a dark image, while large values imply a light image. The standard deviation as given in Equation (3.23) gives an estimate of the average graylevel variation within the image. Small standard deviation values imply a low contrast image, while large values imply a high contrast image.

```
for y = 0 to M-1
    for x = 0   to N-1
        hist{Image[x, y]}=
            hist{Image[x, y]} + 1
    next x
next y
for g = 0 to Gmax
    hist[g] = hist[g]/(N*M)
next g
```

Figure 3.17: Pseudo code to compute the histogram of an image.

Figure 3.18 gives four example histograms representing low and high brightness images, as well as low and high contrast images. Along the horizontal axis are the graylevels of the image, ranging from 0 to G_{max}. The height of each line gives the histogram value (percentage of pixels) at graylevel i. Figures 3.18(a) and (b) show the difference in the histograms for a dark and a light image. The histogram in Figure 3.18(a) shows that most of the pixels within the image take on graylevels that are dark. In Figure 3.18(b), on the other hand, most of the pixels have graylevels that are on the light side. The centroid of the histogram gives an approximate perceived brightness for the image. The width of the histogram describes the graylevel variations within an image. The larger the graylevel variations are, the higher the perceived contrast is. Both histograms in Figures 3.18(a) and (b) are about the same width, implying that the images represented have about the same contrast. The image given by the histogram in Figure 3.18(c) indicates that this image has a low contrast and a high brightness, while the image given by the histogram in Figure 3.18(d) contains a high contrast and a medium brightness. In fact, the contrast of this image is the best that can be obtained. The distribution of graylevels throughout covers the whole range of possible graylevels, and the pixels within this image are uniformly distributed about the total range of graylevels. An image's histogram can also be used to segment an object's pixels from the background pixels. Further discussion of using an image's histogram will be given in Chapter 8 in the discussion of image thresholding.

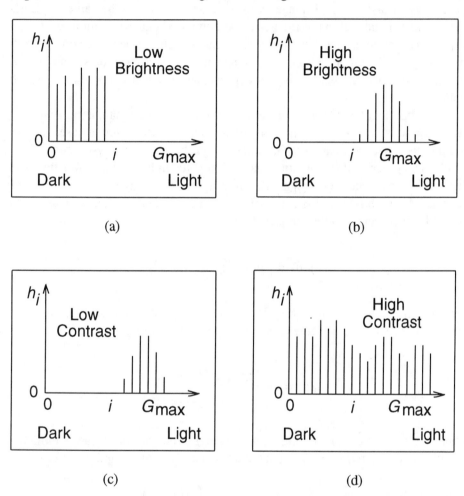

Figure 3.18: An example of four different graylevel histograms: (a) a histogram of a dark image, (b) a histogram of a light image, (c) a histogram of a low contrast image, and (d) a histogram of a high contrast image.

One histogram technique that is used to enhance the brightness and contrast of an image is *histogram equalization*. The goal of histogram equalization is to distribute the graylevels within an image so that every graylevel is equally likely to occur. In other words, histogram equalization takes an image's histogram and produces a new image with a histogram that is uniformly distributed, which is similar to the histogram shown in Figure 3.18(d). Histogram equalization will increase the brightness and contrast of a dark and low contrast image, making features observable that were not visible in the original image. Histogram equalization is also used to standardize the brightness and contrast of images. Since histogram equalization distributes an image's graylevels uniformly about the range of graylevels, all images will have approximately the same brightness and contrast, hence allowing images to be compared equally without a bias due to perceived contrast and brightness differences. This process is similar to the example given in Figure 3.13 of removing the perceived brightness and contrast bias from the two images of the tea cup so that an equal comparison could be made.

The approach taken here will be to derive histogram equalization first using the continuous case and then to convert this result to the discrete case. For the continuous case, the histogram becomes the probability density function (PDF) $p_a(a)$. The process of histogram equalization is to find a mapping function $M(a)$ that maps an input PDF $p_a(a)$ to an output PDF $p_b(b)$ that is uniformly distributed, as shown in Figure 3.19. Given the mapping function $M(a)$, the output PDF is related to the input PDF as

$$p_b(b) = \left\{ p_a(a)\frac{da}{db} \right\} a = M^{-1}(b) \quad . \tag{3.24}$$

For histogram equalization, the desire is to have the output PDF follow a uniform PDF so that each graylevel has an equal probability:

$$p_b(b) = \begin{cases} \dfrac{1}{Z-W} & \text{for } W \leq b < Z \\ 0 & \text{elsewhere} \end{cases} , \tag{3.25}$$

where the parameter Z is the maximum desired graylevel that b can obtain, W is the minimum desired graylevel it can obtain, and $Z - W$ is its graylevel range .

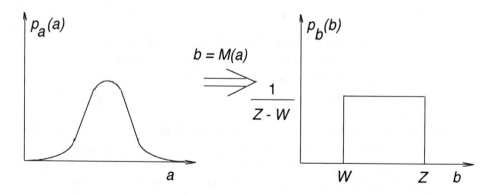

Figure 3.19: The mapping of a PDF to a uniform PDF using histogram equalization.

Equation (3.24) can be rewritten so that the differential probabilities can be equated:

$$p_a(a) \, da = p_b(b) \, db \quad . \tag{3.26}$$

Integrating both sides of Equation (3.26) yields

$$\int_{h \in A} p_a(h) \, dh = \int_{v \in B} p(v) \, dv \quad , \tag{3.27}$$

where A and B are the ranges of values over which $p_a(a)$ and $p_b(b)$ are defined. Substituting Equation (3.25) into Equation (3.27) and performing the integration,

$$\frac{b - W}{Z - W} = \int_0^a p_a(h) \, dh \quad \text{for } W \le b < Z \quad , \tag{3.28}$$

and finally solving for b gives the required mapping function:

$$b = M(a) = (Z - W) \int_{h \in A} p_a(h) \, dh + W \quad \text{for } W \le b < Z \quad . \tag{3.29}$$

For electronic imaging, the range over which b is defined is between 0 and the maximum graylevel $G_{\max}$. Likewise, the input graylevels that define the range of h are also between 0 and $G_{\max}$:

$$b = M(a) = G_{\max} \int_0^a p_a(h) \, dh \quad \text{for } 0 \le b < G_{\max} \text{ and } 0 \le a < G_{\max} . \tag{3.30}$$

Equation (3.30) gives the mapping function $M(a)$ in terms of an integral of the input PDF. For electronic image processing, Equation (3.30), which was derived for the continuous case, must be replaced by its equivalent discrete version. The input PDF becomes the histogram of the original image and the integral is replaced by a summation:

$$b_k = M(a_k) = G_{\max} \sum_{i = 0}^k h_i \quad \text{for } 0 \le k \le G_{\max} \quad , \tag{3.31}$$

where the histogram h_i is defined by Equation (3.20). To perform histogram equalization on an image, the histogram of the image is computed and Equation

(3.31) is used to defined the mapping function. Next, the input image is scanned pixel by pixel, applying this mapping function to the graylevels of each pixel to produce a new histogram equalized image.

Table 3.1 gives the graylevel distribution for an eight graylevel 128×128 image, with graylevel 0 representing black and graylevel 7 representing white. The first column gives the eight graylevels, with the adjacent column giving the number of pixels within the images at that graylevel. This image is of low brightness and contrast, since 80% of the pixels have graylevels less than half the full graylevel scale. The next column gives the histogram values for each graylevel, computed from Equation (3.20). Application of Equation (3.31) produces non-integer values for the mapping function. The last column gives the mapping function rounded to the closest realizable graylevel. The mapping function is determined by reading the input graylevel given in the first column and then reading the output graylevel given in last column. For example, graylevels 5, 6, and 7 in the input image are all mapped to graylevel 7 in the histogram equalized image.

Table 3.1: An example of histogram equalization of an eight graylevel image.

Graylevels	n_i	h_i	Equation (3.31)	$M(a)$
0	1116	0.0681	0.476	0
1	4513	0.2754	2.405	2
2	5420	0.3308	4.720	4
3	2149	0.1312	5.638	6
4	1389	0.0848	6.232	6
5	917	0.0560	6.624	7
6	654	0.0399	6.903	7
7	226	0.0138	7	7
n_t	16384			

Figure 3.20 gives an example of using histogram equalization to improve the overall brightness and contrast of an image. Figure 3.20(a) shows an image of one of Thomas Edison's cars taken at the Edison Museum in Fort Myers, Florida. Figure 3.20(c) is its corresponding graylevel histogram, showing that the pixels within this image range in graylevel between 0 and 80, resulting in a dark and low contrast image. Figure 3.20(b) is the resulting image after histogram equalization, showing an improved brightness and contrast. Details of the car which were not observable in the original image now are clearly visible. The graylevel histogram of the equalized image, Figure 3.20(d), shows that histogram equalization has increased the range of graylevels and has tended to flatten the histogram, making it more uniform. The tendency of histogram equalization to stretch the graylevel range and to flatten the image's histogram has led to the alternative name of *histogram flattening*. Even though the

(a)

(b)

Figure 3.20: An example of using histogram equalization to enhance
a low brightness and contrast image: (a) the original
poor quality image and (b) the enhanced image.

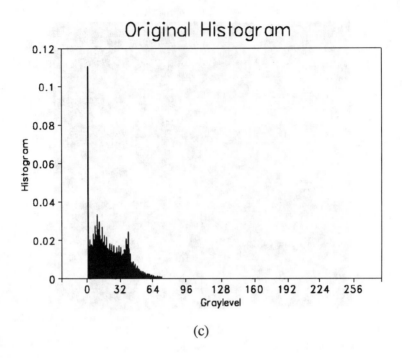

(c)

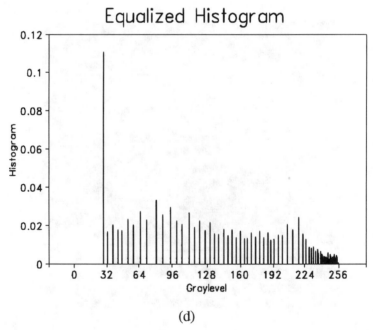

(d)

Figure 3.20: An example of using histogram equalization to enhance a low brightness and contrast image: the graylevel histogram of (c) the original image and (d) the histogram equalized image.

histogram of the equalized image is more uniform than that of the original image it is not perfectly uniform. The reason it deviates from the expected uniform distribution is that the derivation presented here assumed that the graylevels continuously varied between a minimum graylevel and a maximum graylevel. The digitization of an image produces only a finite number of graylevels, and hence Equation (3.31) is only an approximation to the exact solution given by Equation (3.30).

In many instances, it is desired to map an image's graylevels to a new set of graylevels that produce a desired histogram. Unfortunately, the process of histogram equalization produces a mapping function that yields an output image graylevel distribution that is approximately uniformly distributed. The process of *histogram specification* takes the concept of histogram equalization one step further, allowing for the specification of the output graylevel histogram. Given a desired histogram d_i, the first step is to produce a mapping function $D(d_k)$ that histogram equalizes the desired histogram:

$$D(d_k) = \sum_{i=0}^{k} d_i \qquad \text{for } 0 \leq k \leq G_{\max} \qquad . \tag{3.32}$$

The next step is to histogram equalize the input image using Equation (3.31),

$$b_k = M(a_k) = G_{\max} \sum_{i=0}^{k} h_i \qquad \text{for } 0 \leq k \leq G_{max} \qquad . \tag{3.33}$$

Since the desired histogram and the input image have been equalized, producing approximately uniformly distributed histograms, the output graylevels of the histogram equalized input image can be equated to the mapping function that was used to histogram equalize the desired histogram:

$$b_k = D(d_k) \qquad . \tag{3.34}$$

The new graylevel mapping function that maps the equalized image to the desired histogram is the inverse of Equation (3.34):

$$d_k = D^{-1}(b_k) \qquad . \tag{3.35}$$

Figure 3.21(a) is a high brightness and low contrast image of Edison's chemical laboratory taken at the Edison Museum in Fort Myers, Florida. This image has an overall brightness that is too light and also lacking in good contrast. Histogram equalization of this image would tend to darken it but also would increase the contrast more than desired. Figure 3.21(c) gives the histogram of the image showing the graylevel distribution of the pixels. This histogram shows that the image needs to be darkened and the range of graylevels

(a)

(b)

Figure 3.21: An example of using histogram specification to reduce
the brightness and increase the contrast of an image:
(a) the original image and (b) the enhanced image.

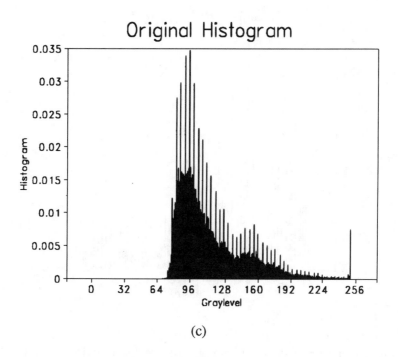

(c)

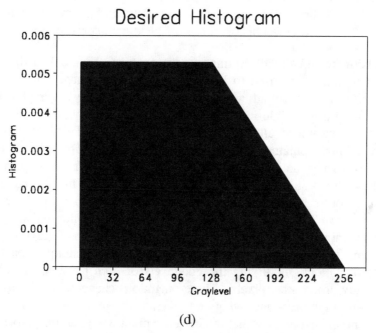

(d)

Figure 3.21: An example of using histogram specification to reduce
the brightness and increase the contrast of an image: (c) the
graylevel histogram of the original image and (d) the desired
histogram that is used in the histogram specification algorithm.

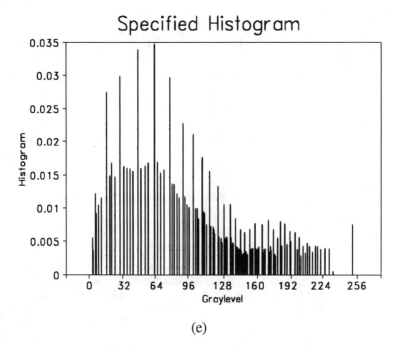

(e)

Figure 3.21: An example of using histogram specification to reduce the
brightness and increase the contrast of an image: (e) the
graylevel histogram of the histogram specified image.

needs to be increased slightly to improve the contrast. Figure 3.21(d) gives the
desired histogram to be used to histogram specify the image. It was chosen
because it biases the output graylevel distribution to lower graylevels, thus
darkening the image, while increasing the range of graylevels. Figure 3.21(b)
gives the output image after histogram specification. The image has been
darkened and now contains the desired contrast. Figure 3.21(e) gives the output
histogram, showing the increase in contrast and the lowering of the brightness.
The choice of the desired histogram given in Figure 3.21(d) has for the most part
left the shape of the histograms of the histogram specified image and the original
image the same. This is unlike histogram equalization, which redistributes the
graylevels within an image to produce a uniform distribution of graylevels.

   This chapter covered several image enhancement techniques based upon
pointwise operations, modifying the graylevels of each pixel within an image
independently of the other pixels. These operations changed both the perceived
contrast and brightness improving the overall image quality. Other image
enhancement operations exist that modify each pixel within an image based upon
the graylevels of its neighboring pixels. These operations are known as spatial
filtering and will be discussed in the next chapter.

CHAPTER 4

Spatial Filtering and Fourier Frequency Methods

Filtering of an image is an important aspect of electronic image processing. It provides a means of reducing the noise present in an image and sharpening of a blurred image. Filtering of an image can be implemented in either the spatial or frequency domain. This chapter first discusses several types of noise that occur commonly within images, followed by several types of spatial filters that can be used to reduce the noise. Next, Fourier frequency filtering methods are presented and then used to restore images that have been degraded by a known degradation process. Both inverse filtering as well as Wiener filtering methods are presented.

4.1 Various Types of Noise that Appear in Images

The reduction of noise present in images is an important aspect of electronic image processing. Noise applied to a noise free image can degrade the image to such a point that important features are no longer observable. Images acquired with an electronic camera are typically corrupted with noise due to the camera's sensor and its associated electronics. Photographs of images contain noise due to the finite size of the silver halide grains that are part of the chemical photographic process. This type of noise is very predominant in images acquired with high speed film due to the film's large silver halide grain size. Another source of photographic noise is due to dust that collects on the optics and the negatives during the development process. Fortunately, this type of noise is easily removed using a median filter that will be discussed later in this chapter. The point-to-point transmission of video images is another common source of noise. Depending on environmental conditions, these images are very often degraded by noise. Video images transmitted via satellites orbiting the earth are very susceptible to the electromagnetic interference due to the sunspot activity on the sun.

The classification of noise is based upon the shape of the probability density function (PDF) or histogram for the discrete case of the noise. Since the topic of this book is electronic image processing of sampled and digitized images, the presentation of the various types of noise that are commonly encountered within images will be presented using the histogram instead of the PDF. Also, the discussion given in this section will be limited to uncorrelated noise. For a discussion on correlated noise, the reader is referred to the many signal processing and random processing books that are available. In terms of a spatially sampled image, uncorrelated noise is defined as the random graylevel variations within an image that have no spatial dependence from pixel to pixel. In other words, the graylevel of a pixel $f(x, y)$ due to uncorrelated noise does not depend on the graylevels of its neighboring pixels. The first type of noise that will be presented is uniform noise. The histogram of this noise is

$$h_i(a_k) = \begin{cases} \dfrac{1}{Z - W} & \text{for } W \le a_k < Z \\ 0 & \text{otherwise} \end{cases} \tag{4.1}$$

Uniform noise produces noise values with equal probability in the range of W to Z. Figure 4.1 gives the histogram for a uniform noise image as it might appear in an image with $G_{max} + 1$ different graylevels. Plotted along the horizontal axis are the graylevels of the image and along the vertical axis is the probability distribution of the noise.

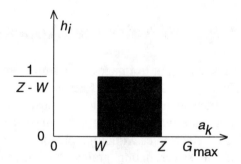

Figure 4.1: A histogram of uniform noise.

The mean and variance are important parameters used to characterize the noise present in an image. The mean is defined in terms of its histogram as

$$mean = M = \sum_{i=0}^{G_{max}} i \cdot h_i \qquad (4.2)$$

and its variance is

$$\sigma^2 = \sum_{i=0}^{G_{max}} i^2 \cdot h_i - M^2 . \qquad (4.3)$$

The standard deviation can be derived from the variance as

$$\sigma = \sqrt{\sum_{i=0}^{G_{max}} i^2 \cdot h_i - M^2} . \qquad (4.4)$$

The mean gives the average brightness of the noise and the standard deviation gives the average peak-to-peak graylevel deviation of the noise. For uniform noise as described by Equation (4.1) the mean and standard deviation are

$$M = \frac{Z + W}{2} \qquad (4.5)$$

and

$$\sigma = \frac{Z - W}{\sqrt{12}} . \qquad (4.6)$$

Figure 4.1 shows that the peak-to-peak graylevel deviation is simply $Z - W$. Rewriting Equation (4.6) gives the peak-to-peak graylevel deviation of the noise in terms of the standard deviation as

$$peak\text{-}to\text{-}peak = \sqrt{12} \cdot \sigma. \qquad (4.7)$$

The most common type of noise that is found within images is Gaussian noise as the result of electronic noise present in electronic cameras and sensors. Gaussian noise is also used to model unknown additive noise sources. By the *central limit theorem*, the sum of a large number of independent noise sources approaches Gaussian distributed noise. In many situations, the noise that is present within an image can be modeled as the sum of many independent noise sources. Under this situation, Gaussian noise can be used to model the noise that is present within an image. An unique feature of Gaussian noise is that linear filtering of this noise also produces Gaussian noise but with a different mean, variance, and correlation. Linear filtering of noise typically changes the histogram of the noise. For example, the lowpass linear filtering of negative exponential noise changes the noise to gamma distributed noise.

Gaussian noise can be expressed in terms of its mean m and its variance σ^2 as

$$h_i(a_k) = \frac{e^{-(a_k - m)^2 / \sigma^2}}{\sigma\sqrt{2\pi}} \qquad \text{for } -\infty < a_k < \infty \ , \tag{4.8}$$

where m is the mean and σ is the standard deviation. Figure 4.2 shows the histogram of this noise. The mean is located at the peak of the histogram and the width is determined by the standard deviation. Figure 4.2 also shows that the Gaussian histogram is symmetric about the mean and the highest probability of an occurrence of Gaussian noise is at the mean. The probability of occurrence decreases the farther the graylevels are from the mean. In fact, 99.7% of the noise graylevels lie within $\pm 3\sigma$ of the mean.

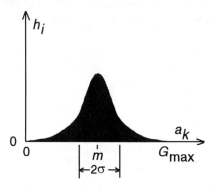

Figure 4.2: A histogram of Gaussian noise.

The histogram for Gaussian noise as defined by Equation (4.8) can only approximate the noise present within a digitized image. Gaussian noise is defined over an infinite range of a_k, but a digitized image has only a finite range of graylevels (0 to G_{max}). If Gaussian noise is added to an image, noise values that exceed the graylevel dynamic range of the image are saturated at 0 and G_{max}. This effectively modifies the first and last bins of the histogram producing values within these two bins that are higher than predicted by Equation (4.8). Care must be taken when comparing these two bins of the image histogram to that predicted by Equation (4.8). For 99.7% of the graylevels that are due to Gaussian noise, the average peak-to-peak graylevel deviation about the mean is

$$peak\text{-}to\text{-}peak = 6 \cdot \sigma \ . \tag{4.9}$$

For example, consider an image containing Gaussian noise with a mean graylevel of 128 and a standard deviation of 10. For 99.7% of the pixels within this image, the peak-to-peak graylevel deviation will be 60. This results in the image's graylevels varying between 98 and 158, for 99.7% of the pixels.

Negative exponential noise appears in images that have been acquired using a laser as the illumination source. The appearance of negative exponential noise in an image is often referred to as *laser speckle*. The histogram for negative exponential noise is

$$h_i(a_k) = \frac{e^{-a_k/a}}{a} \qquad \text{for } 0 \leq a_k < \infty \, , \qquad (4.10)$$

where the mean and standard deviation are equal to a. Figure 4.3 shows the histogram for negative exponential noise. The highest probability of occurrence is at zero and drops off exponentially as the graylevels increase. Like Gaussian noise, negative exponential noise is defined beyond the dynamic range of a typical digitized image. Noise values above the maximum graylevel of the image become saturated, resulting in additional pixels within the image taking on the maximum graylevel. As a result, the value in the last bin of the histogram for a digitized image is higher than predicted by Equation (4.10).

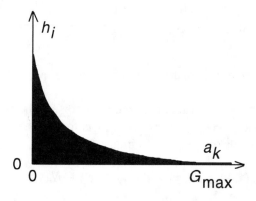

Figure 4.3: A histogram of negative exponential noise.

Another noise that commonly appears in images is that of salt-and-pepper noise. This noise is named for the salt and pepper appearance an image takes on after being degraded by this type of noise. Typically, the cause of this noise can be traced to two sources. For images captured with electronic cameras, malfunctioning pixels usually produce pixels with graylevels of either white or black. This adds black and white noise pixels to an image, giving the image a salt-and-pepper appearance. The other common source of salt-and-pepper noise is dust and lint that appears on the optics during the acquisition of an image. The histogram for salt-and-pepper noise is defined as

$$h_i(a_k) = \begin{cases} p \text{ (pepper)} & \text{for } a_k = G_p \\ p \text{ (salt)} & \text{for } a_k = G_s \\ 0 & \text{otherwise} \end{cases} \quad , \tag{4.11}$$

where the probability of occurrence for the salt-and-pepper noise is $2p$, G_p is the graylevel of the pepper noise, and G_s is the graylevel of the salt noise. In other words, salt-and-pepper noise with a probability of occurrence of $2p$ will occupy $2p\%$ of the pixels within a graylevel image. There will be $p\%$ of the pixels at graylevel G_p and $p\%$ at graylevel G_s. Salt-and-pepper noise belongs to the family of noise called *outlier noise*. This type of noise is so named because the noise values that are generated deviate far beyond the values that are normally expected. Figure 4.4 is the histogram for salt-and-pepper noise showing that it takes on two different graylevels, that of G_p and G_s.

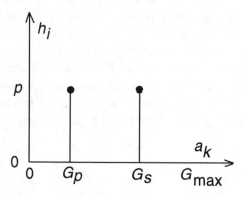

Figure 4.4: A histogram of salt-and-pepper noise.

There are two ways to corrupt an image with noise. The first way to produce a noise degraded image $g(x, y)$ of size $N \times M$ is to add the noise values $n(x, y)$ directly to the graylevels of a noise free image $f(x, y)$:

$$g(x, y) = f(x, y) + n(x, y) \quad \text{for } x \in N, y \in M \tag{4.12}$$

producing *additive noise*. The second method is to multiply the noise values by the graylevels of each pixel within the image :

$$g(x, y) = f(x, y) \cdot n(x, y) \quad \text{for } x \in N, y \in M \; . \tag{4.13}$$

Of the two types of noise, additive noise is much easier to remove from an image than *multiplicative noise*. Nonlinear image processing methods are needed to remove multiplicative type noise. Linear spatial and Fourier frequency domain methods can be used to remove additive noise. The generation of salt-and-pepper noise is neither additive nor multiplicative. Salt-and-pepper noise is generated by replacing $p\%$ of the pixels within an image with graylevels G_p and $p\%$ of the pixels with graylevels G_s.

Figures 4.5(a) - (d) show example images of uncorrelated uniform, Gaussian, negative exponential, and salt-and-pepper noise. Both the uniform and Gaussian noise images were generated using a variance of 400 and a mean value of 128. The negative exponential noise image was generated with a mean value of 50, while the salt-and-pepper noise image was created using a constant 128 graylevel image and a probability of occurrence of 20%. Figure 4.6(a) gives an image of a girl that has been corrupted with additive Gaussian noise with a mean of 0 and a variance of 400. Figure 4.6(b) is the same image of the girl corrupted with 20% salt-and-pepper noise. making it appear as if salt-and-pepper granules have been added to the image. Note how the addition of noise to the image has made it difficult to distinguish features that are present and has also given it a *grainy* appearance.

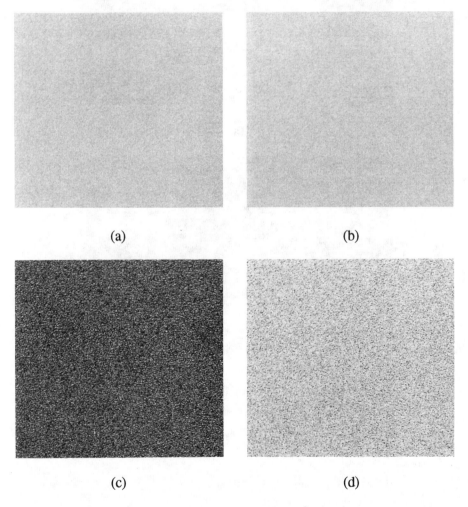

(a)   (b)

(c)   (d)

Figure 4.5: An example of (a) a uniform noise image, (b) a
Gaussian noise image, (c) a negative exponential
noise image, and (d) a salt-and-pepper noise image.

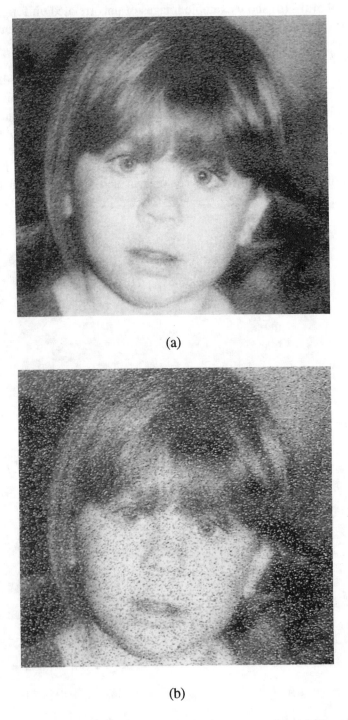

(a)

(b)

Figure 4.6: An example of degrading an image with (a) Gaussian
noise and (b) salt-and-pepper noise.

The four types of noise presented here are the most common noise that appear in images. Knowing the type of noise present within an image makes the selection of what type of image processing filter to use much easier. For example, a median filter performs the best at removing salt-and-pepper type noise. If the noise type is unknown before hand, it may be determined from a homogeneous or background portion of an image. Within this portion of the image, the histogram of the graylevels should give a good approximation of the type of noise that is present. For a further discussion of noise and its effects, the reader is referred to the many texts on signal processing and random processes.

4.2 Spatial Filtering

Spatial filtering of images is an important aspect of electronic image processing, providing a means of removing noise from images and sharpening blurred images. Many types of spatial filtering operations are available, some of which are the arithmetic mean or average, the geometric mean, the harmonic mean, and the median filters. Some of these spatial filters are linear and some are nonlinear. Linear spatial filtering is based upon two-dimensional convolution of an image $f(x, y)$ with the impulse response of a filter $h(x, y)$. The two-dimensional convolution of an $N \times M$ image with a filter impulse response defined over the set of elements given by H is defined as

$$g(x, y) \quad = \quad \sum_{i, j \in H} \sum f(x - i, y - j) \cdot h(i, j) , \qquad (4.14)$$

where $0 \le x - i < N$ and $0 \le y - j < M$. The values and the set of elements defining $h(i, j)$ determine the shape and orientation of the impulse response of the filter. For every pixel within an image, Equation (4.14) is just the weighted sum of the pixel $f(x, y)$ with its neighboring pixels, with the weights chosen by the impulse response of the filter $h(x, y)$.

Figure 4.7 is a graphical illustration showing the pixel $f(x, y)$ and its surrounding eight neighbors. Also shown in Figure 4.7 is a 3×3 filter $h(x, y)$ with its impulse response given by the weights A_i. The center of the filter is given by the coordinate 0, 0 and is defined over the range of $x, y \in -1, 0, 1$. It is common practice to refer to the impulse of the filter as the filter mask. For the 3×3 filter mask given in Figure 4.7, spatial convolution as defined by Equation (4.14) reduces to

$$\begin{aligned}
g(x, y) = \; & A_0 f(x - 1, y - 1) + A_1 f(x, y - 1) \; + A_2 f(x + 1, y - 1) \; + \\
& A_3 f(x - 1, y) \quad + A_4 f(x, y) \quad \;\; + A_5 f(x + 1, y) \quad \;\; + \\
& A_6 f(x - 1, y + 1) + A_7 f(x, y + 1) \; + A_8 f(x + 1, y + 1) \qquad (4.15)
\end{aligned}$$

or

$$g(x, y) = \sum_{i=0}^{2} \sum_{j=0}^{2} f(x - 1 + j, y - 1 + i) A_{(j + 3i)} \cdot \tag{4.16}$$

Equations (4.15) and (4.16) are simply the multiplication of each pixel under the mask with the corresponding filter weight and the replacement of this value at the point with coordinates (x, y). Spatial filtering of an image using a 3×3 filter mask is performed by scanning the image pixel by pixel and then using either Equation (4.15) or (4.16) to perform the spatial filtering operation. The concept of spatial filtering using a filter mask can be expanded to any size or shape desired. The only requirements are that the element positions be defined and the corresponding weights be given. Arbitrarily shaped masks play an important role in the area of morphological filtering. Typically, square or rectangular shaped masks of odd sizes (3×3, 5×5, *etc.*) are used because of their ease in programming and hardware implementation. The use of odd sizes guarantees that the mask can be centered about the pixel being filtered.

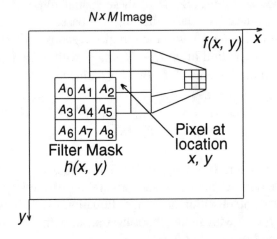

Figure 4.7: An illustration of spatial filtering using a neighborhood mask.

The simplest linear spatial filter is the *mean*, or *average*, filter. This is a lowpass filter that removes high spatial frequencies from an image and is also good at reducing noise present in an image, and for Gaussian noise this filter is the best one for removing Gaussian noise from an image. The negative attribute of this filter is that it blurs images by reducing the sharp edges located within them. This filter is implemented by setting the filter mask coefficients to one over the number of elements defined in the filter. Figure 4.8 shows the 3×3 filter mask, with all of its coefficients set to 1/9, that implements the 3×3 mean filter. Figure 4.9 gives the corresponding nine pixels that lie under the mask at the pixel coordinate of x, y. Substituting the value of 1/9 into Equation (4.16) yields the filtering operation for the 3×3 mean filter as

$$g(x, y) = \frac{1}{9} \sum_{i = 0}^{2} \sum_{j = 0}^{2} f(x - 1 + j, y - 1 + i) \ , \qquad (4.17)$$

where $0 \leq x - i < N$ and $0 \leq y - j < M$. A close inspection of Equation (4.17) shows that the output of the filtering operation is the average of the eight neighboring pixels with the center pixel at the coordinate x, y. Equation (4.17) can be expanded in general for any shaped mean filter as

$$g(x, y) = \frac{1}{N_t} \sum_{i, j \in H} \sum f(x - i, y - j) \ , \qquad (4.18)$$

where $0 \leq x - i < N$ and $0 \leq y - j < M$ and N_t is the number of elements that are defined by the filter mask H.

$\dfrac{1}{9}$	$\dfrac{1}{9}$	$\dfrac{1}{9}$
$\dfrac{1}{9}$	$\dfrac{1}{9}$	$\dfrac{1}{9}$
$\dfrac{1}{9}$	$\dfrac{1}{9}$	$\dfrac{1}{9}$

Figure 4.8: The filter mask coefficients for a 3×3 mean filter.

Careful inspection of Equation (4.17) shows that x can be defined only between 1 and $N - 2$, and likewise y can be defined only between 1 and $M - 2$. This ignores the top and bottom rows of the images as well as the first and last columns. The problem that exists when attempting to filter the border pixels of an image is that not all of the neighboring pixels are defined. For example, consider applying a 3×3 mean filter to the first column of pixels. The difficulty is that the leftmost three neighbors do not exist. There are two approaches to solving this problem. The first is to leave the top and bottom rows and the first and last columns unfiltered. This is not a major limitation, since most of an image's information is located at its center and not at its border. The second method replicates the bordering pixels so that the eight neighbor set can be defined and the mean value computed for each border pixel. The second method is an elegant solution that is often used, but the author believes that it is not worth the extra programming complexity and typically chooses the first solution.

$f(n\text{-}1, m\text{-}1)$	$f(n, m\text{-}1)$	$f(n\text{+}1, m\text{-}1)$
$f(n\text{-}1, m)$	$f(n, m)$	$f(n\text{+}1, m)$
$f(n\text{-}1, m\text{+}1)$	$f(n, m\text{+}1)$	$f(n\text{+}1, m\text{+}1)$

Figure 4.9: The pixels lying under a 3×3 filter mask.

Figure 4.10(a) shows the noise free 256 graylevel image of a young girl and Figure 4.10(b) shows this image degraded with zero mean uncorrelated Gaussian noise with a standard deviation of 24.5. Figure 4.10(d) is the same image as Figure 4.10(a) but corrupted with 15% salt-and-pepper noise. The graylevel for the salt noise was assigned 255, while the graylevel of pepper noise was assigned 0. Figure 4.10(c) is the 7×7 mean filter output image for the Gaussian degraded image. Note the reduction of the noise in comparison to Figure 4.10(b). In fact, the standard deviation of the noise is reduced by the rate of one over the square root of the number of elements in the filter. For the case of an $n \times n$ mean filter, the noise standard deviation drops by n. For Figure 4.10(c) the filtered Gaussian noise standard deviation was reduced to 3.5, or a factor of 7 reduction in the amount of noise present in the image. The reduction of this noise via the 7×7 mean filter was obtained at the price of blurring the image. A smaller filter window size could have been used, reducing the amount of blurring, but less noise would also have been removed. The general rule is, the bigger the filter, the more noise that is removed but the more blurred the filtered image will be.

Figure 4.10(e) is the 7×7 mean filtered image of the salt-and-pepper noise degraded image. In comparing this image with the Gaussian degraded filtered image, the amount of noise present in this image is much greater. The 7×7 mean did not perform as well at removing the salt-and-pepper noise from the image as it did in removing the Gaussian noise from the image. The difficulty is that the mean filter is poor at removing outlier type noise such as salt-and-pepper noise. As mentioned earlier, outlier noise is defined as noise that produces values that drastically deviate from the normal value. For example, consider the five test scores of 100, 99, 99, 3, and 100. The score of 3 would be classified as an outlier, since it deviates drastically from the other four scores. The mean filter is very susceptible to outlier noise since the averaging is biased by the presence of outliers. The best filter at removing outlier noise is the median filter that will be discussed later in this chapter. This filter is very robust in the presence of outlier type noise. In particular, it does an excellent job at removing salt-and-pepper noise from images and does a better job of preserving edges than the mean filter.

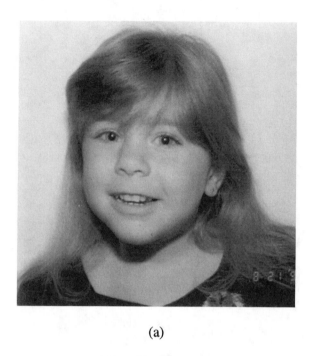

(a)

(b)

Figure 4.10: Examples of using the mean filter: (a) the original image
and (b) Gaussian degraded image with a variance of 600.

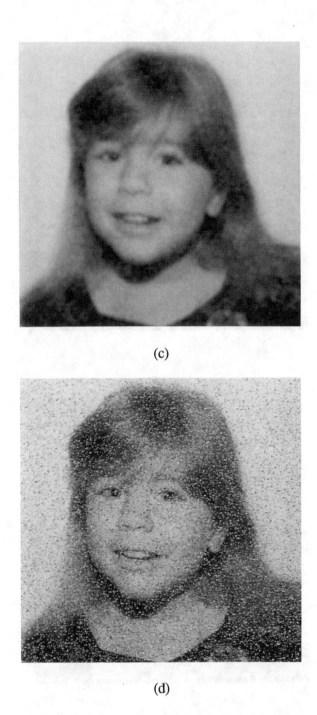

(c)

(d)

Figure 4.10: Examples of using the mean filter: (c) the 7×7 mean
              filtered image of Figure 4.10(b) and (d) the salt-and-
              pepper noise degraded image of Figure 4.10(a) with a
              probability of occurrence of 15% .

(e)

Figure 4.10: Examples of using the mean filter: (e) the 7×7 mean
filtered image of Figure 4.10(d).

To show that the mean filter behaves as a lowpass filter attenuating high
frequency components, the Fourier magnitude spectrum of the mean filter can be
obtained from the DFT of its impulse response using:

$$\mathcal{F}\{h(x, y)\} = H(n, m) = \frac{1}{N^2} \sum_{y=0}^{N-1} \sum_{x=0}^{N-1} h(x, y) \cdot e^{-j2\pi(nx + my)/N} , \quad (4.19)$$

Letting $h(x, y) = 1/N_t$, which is a constant equal to one over the number of points
included in the mean filter operation, and then using the shifting property of the
Fourier transform as defined in Equation (2.55), the Fourier transform of the
impulse response of the mean filter $h(x, y)$ is evaluated as

$$H(n, m) = \frac{1}{N_t} \sum_{x, y \in H} \sum e^{-(j2\pi xn + j2\pi ym)/N} , \quad (4.20)$$

where n and m are the spatial frequencies in the x and y directions, respectively.
Figure 4.11 gives the scaled (between 0 and 255) magnitude spectrum for the
impulse response of a 7×7 mean filter. Notice that this filter attenuates high
spatial frequencies while preserving the low spatial frequencies present in an

image. Also notice how this filter resembles the two-dimensional sinc function. Another way of computing the Fourier magnitude spectrum of the 7×7 mean filter is to the take the DFT of the two-dimensional rect function with a width and height equal to 7 and then scale the result by 49. In Chapter 2 it was shown that the DFT of the two-dimensional rect function is the two-dimensional sinc function. The larger the mean filter becomes, the smaller in width the sinc function becomes and the fewer spatial frequencies the filter preserves. The net effect is that more of the high spatial frequencies are attenuated, increasing the blurring of the image. Note the ringing in the Fourier magnitude spectrum shown in Figure 4.11. This is as expected due to the finite discontinuity that is present in the impulse response of the mean filter.

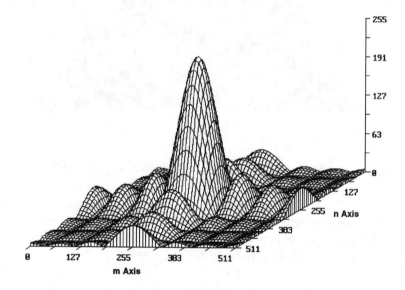

Figure 4.11: The Fourier magnitude spectrum of a 7×7 mean filter.

Another common linear spatial filter is the Gaussian filter. This filter is circularly symmetric and uses filter mask weights that vary according to the Gaussian function:

$$h(x, y) = e^{-\pi(x^2 + y^2)/a^2} \qquad , \qquad (4.21)$$

where the parameter a determines the width of the Gaussian filter. The Fourier magnitude spectrum of the Gaussian filter is also Gaussian in shape and is defined from Equation (2.52) as

$$|\mathcal{F}\{h(x, y)\}| = e^{-\pi a^2 (n^2 + m^2)/N} \qquad , \qquad (4.22)$$

where n and m are the spatial frequencies in the x and y directions, respectively. Equation (4.22) shows that the Gaussian filter is also a lowpass filter, attenuating the high spatial frequencies present in an image. Figure 4.12 gives the 5×5 filter mask for a Gaussian filter as defined by Equation (4.21) for $a = 3.32$. To maintain the same average brightness in the filtered image as that of the original image, this mask must also be scaled so that the sum of all of its coefficients equals one. Summing all of the coefficients given in Figure 4.12 yields a value of 9.96, which is then used to scale each of the 25 coefficients. Actually, what is typically done is to perform the filtering operations using only the numerator values and then divide the final filter output by 249. This allows for integer arithmetic to be used during the convolution part of the filtering algorithm.

The 5×5 Gaussian filter as described by the filtering mask given in Figure 4.12 and the scaling factor of 9.96 were applied to only the right half of the image given in Figure 4.10(a) to make it easier for the reader to observe the blurring caused by the filtering operation. Figure 4.13 is a zoomed-in version of the Gaussian filtered image showing in detail the eyes of the girl. Note the blurring of the eye on the right side of the image in comparison to the left side of the image due to application of the 5×5 Gaussian filter.

$\dfrac{3}{25}$	$\dfrac{6}{25}$	$\dfrac{8}{25}$	$\dfrac{6}{25}$	$\dfrac{3}{25}$
$\dfrac{6}{25}$	$\dfrac{14}{25}$	$\dfrac{19}{25}$	$\dfrac{14}{25}$	$\dfrac{6}{25}$
$\dfrac{8}{25}$	$\dfrac{19}{25}$	$\dfrac{25}{25}$	$\dfrac{19}{25}$	$\dfrac{8}{25}$
$\dfrac{6}{25}$	$\dfrac{14}{25}$	$\dfrac{19}{25}$	$\dfrac{14}{25}$	$\dfrac{6}{25}$
$\dfrac{3}{25}$	$\dfrac{6}{25}$	$\dfrac{8}{25}$	$\dfrac{6}{25}$	$\dfrac{3}{25}$

Figure 4.12: The filter mask coefficients for a 5×5 Gaussian mask filter.

The median filter is a nonlinear spatial filter that is good at removing outlier type noise. It does a better job than the mean filter at preserving edges within an image. The calculation of the median filter is different from that of the mean filter. The convolution operation as defined by Equation (4.14) is not used. The filter mask in this case simply defines what pixels are to be included in the median calculation. The calculation of the median filter begins by ordering the N pixels included in the filter operation as defined by the filter mask from their minimum to maximum values:

$$F_{(0)} \leq F_{(1)} \leq F_{(2)} \cdots \leq F_{(N-2)} \leq F_{(N-1)} \ , \qquad (4.23)$$

where $F_{(0)}$ is the minimum value and $F_{(N-1)}$ is the maximum value of all the pixels included in the filter calculation. The output of the median filter is the median of these values and is found using

$$F_{\text{med}} = \begin{cases} \dfrac{F_{(N/2)} + F_{(N/2-1)}}{2} & \text{for } N \text{ even} \\ F_{((N-1)/2)} & \text{for } N \text{ odd} \end{cases} \qquad (4.24)$$

Typically, an odd number of filter elements are used to avoid the additional averaging of the middle two pixels of the order set as required for an even number of elements. Filters based upon Equation (4.23) are known as *order statistic filters*. There have been many filters derived from Equation (4.23), some of which will be discussed in Chapter 5.

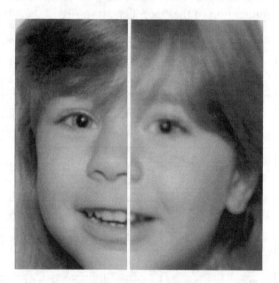

Figure 4.13: An example of filtering the right side of Figure 4.10(a)
with a 5×5 Gaussian filter.

Figure 4.14(a) gives a 5×5 sub-image of a vertical edge separating two graylevel regions of 0 and 9. Also contained within the graylevel region of 9 is one outlier noise pixel at graylevel 18. In computing the 3×3 mean and median filters of this sub-image, it is assumed that the graylevel regions of 0 and 9 extend far beyond just the 5×5 region shown in Figure 4.14(a). The 5×5 median filtered sub-image is shown in Figure 4.14(b). The sharp transition of the vertical edge has been preserved and the outlier pixel has been completely removed. In fact, the median filter can successfully remove an outlier pixel provided that the number of outlier pixels is less than 50% of the total number of pixels used in the calculation. The median filter is known to have a 50%

breakdown value. The breakdown value of a filter is defined as the percentage of outlier pixels in which the filter is no longer able to remove the outlier noise. Figure 4.14(c) is the same 5×5 sub-image filtered this time using a 3×3 mean filter. The mean filter was unsuccessful at completely removing the outlier pixel. The only thing the mean filter did was reduce the height of the outlier and increase its width. Figure 4.14(c) also shows the poor edge preserving quality of the mean filter. The sharp vertical edge in Figure 4.14(a) has been changed to a gradually sloped edge.

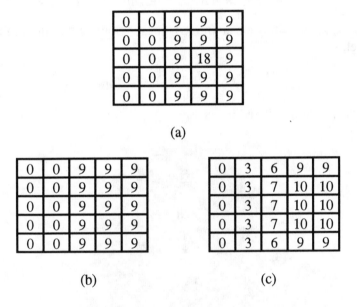

Figure 4.14: A 5×5 image of a vertical edge with one outlier pixel:
(a) the original image, (b) the median filtered image,
and (c) the mean filtered image.

Figure 4.15 is the output image after 3×3 median filtering of the 15% salt-and-pepper image given in Figure 4.10(d). The median filter was able to remove most of the salt-and-pepper noise. The pixel locations where the salt-and-pepper noise is still present are where more that 50% of the pixels within the filter calculation were due to salt-and-pepper noise. At these locations the breakdown value of the median filter was exceeded and the filter was unable to remove the salt-and-pepper noise from the image. In comparing the 7×7 mean filtered version of this image, given in Figure 4.10(e), to the 3×3 median filtered image, the median filter far outperformed the mean filter in the presence of outlier type noise. The mean filtered image was unable to remove the salt-and-pepper noise and also blurred the image.

Another spatial filter of interest is the *highpass*, which is sometimes referred to as an *unsharp* filter. This filter sharpens a blurry image by enhancing the high frequencies present in the image. The spectral components of an image can be divided into low frequency $f_{lowpass}(x, y)$ and high frequency $f_{highpass}(x, y)$ components:

$$f(x, y) = f_{lowpass}(x, y) + f_{highpass}(x, y) \ . \qquad (4.25)$$

Given that the mean filter is a lowpass filter and effectively removes high frequency components from an image, Equation (4.25) can be rewritten as

$$f(x, y) = f_{mean}(x, y) + f_{highpass}(x, y) \ , \qquad (4.26)$$

where $f_{mean}(x, y)$ is the mean filtered image. Solving for the highpass filter output yields

$$f_{highpass}(x, y) = f(x, y) - f_{mean}(x, y) \ , \qquad (4.27)$$

which is the highpass filtered image.

Figure 4.15: An example of using a 3×3 median filter on the salt-and-pepper image of Figure 4.10(d).

In many instances, it is desired to just enhance the high frequency components of an image without attenuating the low frequency components to sharpen a blurred image. In this situation, it is desired to add a proportional amount of the high frequency components to the original image $f(x, y)$. Multiplying Equation (4.27) by a gain term G and adding this quantity to $f(x, y)$ yields

$$f_h(x, y) = (1 + G) \cdot f(x, y) - G \cdot f_{mean}(x, y). \qquad (4.28)$$

For $G = 0$, Equation (4.28) reduces to the original image, and as G increases, the high frequency components are enhanced, improving the overall sharpness of the image. Typical values for G are in the range of 0 to 2.

Equations (4.27) and (4.28) can be implemented first by computing the mean filtered image and then using either Equation (4.27) or (4.28) for every pixel within the image, or the linearity property of spatial convolution can be used. Given two filter masks $h_1(x, y)$ and $h_2(x, y)$ that are defined by the same regions H but with different weights, the difference of the two filter operations can be written as

$$g(x, y) = g_1(x, y) - g_2(x, y) \ , \tag{4.29}$$

or

$$g(x, y) \quad = \quad \sum\sum_{i,\,j \in H} f(x - i, y - j) \cdot h_1(i, j) \ - \ \sum\sum_{i,\,j \in H} f(x - i, y - j) \cdot h_2(i, j) \ . \tag{4.30}$$

Since both summations on the right-hand side of Equation (4.30) are over the same filter regions, Equation (4.30) becomes the convolution of the difference of the filter masks with the input image $f(x, y)$:

$$g(x, y) \quad = \quad \sum\sum_{i,\,j \in H} f(x - i, y - j) \cdot [\, h_1(i, j) - h_2(i, j)\,] \ . \tag{4.31}$$

The concepts used to generate Equation (4.31) can also be used to derive the filter masks required to implement Equations (4.27) and (4.28). Consider the generation of the 3×3 filter masks that implement Equations (4.27) and (4.28). The first mask required is one that produces the original image, effectively performing no filtering. Inspection of Equation (4.15) for the 3×3 spatial filtering operation yields the filter mask coefficients of zero except for A_4, which must be one. Figure 4.16 gives the filter mask that yields the original image $f(x, y)$. The second mask required is that of the 3×3 mean filter mask as given in Figure 4.8. Taking the difference of these two masks produces the filter mask given in Figure 4.17, which implements the unsharp filter as defined by Equation (4.27). Figure 4.18 gives the filter mask that implements Equation (4.28). This mask was generated by the same procedures used for the filter mask for Equation (4.27) except the mean filter mask was multiplied by the gain coefficient G prior to taking the difference.

Figure 4.19(a) is an image of the lunar landing module located at the Kennedy Space Center. Figure 4.19(b) is the highpass filtered image using Equation (4.27) with the 3×3 unsharp filter mask as defined by Figure 4.17.

0	0	0
0	1	0
0	0	0

Figure 4.16: The filter mask coefficients to produce a filtered image $f(x, y)$.

$\frac{-1}{9}$	$\frac{-1}{9}$	$\frac{-1}{9}$
$\frac{-1}{9}$	$\frac{8}{9}$	$\frac{-1}{9}$
$\frac{-1}{9}$	$\frac{-1}{9}$	$\frac{-1}{9}$

Figure 4.17: The filter mask coefficients that implement Equation (4.27).

$\frac{-G}{9}$	$\frac{-G}{9}$	$\frac{-G}{9}$
$\frac{-G}{9}$	$\frac{9 + 8G}{9}$	$\frac{-G}{9}$
$\frac{-G}{9}$	$\frac{-G}{9}$	$\frac{-G}{9}$

Figure 4.18: The filter mask coefficients that implement Equation (4.28).

(a)

(b)

Figure 4.19: An example of using Equation (4.27) with a 3×3
filter mask: (a) the original image and (b) the
unsharp mask image.

This image displays the edges associated with the original graylevel image. The removal of the low frequency components has set the large homogeneous regions to zero and has enhanced the edges within the image. This implies that large homogeneous regions of an image are associated with the low spatial frequencies, while the edges are associated with the higher spatial frequencies. Edge detection of an image is the removal of the low spatial frequencies present and will be further discussed in Chapter 8. Figure 4.20(a) is a blurred image of a young girl. The lack of sharpness is easily observed in viewing the pupils of the girl's eyes. Figure 4.20(b) is the highpass enhanced image obtained using Equation (4.28) with $G = 1.0$ and for an 11×11 filter mask. This mask was applied twice to the image to obtain the desired sharpness.

4.3 Spatial Frequency Filtering

Another common method of image filtering is to manipulate an image's real and imaginary Fourier components computed via the DFT. Actually, the FFT algorithm is used because of its computational improvement over the DFT. Equation (4.32) relates a linear filtered image $g(x, y)$ to the filter's impulse response or filter mask $h(x, y)$ and the input image $f(x, y)$:

$$g(x, y) \quad = \quad \sum_{i, j \in H} \sum f(x - i, y - j) \cdot h(i, j) , \qquad (4.32)$$

where $0 \le x - i < N$ and $0 \le y - j < M$. Equation (4.32) is simply the convolution of the filter mask $h(x, y)$ with the input image $f(x, y)$. The type of filtering that is performed depends completely on the values used in the filter mask.

Instead of implementing Equation (4.32) directly, the two-dimensional DFT of both sides can be obtained using the Fourier transform property given in Equations (2.65) and (2.66), which relates spatial convolution to multiplication in the Fourier domain:

$$G(n, m) = F(n, m) \cdot H(n, m) , \qquad (4.33)$$

where n and m are the spatial frequencies in the x and y directions and $H(n, m)$, $F(n, m)$ and $G(n, m)$ are the Fourier components of $h(x, y)$, $f(x, y)$, and $g(x, y)$, respectively. The DFT has changed the linear filter operation of convolution in the spatial domain to multiplication in the frequency domain. The filtered image is found as the inverse DFT of Equation (4.33)

$$g(x, y) = \mathcal{F}^{-1}\{G(n, m)\} . \qquad (4.34)$$

The advantage of performing image filtering in the Fourier domain rather than the spatial domain is that in many instances the frequency response of the filter

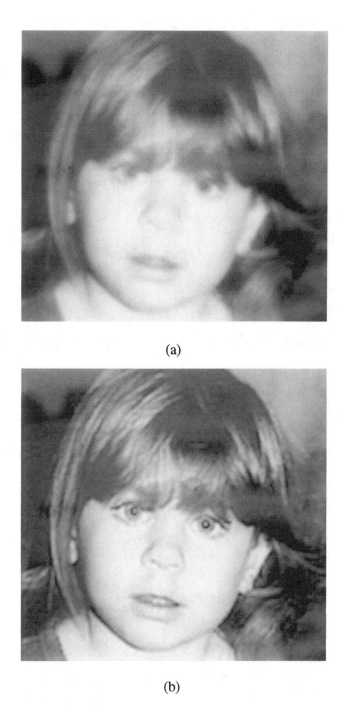

(a)

(b)

Figure 4.20: An example of using an 11×11 highpass filter to
sharpen an image: (a) the blurred image and (b) the
sharpened image.

is known, but its impulse response is unknown, or it is desired to filter a given spatial frequency. In addition, for large sized filter masks, it is faster to compute the DFT of the filter mask and the input image, followed by the application of Equation (4.33) to perform the filtering operation, and then to take the inverse DFT of the filtered spectral components to obtain the filtered image. Figure 4.21 summarizes the steps required to filter an image using Fourier frequency methods.

In general, $H(n, m)$ can be complex valued in the form

$$H(n, m) = H_R(n, m) + j\, H_I(n, m) \;, \tag{4.35}$$

as can the input image spectral components,

$$F(n, m) = F_R(n, m) + j\, F_I(n, m) \;. \tag{4.36}$$

Substituting Equations (4.35) and (4.36) in Equation (3.33) and performing the multiplication yields

$$G_R(n, m) \;=\; [F_R(n, m)\, H_R(n, m) - H_I(n, m)\, F_I(n, m)] \tag{4.37}$$

and

$$G_I(n, m) \;=\; j\, [F_I(n, m)\, H_R(n, m) + F_R(n, m)\, H_I(n, m)] \;, \tag{4.38}$$

where $G_R(n, m)$ and $G_I(n, m)$ are the real and imaginary frequency components of the filtered image $g(x, y)$.

The magnitude spectrum of the filtered image is

$$|G(n, m)| = \sqrt{G_R(n, m)^2 + G_I(n, m)^2} \tag{4.39}$$

and its phase spectrum is

$$\theta(n, m) = \tan^{-1}\!\left[\frac{G_I(n, m)}{G_R(n, m)}\right] \;, \tag{4.40}$$

or

$$\theta(n, m) = \tan^{-1}\!\left[\frac{F_I(n, m)\, H_R(n, m) + F_R(n, m)\, H_I(n, m)}{F_R(n, m)\, H_R(n, m) - H_I(n, m)\, F_I(n, m)}\right] \;. \tag{4.41}$$

Inspection of Equation (4.41) reveals that in general, a complex filter function $H(n, m)$ modifies the phase components of the original unfiltered image. In the filtering of images, as stated in Section 2.2 of Chapter 2, the phase of the original spectrum must be preserved during the filtering operation. This is accomplished if $H(n, m)$ is real, containing no imaginary components. In this situation, Equation (4.41) reduces to

$$\theta(n, m) = \tan^{-1}\left[\frac{F_I(n, m)\, H_R(n, m)}{F_R(n, m)\, H_R(n, m)}\right] = \tan^{-1}\left[\frac{F_I(n, m)}{F_R(n, m)}\right] , \qquad (4.42)$$

which is the phase spectrum of the unfiltered image. Filters of this type are known as *zero phase* or *phase preserving* filters. The only requirement for $H(n, m)$ to be a phase preserving filter is to have its values be real. A phase preserving filter also implies a symmetric impulse response:

$$h(x, y) = h(-x -y) . \qquad (4.43)$$

Fourier frequency filtering using phase preserving filters is accomplished by scanning every spatial frequency component of the $F(n, m)$ and then multiplying both the real and imaginary parts of each of these components by the corresponding $H(n, m)$, thereby producing a new filtering spectrum $G(n, m)$.

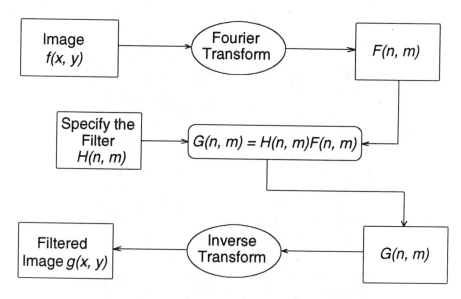

Figure 4.21: A flow chart showing the steps required to use Fourier frequency techniques to filter an image.

Figure 4.22(a) is an image of a young girl taken with a faulty charge coupled device (CCD) camera. Close inspection of this image shows that every other line is at a different brightness level. This is due to a gain misalignment of the amplifier stages in the CCD camera between the two different fields that compose one frame of NTSC video. This noise is readily observable as

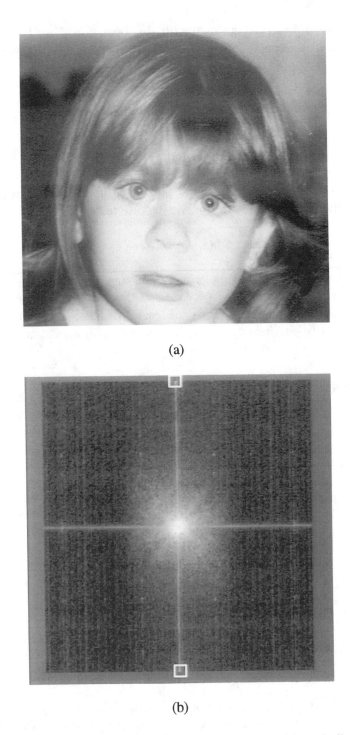

(a)

(b)

Figure 4.22: An example of an image corrupted with periodic noise:
(a) the original image and (b) its Fourier magnitude
spectrum.

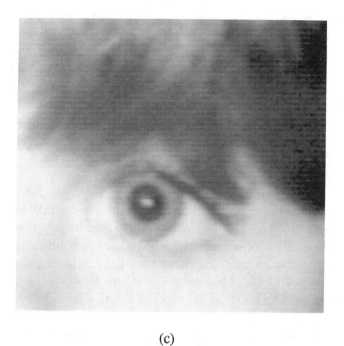

(c)

(d)

Figure 4.22: An example of an image corrupted with periodic noise:
(c) the zoomed image of the left eye showing the
periodic noise, and (d) the enhanced image
of the left eye after Fourier frequency filtering.

horizontal streaks predominant to the right of the girl's hair. Figure 4.22(c) is a zoomed-in version of the left eye, highlighting the periodic interference present in Figure 4.22(a). The horizontal streaks are quite prevalent near the pupil of the eye. The uneven widths of the line streaks are due to the sampling process of the electronic printer used to produce the hard copy photographs for this book.

This degradation can be modeled as additive periodic noise $n_p(x, y)$, with the period of the noise equal to exactly the vertical Nyquist sampling frequency

$$g(x, y) = f(x, y) + n_p(x, y) \quad , \qquad (4.44)$$

where $f(x, y)$ is the noise free image and $g(x, y)$ is the degraded image. The Fourier transform of the degraded image yields the original Fourier spectral components of the noise free image plus two impulses located at $n = 0$ and $M = \pm M/2$. Figure 4.22(b) shows the brightened scaled magnitude spectrum of the degraded image. Two white boxes have been used to highlight the location of these two impulses at the coordinates $n = 0$ and $M = \pm M/2$. Because of the impulse removing capability of the median filter, a 3×3 median filter was applied at these two frequency locations to remove the two impulses due to the periodic noise $G(0, M/2)$ and $G(0, -M/2)$. The median filter was applied separately to both the real and imaginary components of $G(0, M/2)$ and $G(0, -M/2)$. Inverse Fourier transforming the filtered image yields a restored image of the girl's eye as shown in Figure 4.22(d). Note that the horizontal streaks are no longer visible near the pupil of the eye.

The filter function $H(n, m)$ is a two-dimensional function that depends on the spatial frequencies n and m. The values for $H(n, m)$ can be chosen in two ways. The first method is to specify a two-dimensional function that depends on n and m directly. The other method starts with a one-dimensional filter and expands this filter to two dimensions. The benefit of the second approach over the first is the vast knowledge of one-dimensional filters that is available. A common method of writing $H(n, m)$ using one-dimensional filters is to separate the filter into the multiplication of two one-dimensional filters along the n and m spatial frequency directions:

$$H(n, m) = H_1(n) \cdot H_2(m) \quad . \qquad (4.45)$$

Another common method is to use circular symmetry and only a single one-dimensional filter to describe $H(n, m)$:

$$H(n, m) = H_1(\rho) \quad , \qquad (4.46)$$

where

$$\rho^2 = n^2 + m^2 \quad \text{for } \rho \geq 0 \qquad (4.47)$$

and

$$\phi = \tan^{-1}\left[\frac{m}{n}\right] \, .\qquad(4.48)$$

Equation (4.46) takes a one-dimensional filter and rotates it 360° about $n = m = 0$, producing a circularly symmetric filter that is two-dimensional in shape. As stated earlier, use of either Equation (4.45) or (4.46) to Fourier filter images requires that the one-dimensional filters also be phase preserving.

Figure 4.23(a) gives the magnitude spectrum for a typical lowpass filter. This filter passes the frequencies below the cutoff frequency ω_c and attenuates the frequency components above the cutoff frequency. One way of converting this one-dimensional filter to two dimensions is accomplished by setting both $H_1(n)$ and $H_2(m)$ to $H_L(f)$, as is given in Figure 4.23. The other way is to use Equation (4.46) with f set equal to ρ. Figure 4.23(b) gives a typical one-dimensional highpass filter. This filter passes the high frequency components above the cutoff frequency ω_c and attenuates the low frequency components. In many cases when using a highpass filter to filter an image, it is desired not to change the average brightness of the image. This is accomplished by leaving $F(0, 0)$, the average value of the image, unchanged during the Fourier filtering operation.

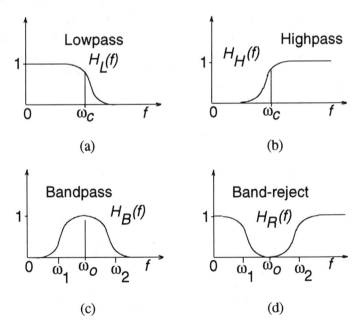

Figure 4.23: Four common one-dimensional filters used in Fourier filtering of images: (a) the lowpass filter, (b) the highpass filter, (c) the bandpass filter, and (d) band-reject filter.

Figure 4.23(c) gives the magnitude spectrum for a bandpass filter. This filter passes frequency components in the range of ω_1 to ω_2, with the center frequency of the pass band given by ω_0. The final filter is the band-reject filter, given in Figure 4.23(d). This filter attenuates the frequency components in the range of ω_1 to ω_2. Like the lowpass filter, these other three filters can also be converted to a two-dimensional filter using either Equation (4.45) or (4.46).

An ideal lowpass filter is defined as one that contains no transition set of frequencies in which the filter goes from the passing of frequency components to the attenuation of frequency components:

$$H(f) = \begin{cases} 1 & \text{for } f \leq \omega_c \\ 0 & \text{for } f > \omega_c \end{cases} . \tag{4.49}$$

Figure 4.24(a) shows a plot of its magnitude spectrum. This filter has a finite discontinuity at the cutoff frequency ω_c. In a similar way, the ideal highpass filter is defined as

$$H(f) = \begin{cases} 1 & \text{for } f \geq \omega_c \\ 0 & \text{for } f < \omega_c \end{cases} . \tag{4.50}$$

This ideal filter also has a finite discontinuity. Figure 4.24(b) gives the highpass filter magnitude spectrum. Care must be taken when using the ideal lowpass and highpass filters, since these filters have a finite discontinuity in their magnitude spectrum. The truncation of the Fourier spectral components at the cutoff frequency introduces ringing artifacts that are observable in the filtered image, due to the Fourier transform property that truncation in one domain, which produces a discontinuity, results in ringing in the other domain.

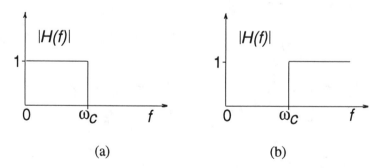

(a)   (b)

Figure 4.24: The magnitude spectrums for (a) the ideal lowpass filter and (b) the ideal highpass filter.

There are many types of one-dimensional filters that are available. Many of these filters come directly from electronic circuit designs that have existed for many decades. Some of the most common ones are the *Butterworth, elliptical, Chebyshev,* and *Bessel* filters as well as the simple first order resistor and capacitor network and the second order resistor, capacitor, and inductor network. Unfortunately, because of the phase preserving requirements of the Fourier filtering of images, only the magnitude spectrum of these filters can be used to generate $H(n, m)$ using either Equations (4.45) or (4.46). The Butterworth filter offers the ability to control the rate at which the filter changes from passing frequency components to attenuating them. In addition, this filter linearly changes from the pass band to the attenuation band without the presence of the ripples that are typically found in some of the other filters.

The one-dimensional Butterworth lowpass filter is given by

$$H(f) = \frac{1}{\sqrt{1 + \left(\dfrac{f}{\omega_c}\right)^{2N}}} \, , \tag{4.51}$$

where N is the order of the filter and ω_c is the cutoff frequency or 3 dB attenuation point ($20 \cdot \log_{10}[|H(f)|]$). Figure 4.25(a) gives three plots of the magnitude spectrum for $N = 1$, $N = 4$, and $N = \infty$. As N increases, the rate at which the filter's frequency response changes from the pass band to the attenuation band decreases. For $N = \infty$, the lowpass Butterworth filter becomes the ideal lowpass filter. At the cutoff frequency, this filter takes on the value of 0.707 for all values of N. Similar to the Butterworth lowpass filter, the Butterworth highpass filter is defined as

$$H(f) = \frac{1}{\sqrt{1 + \left(\dfrac{\omega_c}{f}\right)^{2N}}} \, . \tag{4.52}$$

Figure 4.25(b) gives the magnitude response of this filter for $N = 1$, $N = 4$, and $N = \infty$.

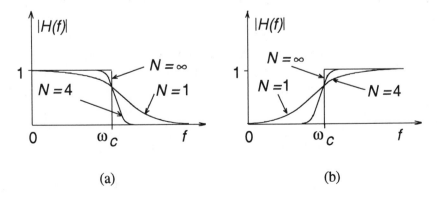

(a)   (b)

Figure 4.25: The Butterworth (a) lowpass and (b) highpass filters.

Like the lowpass Butterworth filter, the order of the filter determines the rate at which the filter's frequency response changes from the attenuation band to the pass band and for $N = \infty$, this filter becomes the ideal highpass filter. Both the lowpass and highpass Butterworth filters are easily converted to a two-dimensional filter by using either Equation (4.45) or (4.46). For the circularly symmetric case, ρ is substituted for f in Equations (4.51) and (4.52).

Figure 4.26(a) shows an example of the circularly symmetric Butterworth lowpass filter generated using Equations (4.51) and (4.46), with $N = 4$, $\omega_c = 100$, and with an image size of 512×512. Figure 4.26(b) is the corresponding circularly symmetric highpass filter. Assuming a normalized sample spacing in the x and y directions of one, the sampling frequency for an $N \times N$ image becomes unity. The maximum normalized frequency as defined by the Nyquist theorem is then in the range of -0.5 to 0.5 and the normalized cutoff frequency relative to the sampling frequency is given by ω_c/N. For Figures 4.26(a) and (b), the normalized cutoff frequency is $100/512 = 0.195$. In generating both the lowpass and highpass Butterworth magnitude spectrums, the radial DC ($\rho = 0$) frequency was placed in the center, as shown in the figures. The Butterworth lowpass filter starts with a response of one and decreases in value radially as the spatial frequencies are increased. On the other hand, the highpass filter starts with a response of 0 at DC and increases in value as the spatial frequencies increase radially. Both filters show that the response of circularly symmetric filters is constant for a fixed radial spatial frequency.

To filter an image using these two filters, the Fourier components of the image must be centered within the arrays used to store the real and imaginary spatial frequency components. This is easily accomplished using the centering property of the DFT given by Equation (2.64) from Chapter 2:

$$f(x, y) \Rightarrow f(x, y) \, (-1)^x \, (-1)^y \; . \tag{4.53}$$

Prior to taking the DFT of an image, every pixel in the image is multiplied by $(-1)^{x+y}$. Figure 4.27(a) shows the uncentered magnitude spectrum of an image containing a white square object. Figure 4.27(b) shows the DFT of the same image after application of the centering property of the DFT as defined from Equation (4.53). Notice how the DC frequency is now located at the center of the discrete Fourier space. Since application of circularly symmetric filters requires the use of the centering property of the DFT, the radial spatial frequency for the DFT of an $N \times N$ image is then defined as

$$\rho = \sqrt{(n - N/2)^2 + (m - N/2)^2} \quad \text{for } 0 \le n, m \le N - 1 , \tag{4.54}$$

where n and m in this case are the array indices that access the real and imaginary spatial frequency components. Figures 4.28(a) and (b) give an example of using a circularly symmetric Butterworth lowpass filter to remove

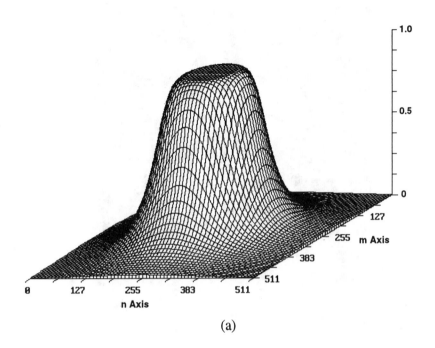

(a)

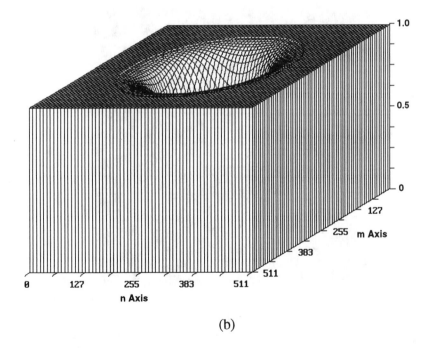

(b)

Figure 4.26: An example of circularly symmetric filters: (a) the
Butterworth lowpass and (b) the highpass Butterworth
filters with $\omega_c = 100$ and $N = 4$.

(a)

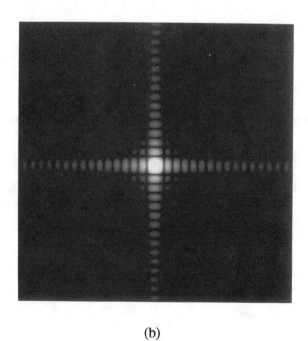

(b)

Figure 4.27: An example of using the centering property of the
DFT: (a) the uncentered spectrum and (b) the
centered spectrum.

(a)

(b)

Figure 4.28:   An example of using the lowpass Butterworth filter to blur an image: (a) the original image and (b) the lowpass Butterworth filtered image with $\omega_c = 6$ and $N = 1$.

the high spatial frequencies from an image. Figure 4.28(a) is the original 512×512 image and Figure 4.28(b) is the Butterworth lowpass filtered image with $N = 1$ and $\omega_c = 6$ (normalized cutoff frequency of 0.012). This filter has removed most of the high spatial frequency components, blurring the filtered image. As shown before, the sharpness of an image is directly related to the amount of high spatial frequency components present.

4.4 Image Restoration

Several image enhancement techniques based on point and neighborhood operations were presented in Chapter 3 and the first part of this chapter. The goal of image enhancement is to improve the overall quality of an image for human viewing or for input to an automatic image recognition system. The enhancement of images does not require any knowledge of how an image was degraded. In some noise degradation instances, only an estimate of the type of noise is desired so that the best enhancement filter can be chosen. The goal of image restoration, on the other hand, is to modify a degraded image to produce the original undegraded image. The use of image restoration techniques requires knowledge about how an image was degraded. In other words, a model for the degradation must exist and be known.

A short analogy will illustrate the difference between enhancement and restoration. For example, consider the repair of an antique car in which the generator has gone bad due to a poor engineering design. Upon visiting the local automotive store, it was discovered that the original generator is not available and it is recommended that a similar generator from another manufacturer's car be purchased because of its better quality and design. Purchase and installation of this generator makes the antique car operational again. This is an example of enhancement, not restoration. Restoration requires placing the car back into its original manufactured condition for better or worse. Taking restoration to the extreme would be to obtain the original manufacturer's blueprints for the generator and have one generator custom fabricated, a costly procedure given that this generator will still be of lower quality than the replacement generator recommended by the automotive parts store. Yet, this is restoration.

Restoration of images follows the same philosophy, returning a degraded image to its original condition in hope that the restored image will be of better quality than the degraded image. A typical example is the restoration of a blurred image, given that the blurring function is known. Image enhancement, on the other hand, might be the brightening of a dark image, making the enhanced image more acceptable for viewing. Figure 4.29 gives a block diagram of a common electronic imaging system that contains a degradation as a result of a lens aberration or due to an improperly focused lens. An object is imaged onto a camera's sensor via the lens. The camera's sensor converts the irradiance variations across its surface to electrical voltages. At this stage, electronic noise is added to the image, further degrading the image quality. This image is then sampled and digitized by an image acquisition system. The blurring of the image as a result of lens aberrations or improper focus can be reduced by image restoration.

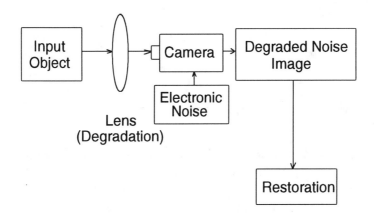

Figure 4.29: A block diagram showing the degradation of an image with the addition of electronic noise.

Figure 4.30 gives the model that describes the degradation of an image by an improperly focused or aberrated optical system. The image blurring by an optical system is modeled by $h(x, y)$ and the additive noise due to the camera's electronics is given by $r(x, y)$. The goal is to restore the degraded image $g(x, y)$ back to the original image $f(x, y)$. The degraded output image can be written as

$$g(x, y) = h[f(x, y)] + r(x, y) , \tag{4.55}$$

where $h[f(x, y)]$ defines the image degradation operation. For image restoration to be possible, two assumptions have to be made. The first assumption is that the degradation model $h(x, y)$ is linear,

$$h[k_1 f_1(x, y) + k_2 f_2(x, y)] = k_1 h[f_1(x, y)] + k_2 h[f_2(x, y)] , \tag{4.56}$$

and the second is that the degradation model is homogeneous and space invariant. If

$$g(x, y) = h[f(x, y)] , \tag{4.57}$$

then

$$g(x - a, y - b) = h[f(x - a, y - b)] . \tag{4.58}$$

The first assumption, given in Equation (4.56), states that the response to a sum of inputs is equivalent to the sum of the responses from each input. The second assumption, that the image degradation be homogeneous and space invariant, requires that the degradation be the same throughout the image. In other words, the topmost left part of the image is degraded the same as the bottom most right part of the image.

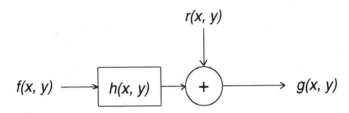

Figure 4.30: A degradation model that will be used in the process of image restoration.

These two assumptions allow the use of spatial convolution to model the degradation process. For an $N \times M$ image, Equation (4.55) becomes

$$g(x, y) = \sum_{m=0}^{y} \sum_{n=0}^{x} f(n, m)\, h(x - n, y - m) + r(x, y) \qquad (4.59)$$

or

$$g(x, y) = f(x, y) * h(x, y) + r(x, y), \qquad (4.60)$$

where * defines spatial convolution. Taking the DFT of Equation (4.60) and using the fact that convolution in the spatial domain is equivalent to multiplication in the Fourier domain yields

$$G(n, m) = H(n, m)\, F(n, m) + R(n, m), \qquad (4.61)$$

where $H(n, m)$, $F(n, m)$, $R(n, m)$, and $G(n, m)$ are the Fourier frequency components of the image degradation $h(x, y)$, the original undegraded image $f(x, y)$, the electronic noise $r(x, y)$, and the degraded image $g(x, y)$, respectively. The goal is to find $F(n, m)$ from Equation (4.61) given $G(n, m)$ and $H(n, m)$.

If the degradation model is not known directly, it can be found through a simple experiment. Consider the optical blurring of an image as given in Figure 4.29. This degradation is modeled by Equations (4.60) and (4.61). To determine $h(x, y)$, an object that is at least ten times smaller than the maximum spatial resolution of the undegraded imaging system is acquired with the degraded imaging system to produce an $N \times M$ sampled and digitized image. If an object is a least ten times smaller than the resolution of the imaging system, it can be approximated as an impulse $\delta(x, y)$. From Equation (2.53), the DFT of an impulse yields a constant Fourier spectrum of $1/(NM)$. Substituting the Fourier transform of the impulse response for the Fourier components of the noise free image in Equation (4.61) gives

$$G(n, m) = H(n, m) \frac{1}{NM} + R(n, m) \; . \qquad (4.62)$$

Applying the inverse DFT to Equation (4.62) gives

$$g(x, y) = h(x, y) \frac{1}{NM} + r(x, y) \; . \qquad (4.63)$$

If possible, several images of this small object should be obtained and averaged together to reduce the noise term $r(x, y)$, using the same approach that was given in the image averaging example given in Chapter 3. Equation (4.63) now reduces to

$$g(x, y) = h(x, y) \frac{1}{NM} , \qquad (4.64)$$

which is the degradation impulse response.

One approach to restoring the degraded image is to divide both sides of Equation (4.61) by the Fourier frequency components of the degradation $H(n, m)$ and solve for $F(n, m)$:

$$F(n, m) = \frac{G(n, m)}{H(n, m)} - \frac{R(n, m)}{H(n, m)} \; . \qquad (4.65)$$

Equation (4.65) is valid and can be used to find the spectral components of the noise free image provided $R(n, m)$ is known and $H(n, m)$ does not evaluate to zero over any spatial frequency.

For the noise free case, Equation (4.65) reduces to

$$F(n, m) = \frac{G(n, m)}{H(n, m)} \; . \qquad (4.66)$$

The division of a degraded image's Fourier components by the Fourier frequency components of the image degradation is known as *inverse filtering*. If $H(n, m)$ is zero at any particular frequency, $G(n, m)$ will also be zero and Equation (4.66) will yield zero over zero, which is undefined. In other words, for those frequencies in which $H(n, m)$ equals zero, the Fourier frequency components from the original undegraded image will be set to zero in the degraded image. Once this occurs there is no way to restore these components during the restoration process. The best that can be done is to estimate these Fourier components from surrounding Fourier spectral values. In summary, the steps involved to perform inverse filtering are to:

- take the centered DFT of the degraded image $g(x, y)$ to obtain $G(n, m)$,
- take the centered DFT of the degradation model $h(x, y)$ to obtain $H(n, m)$,
- compute $F(n, m)$ from Equation (4.66),
- and finally take the inverse DFT to obtain the restored image $f(x, y)$.

The set of images shown in Figure 4.31 gives an example of using inverse filtering to remove image blurring. Figure 4.31(a) is the undegraded noise free image and Figure 4.31(c) is the degraded image. This image has undergone a blurring degradation in the horizontal direction. Figure 4.31(c) was created by applying the centered DFT to the undegraded image to obtain its centered Fourier frequency components and then lowpass filtering these spectral components using the degradation model $H(n)$ given in Figure 4.31(b). Figure 4.31(d) is the result of applying the inverse filtering operation given in Equation (4.66). The inverse filter operation was unable to restore the image because the values of $H(n)$ decreased to zero very rapidly for large spatial frequencies.

Figure 4.31(e) is the inverse filtered image with the added constraint that only the Fourier frequency components in which $H(n) > 0.01$ were included in the restored image. All other frequency components were set to zero in $G(n, m)$. Figure 4.31(e) shows the sharpening of the degraded noise free image using inverse filtering. Figure 4.31(f) is the degraded image given in Figure 4.31(c) with additive zero mean uncorrelated Gaussian noise with a variance of 5. The restored image given in Figure 4.31(g) was generated using the exact same constraint that was used to generate the restored image given in Figure 4.31(e). Note the difficulty the inverse filtering operation had in restoring this image in the presence of noise. A variance of 5 corresponds to an average peak-to-peak graylevel variation due to the noise of 13. In fact, the noise is barely observable in Figure 4.31(f). Figure 4.31(h) is the degraded image from Figure 4.31(c) with additive zero mean uncorrelated Gaussian noise with a variance of 200. The Gaussian noise is now observable in this image. Figure 4.31(i) is the corresponding inverse filtered image, showing that there are very few features present in this image that were present in the noise degraded image. In this case, the inverse filtering operation produced a restored image that is worse than the degraded image. Inverse filtering performs poorly in the presence of noise, as illustrated by the last two examples given in Figure 4.31.

The *Wiener filter* performs better in the presence of noise than does the inverse filter. The use of the Wiener filter assumes that in addition to the degradation model, the second order statistics of the noise and the undegraded image are known. Let $S_r(n, m)$ be the power spectral density (PSD) of the noise and $S_f(n, m)$ be the power spectral density of the original undegraded image. A very well known theorem known as the Wiener-Khintchine theorem states that the PSD of a signal is related to its autocorrelation function through the Fourier transform. The Wiener-Khintchine theorem allows the PSD of the noise and the PSD of the undegraded $N \times M$ image to be written in terms of their autocorrelation functions $R_f(x, y)$ and $R_r(x, y)$ as

$$S_f(n, m) = \frac{1}{NM} \sum_{y=0}^{M-1} \sum_{x=0}^{N-1} R_f(x, y) \cdot e^{-j2\pi(nx/N + my/M)} \tag{4.67}$$

and

$$S_r(n, m) = \frac{1}{NM} \sum_{y=0}^{M-1} \sum_{x=0}^{N-1} R_r(x, y) \cdot e^{-j2\pi(nx/N + my/M)}, \tag{4.68}$$

where the autocorrelation function of the undegraded image is given by

$$R_f(x, y) = \frac{1}{(N-x)(M-y)} \sum_{m=0}^{M-1-y} \sum_{n=0}^{N-1-x} f(n, m) f(x+n, y+m) \tag{4.69}$$

and the autocorrelation of the noise is

(a)

Figure 4.31: An example of inverse filtering: (a) the original image.

Degradation Model

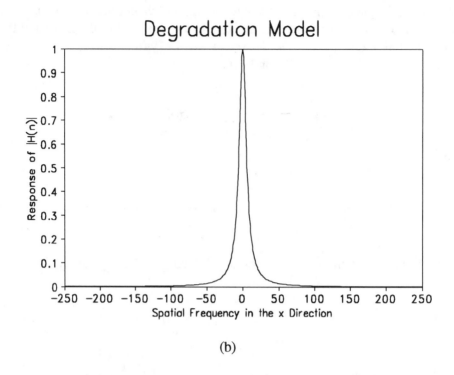

(b)

(c)

Figure 4.31: An example of inverse filtering: (b) the degradation model
and (c) the degraded image with no noise.

(d)

(e)

Figure 4.31: An example of inverse filtering: (d) the inverse filtered image with no constraints and (e) the inverse filtered image with the constraint of excluding all spectral components in the filtering process in which the degradation model is less than 1%.

(f)

(g)

Figure 4.31: An example of inverse filtering: (f) the degraded image
with additive Gaussian noise with a variance of 5 and
(g) the inverse filtered image with the constraints
as used for Figure 4.31(e).

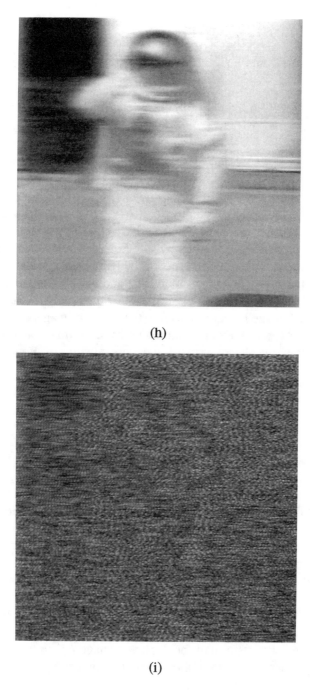

(h)

(i)

Figure 4.31: An example of inverse filtering: (h) the degraded image
with additive Gaussian noise with a variance of 200
and (i) the inverse filtered image with the constraints
as used for Figure 4.31(e).

$$R_r(x, y) = \frac{1}{(N-x)(M-y)} \sum_{m=0}^{M-1-y} \sum_{n=0}^{N-1-x} r(n, m)\, r(x+n, y+m) . \qquad (4.70)$$

Using Equations (4.67) and (4.68) the Wiener filter is defined as

$$F(n, m) = \frac{H^C(n, m)}{|H(n, m)|^2 + \gamma \dfrac{S_r(n, m)}{S_f(n, m)}} G(n, m) , \qquad (4.71)$$

where $H^C(n, m)$ is the complex conjugate of $H(n, m)$ and the parameter γ selects the type of Wiener filter that is implemented. For γ not equal to one, the Wiener filter is referred to as the *parametric Wiener Filter*. The Wiener filter gives the restored frequency components in terms of the degradation spectral components, the degraded image's spectral components, and the PSDs of the noise and undegraded image. The interesting thing about Equation (4.71) is that the PSD of the undegraded image is required to find the restored image.

Usually, the PSDs of the noise and the undegraded image are not known. In this situation, the right-hand term of the denominator of Equation (4.71) is replaced with a constant K,

$$F(n, m) = \frac{H^C(n, m)}{|H(n, m)|^2 + K} G(n, m) , \qquad (4.72)$$

and this parameter is adjusted until the best restored image is obtained. For $K = 0$, the Wiener filter becomes the inverse filter. As the value of K is increased, the Wiener filter performs less restoration and begins to degrade the degraded image with the degradation model. For the case of image restoration of a blurred image, a value of $K = 0$ yields the sharpest image but also enhances any noise that is present in the degraded image. As K increases, the Wiener filter removes less of the blurring from the degraded image. Hence, under this condition, the Wiener filter does not enhance as much of the noise present in the degraded image. In summary, the steps required to compute the Wiener filter as described by Equation (4.72) are to:

- take the Fourier transform of the degraded image $g(x, y)$ to obtain $G(n, m)$,
- take the Fourier transform of the degradation model $h(x, y)$ to obtain $H(n, m)$,
- select a value for the parameter K,
- compute $F(n, m)$ using Equation (4.72),
- take the inverse Fourier transform of $F(n, m)$ to obtain the restored image $f(x, y)$.

(a)

(b)

Figure 4.32: An example of Wiener filtering: (a) the degraded image with no noise and (b) the Wiener filtered image with $K = 0.0002$.

(c)

(d)

Figure 4.32: An example of Wiener filtering: (c) the degraded image
with additive Gaussian noise with a variance of 5 and
(d) the Wiener filtered image with $K = 0.002$.

(e)

(f)

Figure 4.32: An example of Wiener filtering: (e) the degraded image with additive Gaussian noise with a variance of 200 and (f) the Wiener filtered image with $K = 0.008$.

These steps are then repeated changing the value of K until the best restored image is obtained.

Figure 4.32(a) is the noise free degraded image of Figure 4.31(a) obtained using the same degradation model $H(n)$ given in Figure 4.31(b). Figure 4.32(b) is the restored image using the Wiener filter with $K = 0.0002$. The Wiener filter outperformed the unconstrained inverse filter, as seen by comparing 4.31(d) and 4.32(b). On the other hand, the constrained inverse filter and the Wiener filter performed about the same, as seen by comparing Figures 4.31(e) with 4.32(b). Figure 4.32(c) is the degraded image with additive zero mean uncorrelated Gaussian noise with a variance of 5. Figure 4.32(d) is its corresponding Wiener filtered image with $K = 0.002$. Comparing Figure 4.31(g) with Figure 4.32(d) it becomes clear that the Wiener filter is a much better filter to use than the inverse filter in the presence of noise. The image given in Figure 4.32(d) has been deblurred and is no longer objectionable. Figure 4.32(e) is the degraded image with additive uncorrelated zero mean Gaussian noise with a variance of 200. Figure 4.32(f) is the Wiener filtered image with $K = 0.008$. Again the Wiener filtered output image is much better than the inverse filtered image given in Figure 4.31(i). Figure 4.32(f) shows that the addition of noise in the degradation process makes the restoration problem much more difficult.

CHAPTER 5

Nonlinear Image Processing Techniques

The nonlinear image processing techniques discussed in this chapter can be separated into three categories. In the first category are those filters that operate within a small local window as defined by a filter template similar to the spatial filters discussed in Chapter 4. Many of these filters are based on the graylevel ordering of the pixels included in the filter operation from their minimum to their maximum values. The second area of nonlinear filters discussed includes several of the adaptive filters that are commonly used in electronic image processing. These filters change their filtering characteristics depending on the noise and the image features that are being filtered. The last category discussed is homomorphic filtering, used to remove multiplicative noise and image shading from an image.

5.1 Nonlinear Spatial Filters Based on Order Statistics

Linear spatial lowpass filters are commonly used to reduce noise in an image but at the cost of blurring the original image. Another limitation of linear spatial filters is that they perform poorly in the presence of outlier type noise. This was shown in Chapter 4 with the 7×7 mean filtering of a salt-and-pepper noise degraded image, given in Figure 4.10(e). The mean filter did not perform as well as the median filter. The mean filter blurred the image and did not effectively remove the salt-and-pepper noise. The median filter, which is a nonlinear spatial filter, preserved the image sharpness and removed a majority of the salt-and-pepper noise, as was shown in Figure 4.15. Nonlinear filters typically do better at preserving edges within an image, producing an overall sharper image than linear spatial lowpass filters.

Spatial nonlinear filters are based upon a local neighborhood about a pixel $f(x, y)$ as defined by a filter template H. Figure 5.1 shows an $N \times M$ image with a filter mask overlying the pixel $f(x, y)$. The hot spot defines which element of the filter mask overlies the image pixel $f(x, y)$. The hot spot and the elements defined by the filter mask and their values are all required to define a filter mask. Typically, square or rectangular shaped masks of odd lengths are used because of their ease in programming and hardware implementation. In addition, the use of odd sizes for these filter masks guarantees that the mask will be centered about the pixel $f(x, y)$ being filtered. Spatial filtering of the image is then accomplished by scanning the image pixel by pixel and then applying the filtering operation to each pixel within the image, using only those pixels defined by the filter mask.

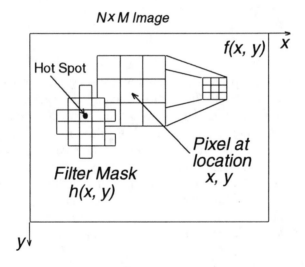

Figure 5.1: An illustration of spatial filtering using a neighborhood mask.

Nonlinear filters based on order statistics require that all the pixels defined in the filtering operation be ordered from their minimum graylevel to their maximum graylevel. Let F be the set of N pixels defined in the filter operation as given by the filter mask H. The first step in finding the order statistics is to order this set of pixels from their minimum to their maximum values:

$$F_{(0)} \leq F_{(1)} \leq F_{(2)} \cdots \leq F_{(N-2)} \leq F_{(N-1)} \; , \qquad (5.1)$$

where $F_{(0)}$ is the minimum value of all the pixels and $F_{(N-1)}$ is the maximum value of the pixels. In Chapter 4, the median filter was defined as the median of this ordered set:

$$F_{\text{med}} = \begin{cases} \dfrac{F_{(N/2)} + F_{(N/2-1)}}{2} & \text{for } N \text{ even} \\ F_{((N-1)/2)} & \text{for } N \text{ odd} \end{cases} \quad , \qquad (5.2)$$

where typically an odd number of pixels are used in the median filter to avoid the extra averaging of the center two values required for an even number of pixels.

Chapter 4 showed that the median filter performs better than a linear filter in preserving edges and removing salt-and-pepper noise and that the breakdown for this filter was 50%. Unfortunately, one of the disadvantages of the median filter is that it removes one pixel wide lines from an image. This is illustrated by a 5×5 sub-image of a one pixel wide horizontal line given in Figure 5.2. The 3×3 median filtering of the center pixel from this sub-image produces the order set of nine pixels as

$$\{0, 0, 0, 0, 0, 0, 5, 5, 5\} \ . \qquad (5.3)$$

The median, which is the center of this ordered set of pixels, is zero. Applying the 3×3 median filter to each pixel of this sub-image produces a median filtered image that is completely zero, removing the one pixel wide line from the image.

0	0	0	0	0
0	0	0	0	0
5	5	5	5	5
0	0	0	0	0
0	0	0	0	0

Figure 5.2: A 5×5 sub-image of a one pixel wide horizontal line.

To solve the elimination of one pixel wide lines by the median filter, the *weighted median* filter was developed. This filter is similar to the median filter, but the filter mask now defines which pixels are to be used in the median calculation, and the value of the mask determines the number of times a pixel's graylevel is repeated in the ordering of the pixels from their minimum to their maximum values. Equation (5.2) is then used on this weighted set of ordered pixels to compute the median value. Like the median filter, it is desired to use only an odd number of pixels to reduce the number of calculations required to compute the median value. This requires that the sum of all the element values of the filter mask be an odd number.

Figure 5.3(a) gives a 3×3 filter mask with mask weights that when used in conjunction with the weighted 3×3 median filter preserves one pixel wide horizontal lines. Applying the 3×3 weighted median filter in conjunction with this mask to the center pixel of the sub-image given in Figure 5.2 produces the following ordered set of 15 pixels:

$$\{0, 0, 0, 0, 0, 0, (5, 5, 5), (5, 5, 5), (5, 5, 5)\} \ . \qquad (5.4)$$

The parentheses in Equation (5.4) show the three horizontal pixels being repeated three times as specified by the filter mask. The median of this value is now 5. Scanning the sub-image pixel by pixel and applying this weighted median filter to each pixel preserves the horizontal line. The question that should be asked now is, "Will this weighted median filter remove one pixel outliers?" Consider the 5×5 sub-image given in Figure 5.2 except the center pixel is at graylevel 5 and all other pixels are at graylevel 0. The ordering of this set of pixels using the 3×3 filter mask given in Figure 5.3(a) gives

$$\{0, 0, 0, 0, 0, 0, (0, 0, 0), (0, 0, 0), (5, 5, 5)\} \ . \qquad (5.5)$$

The median of this ordered set is zero, removing the one pixel outlier with graylevel 5.

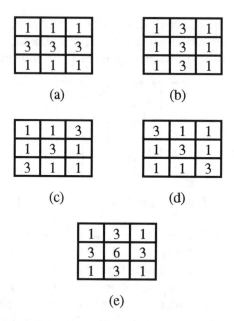

(a) (b)

(c) (d)

(e)

Figure 5.3: An example of various filter masks used by the weighted median: the filter masks preserve one pixel wide (a) horizontal lines, (b) vertical lines, (c) diagonal lines with a slope of 1, (d) diagonal lines with a slope of −1 and (e) horizontal and vertical lines.

Figure 5.3(b) gives the 3×3 weighted median filter mask that preserves one pixel wide vertical lines, while Figures 5.3(c) and (d) give the filter masks that preserve diagonal lines. The mask given in Figure 5.3(e) preserves both one pixel wide vertical and horizontal lines and was created from the filter masks given in Figures 5.3(a) and (b). It also should be mentioned that the weighted median filter has a smaller breakdown value than the median filter, resulting in less noise attenuation than   the median filter. Figure 5.4 shows an example of using the median filter to remove particular oriented lines from an image. Figure 5.4(a) is an image of two coins placed in front of a one pixel wide black grid. It is desired to remove the horizontal lines while leaving the vertical lines. Figure 5.4(b) is the result of applying a standard 3×3 median filter. This median filter has removed all of the black grid except at the intersections of the vertical and horizontal lines. Figure 5.4(c) shows the result of using the 3×3 weighted median filter with the filter mask given in Figure 5.3(b). The weighted median filter correctly preserved the vertical lines. Figure 5.4(d) shows the result of applying the filter mask given in Figure 5.3(a) that preserves horizontal lines. Applying the filter mask given in Figure 5.3(e) produces a weighted filtered image that contains both horizontal and vertical lines.

Another order statistic filter is the *midpoint* filter. This filter is defined as the average of the maximum and minimum graylevels of the order set of pixels that are included in the filter operation:

$$g(x, y) = \frac{F_{(0)} + F_{(N-1)}}{2} \, , \tag{5.6}$$

where $g(x, y)$ is the filtered output pixel and $F_{(0)}$ and $F_{(N-1)}$ are the minimum and maximum graylevels of all of the pixels included in the filter operation. The filter mask for this filter simply defines the pixels within the image that are to be included in the filter operation. This is the best filter to use to remove uniform type noise from an image.

The midpoint filter should not be used with images that contain outlier noise such as salt-and-pepper noise. This filter is very susceptible to outlier noise by the very fact that it uses the maximum and minimum values of the ordered set of pixels in its calculation. In the ordered set of pixels, outlier noise, if present, will usually be located within a few values of the minimum or maximum of the ordered set. Like the mean, median, and weighted median filters, the midpoint filter is an *unbiased* filter in that the average brightness of the filtered image remains the same as the original image. The midpoint filter blurs an image, similar to the mean filter. Consider the 5×5 sub-image containing a vertical edge as shown in Figure 5.5(a). Application of a 3×3 midpoint filter changes the set of vertical pixels on both sides of the vertical edge to 4.5 (the nearest grayscale value is 4), as shown in Figure 5.5(b). The sharp edge has been replaced with a new vertical edge that gradually changes from graylevel 0, 4, 4, to 9 over four pixels. The only difference between this filter and the mean filter is that the step is a linear edge with graylevels ranging from 0, 3, 6, to 9, as shown in Figure 5.5(c). The width of the vertical edge for both filtered images is basically the same.

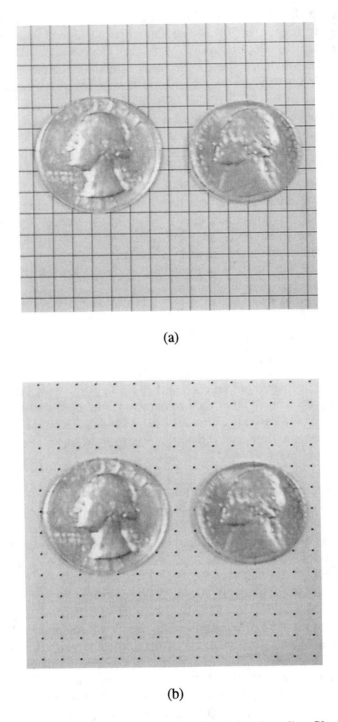

(a)

(b)

Figure 5.4: An example of using a 3 × 3 weighted median filter: (a) the
original image and (b) the 3 × 3 median filtered image.

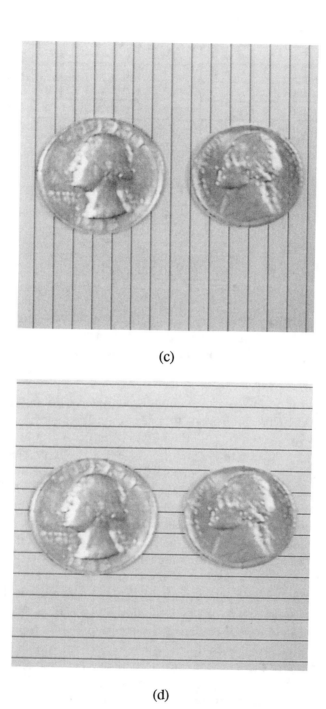

(c)

(d)

Figure 5.4: An example of using a 3 × 3 weighted median filter: the 3 × 3 weighted median filtered image using the filter masks given in (c) Figure 5.3(b) and (d) Figure 5.3(a).

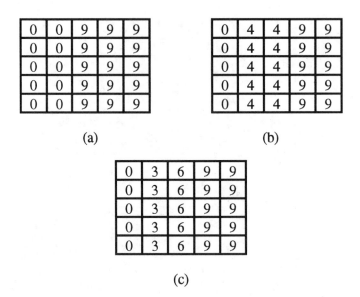

(a)   (b)

(c)

Figure 5.5: A 5 × 5 image of a vertical edge: (a) the original image,
(b) the 3 × 3 midpoint filtered image, and (c) the
3 × 3 mean filtered image.

Figure 5.6(a) is an image of a young boy corrupted with uniform noise with
a variance of 500. This corresponds to a peak-to-peak graylevel variation of 77
due to the noise [Equation (4.7)]. Figure 5.6(b) is the 3 × 3 midpoint filtered
image, showing a reduction in the uniform noise that is observable in the image.
Most noticeable is the reduction of this noise on the face just below the eyes.

The *maximum* and *minimum* filters are ordered statistic filters that are
commonly used to remove either salt-or-pepper noise. The minimum filter is
defined as the minimum graylevel of all of the pixels defined by the filter mask
$F_{(0)}$. If an image contains only salt noise, then the minimum filter effectively
removes this noise. Since the minimum filter selects the first element of the
ordered set of pixels, this filter should not be used to remove pepper noise. In
fact, this filter will enhance pepper noise. The maximum filter is the maximum
graylevel of the ordered set of pixels defined by the filter mask $F_{(N-1)}$.
Likewise, the maximum filter is used to remove pepper noise from an image and
will enhance salt noise. Both filters are also used in morphological image
processing to perform the morphological operations of erosion and dilation. The
concept of shrinking and increasing the geometrical size of an object within an
image is the basic concept behind the morphological filtering operations of
erosion and dilation. In fact, the maximum filter dilates (makes bigger) light
objects while eroding dark objects. The inverse holds true for the minimum
filter. The minimum filter implements the erosion operation, while the
maximum filter implements the dilation operation. The use of arbitrary shaped

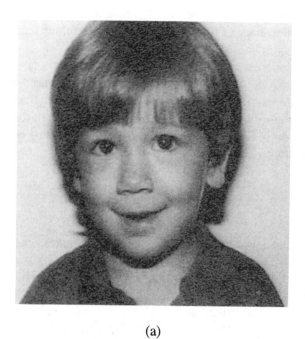

(a)

(b)

Figure 5.6: An example of using the midpoint filter: (a) the original image degraded with uniform noise with a variance of 500 and (b) the 3 × 3 midpoint filtered image.

filter masks is essential in the implementation of morphological image processing. It is the shape of the filter mask that determines the final shape of an object within the morphological filtered image.

Both the minimum and the maximum filters are biased filters. The maximum filter increases the average brightness of the filtered image, while the minimum filter decreases it. Application of the minimum filter to remove salt noise not only removes the salt noise but increases the size of the dark regions of an image while decreasing the size of the light regions. The breakdown values for both of these filters is

$$breakdown\ value = \frac{(N-1)}{N} \times 100\% ,\qquad (5.7)$$

where N is the total number of pixels included in the filter operation. Provided that at least one pixel is of the correct value and the other $(N-1)$ pixels are only one sided outliers, the maximum and the minimum filters will still remove the outlier noise from the image.

Figure 5.7(a) shows an image of a Hewlett Packard model 35 calculator keyboard corrupted with 10% salt noise of graylevel 255. This image was chosen to illustrate the filtering effects of the minimum filter because it contains both light letters on dark keys as well as dark letters on light keys. Figure 5.7(b) is the corresponding 3×3 minimum filtered image. This filter has done remarkably well at completely removing the salt noise from the image. Figure 5.7(c) is the same image of the calculator keyboard but this time corrupted with 10% pepper noise. The 3×3 maximum filter was chosen to remove the pepper noise from the image. Like the minimum filter, the maximum filter did quite well at removing all of the pepper noise from the image, as shown in Figure 5.7(d).

A comparison of the two filter outputs given in Figures 5.7(b) and (d) illustrates some of the filtering effects of these two filters. The minimum filtered image appears darker than the original due to the negative bias of the minimum filter. The maximum filter, on the other hand, has brightened the image as expected due to the positive bias of the maximum filter. The light letters on dark keys and the dark letters on light keys illustrate how the maximum and minimum filters reduce the size of dark and light regions. Figures 5.7(a) and (c) show that both the light and dark keys have lettering of the same width. The minimum filtered image given in Figure 5.7(b) has enlarged the dark letters while reducing the width of the light letters, making the dark lettered keys more visible. The opposite effect occurred in the maximum filtered image, with the width of light letters increasing in size and the width of the dark lettered keys decreasing. If a 5×5 minimum filter were used, the light letters would be completely removed and the darks letters would be further increased in size.

The mean filter is the best filter at removing Gaussian type noise but tends to blur an image. Since the mean of a set of pixels is very susceptible to the

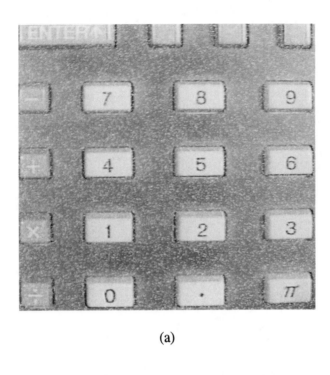

(a)

(b)

Figure 5.7: An example of using the minimum filter: (a) the 10%
salt noise degraded image and (b) the minimum
filtered image.

(c)

(d)

Figure 5.7: An example of using the maximum filter: (c) the 10%
             pepper noise degraded image and (d) the maximum
             filtered image.

presence of outlier noise, the mean filter is a very poor filter to use in the presence of this noise. The median filter, on the other hand, performs quite well in the presence of outlier noise and preserves image sharpness, but performs poorly in the presence of Gaussian noise. A filter that gives a mixture of the mean and median filters is the *alpha-trimmed mean* filter. This filter performs reasonably well in the presence of both Gaussian and outlier noise. The computation of the alpha-trimmed mean filter requires the ordering of pixels that are defined in the filter operation from their minimum to their maximum graylevels producing the ordered set of pixels. For this filter, the filter mask determines which pixels from within the image are used in the filtering of the pixel at the location $f(x, y)$.

The alpha-trimmed mean filter is defined as

$$alpha\text{-}mean[f(x, y)] = \frac{1}{N - 2p} \sum_{i = p}^{N-p-1} F_{(i)} \, , \qquad (5.8)$$

where $p = 0, 1, 2, 3 \cdots \text{int}[N/2]$ and $N - 2p$ is the total number of pixels included in the average calculation. Each pixel within the image is scanned pixel by pixel and Equation (5.8) is evaluated for each pixel, producing a new filtered image. The value of p determines the type of filtering that is performed. For $p = 0$, the alpha-trimmed mean filter reduces to the mean filter. The upper limit for p of $\text{int}[N/2]$ is valid only for an odd number of pixels N that are included in the filter operation. For N even, the upper limit must be changed to $N/2 - 1$. For N even or odd, when p is set to its maximum value the alpha-trimmed mean filter reduces to the median filter as given by Equation (5.2).

The alpha-trimmed mean filter is just a mean or averaging filter that removes a selective number of pixels (depending on p) that are close in graylevel to the maximum and minimum graylevels contained within the filtering operation. If an image contains both outlier and Gaussian type noise, selecting p other than zero removes some of the outlier noise pixels in the calculation of the mean. Since the mean filter is very sensitive to outlier noise, the goal of the alpha-trimmed mean filter is to remove as many outlier pixels as possible and still perform a mean calculation to reduce the Gaussian noise that is present within an image. Because the alpha-trimmed mean filter performs averaging, it does blur an image, but not as much as an equivalent sized mean filter. This is because for a given p not equal to zero, the alpha-trimmed mean filter effectively averages over fewer pixels. Also, because pixels associated with an edge appear as outlier pixels, these pixels are removed from the mean calculation producing less smoothing of the edge.

Figure 5.8(a) is an example image containing 10% salt-and-pepper noise and additive zero-mean Gaussian noise with a variance of 500. For a 5 × 5 alpha-trimmed mean filter, the total number of pixels in the filter operation will

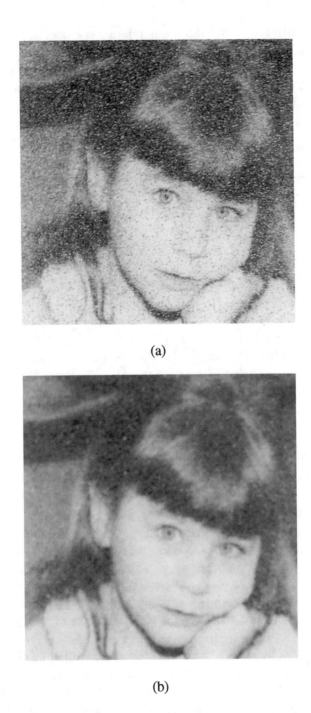

(a)

(b)

Figure 5.8: An example of using the alpha-trimmed mean filter:
(a) the degraded 10% salt-and-pepper and Gaussian
noise (variance = 500) image and (b) the 5×5
alpha-trimmed mean filtered image with $p = 2$.

(c)

(d)

Figure 5.8: An example of using the alpha-trimmed mean filter:
(c) the 5×5 median filtered image and (d) the 5×5
mean filtered image.

be 25, and on the average about 3 of these pixels will be due to outlier noise. Setting $p = 2$ removes the four pixels with graylevels closest to minimum and maximum graylevels from the mean calculation. The goal is to set p just small enough to remove the outlier noise pixels from the mean calculation yet still provide enough pixels for the mean calculation. Recall from Chapter 4 that the larger the filter mask used the better the mean filter performs at reducing the noise. Unfortunately, the larger the filter window size, the blurrier the filtered image becomes.

Figure 5.8(b) gives the 5×5 alpha-trimmed mean filtered image with $p = 2$. The Gaussian noise has been drastically reduced, as well as most of the salt-and-pepper noise. Figure 5.8(c) gives the corresponding 5×5 median filtered image. The salt-and-pepper noise has been completely removed by the median filter operation. Yet, the alpha-trimmed mean filter did not perform as well as the median filter in removing the salt-and-pepper noise from the degraded image. This is due to the fact that fewer outlier pixels are removed by the alpha-trimmed mean filtering calculation than by the median filter. Hence, the alpha-trimmed mean filter has a lower breakdown value than the median filter. The breakdown value for the alpha-trimmed mean filter is given by

$$
breakdown\ value = \begin{cases} \dfrac{p}{2 \cdot \text{int}[N/2]} & \text{for } N \text{ odd} \\[2ex] \dfrac{p}{N-2} & \text{for } N \text{ even} \end{cases} \tag{5.9}
$$

For $p = 0$, which is the mean filter, the breakdown value is 0. For $p = \text{int}[N/2]$ when N is odd or for $N/2 - 1$ when N is even, the breakdown value is 50%, which is the same value as the median filter.

A close inspection of the wall behind the girl on the left side of the image shows that the alpha-trimmed mean filter reduces the Gaussian noise just slightly better than the median filter. This is because the median filter also reduces Gaussian noise only slightly worse than the mean filter. The image sharpness for both images is about the same, slightly blurrier than the original degraded image. Figure 5.8(d) is the mean filtered output image. The noise remaining in this image is much larger, and the image is blurrier than for the median and the alpha-trimmed mean filters. The mean filter performed poorly because of the presence of the salt-and-pepper noise.

Another ordered statistic filter that is commonly used to remove uniform noise is the *complementary-alpha-trimmed mean* filter. Uniform noise belongs to the class of noise known as short tail distributed noise. This noise is given this name because the tails of the PDFs approach zero at an extremely fast rate. Instead of removing the p number of pixels from the minimum and the maximum of the ordered set of pixels, the complementary-alpha-trimmed mean filter averages over these pixel. It is defined as

$$\textit{complementary-alpha mean}[f(x, y)] = \frac{1}{2(p + 1)} \left\{ \sum_{i=0}^{p} F_{(i)} + \sum_{i=N-p-1}^{N-1} F_{(i)} \right\} ,$$

(5.10)

where $p = 0, 1, 2, 3 \cdots \text{int}[N/2] - 1$ and $2(p + 1)$ is the total number of pixels included in the average calculation. For $p = 0$, the complementary-alpha-trimmed mean filter reduces to the midpoint filter. The midpoint filter has been previously discussed as the best filter in removing uniform noise from an image.

The goal of filtering an image containing both Gaussian and salt-and-pepper noise is to remove the outlier pixels from the filter operation and then perform an averaging calculation to reduce the Gaussian noise present within the image. For Gaussian noise, the peak-to-peak graylevel variations due to this noise will be in the range of ± 3 times its standard deviation. In homogeneous regions of an image, any pixels with graylevels outside this range are probably due to outlier pixels. Hence, a measure to determine if a pixel's graylevel is due to an outlier is to use the standard deviation as a means of eliminating pixels from the filter calculation. The goal is to consider only pixels within some $\pm K$ standard deviation about the mean graylevel.

The difficulty is estimating the mean values of a set of pixels that also contains outliers. The calculation of the mean value directly would be biased by any outlier pixels that are present. By definition, the mean and median values are equal for all noise that have symmetric histograms. Since Gaussian noise has a symmetric histogram, the mean and median values are equal. Essentially, this allows the median value to be used as an estimate of the mean value. The advantage of this is that the median value is unaffected by the presence of outlier pixels. In fact, up to 50% of the pixels can be outliers and the median produces the correct value. The *modified trimmed mean* filter uses the median as an estimate for the mean value and removes all pixels in calculating the mean that are beyond $\pm K$ standard deviation from the mean graylevel.

The first step required to implement the modified trimmed mean filter is to compute the median, $\text{med}[f(x, y)]$, of the pixels that are included in the filter calculation as defined by the filter mask H using Equations (5.1) and (5.2). Next, the mean of the pixels is calculated using a similar approach to that for the mean filter

$$g(x, y) = \frac{1}{N_t} \sum_{i, j \in H} \sum f(x - i, y - j) , \qquad (5.11)$$

where only the pixels that are defined by the filter mask and that meet the criteria

$$|f(i, j) - \text{med}[\, f(x, y)]| \le K \quad \text{for } i, j \in H \qquad (5.12)$$

are included in this mean calculation.

The value of K is typically between two and three times the standard deviation of the Gaussian noise. For $K = 0$, the modified trimmed mean filter becomes the median filter and for $K = \infty$, it becomes the mean filter. The modified trimmed mean filter changes from a median to a mean filter, similar to the way the alpha-trimmed mean filter changed from a mean to a median filter. Since the median filter preserves edge sharpness and the mean filter blurs edges, the modified trimmed mean filter produces an image that has an image sharpness somewhere between these two filters. If two different sized filter masks are used to define the pixels for the mean and median calculations, the modified trimmed mean filter becomes the *double windowed modified trimmed mean* filter. Usually the median filter window is of a smaller size than the mean filter window.

Figure 5.9 is the 5×5 modified trimmed mean filtered image of the 10% salt-and-pepper and the Gaussian noise (variance = 500) degraded image given in Figure 5.8(a). The parameter K was chosen to be 45, which is 2.0 times the standard deviation of the Gaussian noise. In comparing Figure 5.9 to the alpha-trimmed mean, median, and mean filtered images given in Figures 5.8(b) through (d), the modified trimmed mean filter definitely outperformed the mean filter and gave about the same performance as the median and alpha-trimmed mean filters. A close inspection of Figure 5.9 does show a slight improvement in filtering the Gaussian noise as compared to the median filtered image of Figure 5.8(c).

Figure 5.9: An example of using the modified trimmed mean filter.

Several other nonlinear ordered statistic filters make great edge detectors. Each of these filters is based upon producing a filter output that is proportional to the variation of graylevels from a set of pixels as defined by the filter mask. The first of these filters is the *range* filter, which is defined as

$$range[f(x, y)] = F_{(N-1)} - F_{(0)} \,, \tag{5.13}$$

which is the maximum minus the minimum graylevels of the pixels that are included in the filter operation. The range filter is implemented by scanning an image pixel by pixel and then using Equation (5.13) to determine the filtered pixel. This filter is capable of detecting edges with small graylevel differences but is very susceptible to noise.

Since outlier noise usually appears as the maximum and minimum graylevels within the ordered set of pixels, any outlier noise that is present during the range filtering operation will produce an output from this filter. Even as little as one outlier will corrupt the output. The range filter cannot tell the difference between outlier noise and an edge. Similar to the range filter is the *quasi-range* filter

$$quasi\text{-}range[f(x, y)] = F_{(N-1-i)} - F_{(i)} \quad \text{for } i = 0, 1, \cdots \text{int}[(N-1)/2] \,. \tag{5.14}$$

The breakdown value of this filter varies from zero for $i = 0$ to 50% as i increases. As the parameter i increases from 0 to $\text{int}[(N-1)/2]$, the quasi-range filter becomes less susceptible to outlier noise but its sensitivity to finding edges decreases. The difficulty is that these filters are based on order statistics. The basis for order statistic filters is to separate normal pixels from outlier pixels. To order statistic filters, outlier pixels and edge pixels are same.

Another nonlinear order statistic filter is the *dispersion edge* filter. This filter is defined as

$$dispersion[f(x, y)] = \frac{1}{\text{int}[N/2]} \sum_{i=0}^{\text{int}[(N-1)/2]} \{F_{(N-1-i)} - F_{(i)}\} \,. \tag{5.15}$$

In comparison to the range and quasi-range filters, the dispersion filter is by far the most sensitive to edges. It can detect gradual edges easier than the other two filters. Unfortunately, this edge sensitivity comes at a price. The dispersion filter is also sensitive to outlier noise because of its use of the maximum and minimum values, although it is less susceptible than the range filter.

Figure 5.10 gives examples of using the range, quasi-range, and dispersion filters to detect edges within an image in the presence of salt-and-pepper noise. Figure 5.10(a) is the original undegraded image and Figure 5.10(b) is the

(a)

(b)

Figure 5.10: Examples of using order statistic filters as edge
              detectors: (a) the original image and (b) the 10%
              salt-and-pepper noise degraded image.

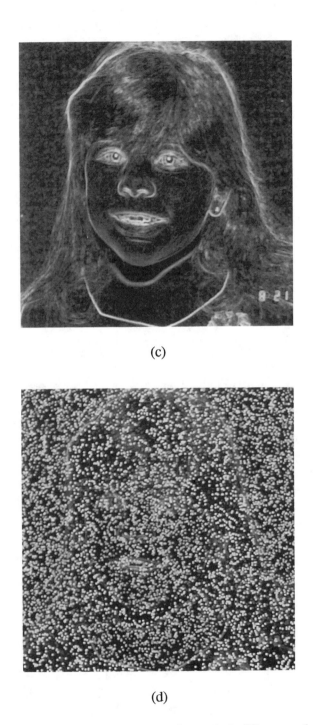

(c)

(d)

Figure 5.10: Examples of using order statistic filters as edge
detectors: the range filtered images of (c)
Figure 5.10(a) and (d) of Figure 5.10(b).

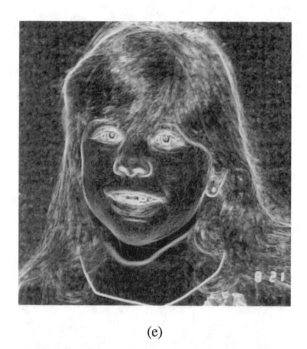

(e)

(f)

Figure 5.10: Examples of using order statistic filters as edge
              detectors: (e) the dispersion filtered image of Figure
              5.10(a) and (f) the quasi-range filtered image of Figure
              5.10(b).

corresponding 10% salt-and-pepper degraded image. The edge image given in Figure 5.10(c) is the 3 × 3 range filtered image of the original noise free image, generated using Equation (5.13). Notice how well the edges are defined in this image. Figure 5.10(d) is the 3 × 3 range filtered image of the salt-and-pepper degraded image, showing the poor robustness of this filter in the presence of outlier noise. The resulting edge image has lost any resemblance to that of the original image.

Figure 5.10(e) is the 3 × 3 dispersion filtered image of the original image given in Figure 5.10(a). Of the range and quasi-range filters, the dispersion filter has found the most edges. This filter is the most sensitive of the three in detecting edges but is not very robust in the presence of outlier noise. Since the minimum and maximum graylevels of the pixels are used in the filtering operation, a single outlier noise pixel will corrupt the output of this filter. Figure 5.10(f) is the 3 × 3 quasi-range filtered image of the degraded salt-and-pepper image with i set to 2 in Equation (5.14). This filter has done a much better job at finding edges in the presence of outlier noise. The only disadvantage, though, is that it has found fewer edges than the range filtered image of Figure 5.10(c). Increasing the parameter i increases the robustness of this filter to noise but decreases its edge finding sensitivity. This can be seen in the number of edges that were found in Figure 5.10(f) in comparison to Figure 5.10(c).

In implementing the order statistic filters given in the section, there is no need to develop a separate algorithm for each filter. Most of these filters can be implemented with one algorithm with different parameters passed to it to determine the type of filter operation that is to be performed. Figure 5.11 gives the basic block diagram used to implement all of the ordered statistic filters discussed here. The first step is to define two arrays to hold the pixels that are to be included in the filter operation and the other is to store the coefficients, which define the type of filter that is to be implemented. The next step in filtering a pixel within an image at the coordinate x, y is to place into one of these arrays the N pixels that are to be used in the filtering operation as defined by filter mask H. A sorting algorithm, such as a bubble sorting algorithm, is then used to sort this array of pixels from its minimum to maximum values. Each element of this array is then multiplied by the filter coefficients A_i as shown in Figure 5.11. The output of this filter is then the sum of all the weighted sorted pixels.

Table 5.1 gives the coefficients required to implement the various order statistic filters that have been discussed in this section. The only two filters that are not implemented using the block diagram given in Figure 5.11 are the weighted median filter and the modified trimmed mean filter. All others have been included in Table 5.1. For example, to compute the range filter, all coefficients are set to zero except for A_0, which is set to −1, and A_{N-1}, which is set to 1. The minimum and the maximum filters are computed with $A_0 = 1$ and $A_{N-1} = 1$, respectively. All other coefficients are set to zero for these two filters. Even the mean filter can be implemented using Figure 5.11. In this case, all of the coefficients are set to $1/N$. This is an inefficient way of computing

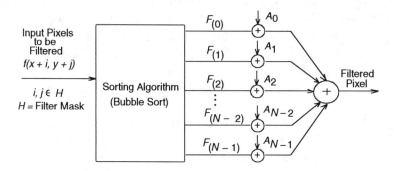

Figure 5.11: A block diagram that illustrates the implementation of order statistic filters (adapted from Pitas and Venetsanopoulos, 1990).

Table 5.1: The coefficients required to implement the various order statistic filters given in this section.

Filter Type	Coefficients
median	$A_{N/2} = 0.5$, $A_{N/2-1} = 0.5$ for N even $A_{(N-1)/2} = 1.0$ for N odd all other coefficients are set to zero
midpoint	$A_0 = 0.5$, $A_{N-1} = 0.5$ all other coefficients are set to zero
maximum	$A_{N-1} = 1.0$ all other coefficients are set to zero
minimum	$A_0 = 1.0$ all other coefficients are set to zero
range	$A_0 = -1.0$, $A_{N-1} = 1.0$ all other coefficients are set to zero
quasi-range (i)	$A_i = -1.0$, $A_{N-1-i} = 1.0$ all other coefficients are set to zero
dispersion	$A_i = -1/\mathrm{int}[N/2]$, $A_{N-1-i} = 1/\mathrm{int}[N/2]$ for $i = 0$ to $\mathrm{int}[(N-1)/2]$
complementary-alpha-trimmed mean filter (p)	$A_i = 1/(2p+2)$ for $i = 0$ to p and $i = N - p - 1$ to $N - 1$ all other coefficients are set to zero
alpha-trimmed mean filter (p)	$A_i = 1/(N - 2p)$ for $i = p$ to $N - p - 1$ all other coefficients are set to zero
mean	$A_i = 1/N$ for $i = 0$ to $N - 1$

the mean filter since sorting of the pixels based on graylevels is not needed for the mean filter. In fact, the advantage of using Figure 5.11 is that only one algorithm is needed. The disadvantage is the computational efficiency that is lost by using a general algorithm to perform the filtering computation. Also, the computational efficiency of the minimum and maximum filters are affected by using Figure 5.11 to compute these filters. It is faster to find the minimum and maximum of a set of numbers individually than by sorting the whole set of numbers. The reader interested in obtaining further information concerning ordered statistic filters is referred to the book *Nonlinear Digital Filters* by Pitas and Venetsanopoulos. This book gives an excellent treatment of order statistical filters.

5.2 Nonlinear Mean Filters

There are four nonlinear spatial mean filters that perform better than the mean filter, with a slight increase in computational cost. Similar to the standard mean filter, these filter a pixel $f(x, y)$ within an image by including a set of neighborhood pixels as defined by the filter mask H. The filter mask can be of any shape but usually takes on an $n \times n$ square shape because of implementation ease. The image is scanned pixel by pixel until every pixel within the image has been filtered, producing a new filtered image. The first of these nonlinear filters is the *geometric mean* filter, defined as the $1/N$ root of the product of the N pixels included in the filtering operation:

$$\textit{geometric mean}[f(x, y)] = \prod_{i, j \in H} \{ f(x+i, \ y+j) \}^{1/N}, \qquad (5.16)$$

where H is the filter mask definition and the coordinates $x + i, y + j$ are defined over the image $f(x, y)$. This filter performs better than the mean filter at reducing uniform and Gaussian type noise from an image. It also has the same edge preserving quality as the mean filter. Because this filter uses multiplication, it is very susceptible to negative outlier noise such as pepper noise. A single pixel with a zero graylevel will set the output of this filter to zero even if all other pixels are nonzero.

The second nonlinear mean filter that will be presented here is the *harmonic mean* filter, defined as

$$\textit{harmonic mean}[f(x, y)] = \cfrac{N}{\displaystyle\sum_{i, j \in H} \cfrac{1}{f(x+i, \ y+j)}}. \qquad (5.17)$$

As with Equation (5.16), H is the filter mask definition and the coordinates $x + i$, $y + j$ are defined over the image $f(x, y)$. If any of the pixels defined within the filter operation contain pixels with a graylevel of zero, the output of this filter is defined as zero. For the same size filter mask, this filter also reduces the variance of uniform and Gaussian noise more than that of the standard mean filter. The harmonic filter is good at removing positive outliers. Outlier graylevels appear as one over their value in computing the summation given in Equation (5.17). Essentially, positive outlier pixels contribute very little to the total summation value. For the same size filter mask, the harmonic filter outperforms the mean and median filters in removing positive outlier or salt noise.

The third nonlinear mean filter is the *contra-harmonic mean* filter, defined as

$$\text{contra-harmonic mean}[f(x, y), K] = \frac{\sum_{i,j \in H} f(x + i, \ y + j)^{K+1}}{\sum_{i,j \in H} f(x + i, \ y + j)^{K}} \quad , \qquad (5.18)$$

where the parameter K determines the order of the filter. If all of the pixels within the filtering operation are zero, then the output of the filter reduces to zero. This filter performs exceptionally well at removing either salt noise or pepper noise but not both. Salt noise is removed for negative values of K, while pepper noise is removed for positive values of K. It is a biased filter, increasing the average brightness of the filtered image for positive values of K while decreasing the average brightness for negative values of K. This filter tends toward the maximum filter for large positive K values, enhancing light regions within the image. For large negative K values, it tends toward the minimum filter, enhancing dark regions. The contra-harmonic filter also does a good job at preserving the edges within the image.

The last nonlinear filter presented in this section is the *power mean* filter. This filter is very similar to the standard mean filter except the graylevel of each pixel is raised to the Kth power before averaging the N pixels as defined by the filter mask. The output of the power mean filter is the Kth root of this averaging operation

$$\text{power mean}[f(x, y), K] = \left\{ \sum_{i,j \in H} \frac{f(x + i, \ y + j)^{K}}{N} \right\}^{1/K} , \qquad (5.19)$$

The power mean filter has filtering characteristics very similar to the contra-harmonic mean filter. For positive values of K, it removes pepper noise and tends toward the maximum filter, while it tends toward the minimum filter for negative values of K, enhancing the dark regions within the filtered image. This filter also does a better job at removing Gaussian noise and preserving edge features than the standard mean filter.

Figures 5.12(a) and (b) show the edge preserving qualities of the nonlinear mean filters in comparison to the standard mean filter. The figures were created using a two graylevel image of 64 and 128 containing a vertical edge. This image was then corrupted with additive uncorrelated zero mean Gaussian noise with a variance of 100. Next, for each filter, a 7×7 square mask was used to define the pixels to be included in the filtering operation. The Gaussian noise degraded image was then filtered using the standard mean, harmonic, contra-harmonic, geometric, and power mean filters. A graylevel profile of the degraded and filtered images as a function of horizontal position was created for a fixed vertical position of $Y = 221$. Figures 5.21(a) and (b) shows the result of the various filtering operations.

All of the filters were capable of reducing the noise that was present in the degraded image, but the edge preserving quality of each of the filters varied greatly. The worst of the filters in terms of edge preservation was the standard mean filter. This filter took the sharp edge in the degraded image and drastically reduced its slope. The geometric mean filter had about the same edge preservation as the standard mean filter. Only slightly better than these was the harmonic mean filter. The slope of the geometric filtered edge is slightly higher than the slope of the standard filtered edge.

Both the contra-harmonic and the power mean filters performed the best at preserving the sharpness of an edge. For positive values of K, both filters produced sharp edges to the left of the original edge. This is due to the maximum filter nature of these two filters for positive values of K. The maximum filter tends to make light regions within an image bigger (dilated) and dark regions smaller (eroded). Negative values of K, on the other hand, produced edges to the right of the original edge. Because of the similarity of these two filters to the minimum filter for negative values of K, these filters tend to enlarge dark regions, while reducing light regions. Having an edge start to the right of the original as shown in Figure 5.12(a) and (b) results in enlarging the dark regions of the image, while decreasing the light regions. For a fixed filter mask, the graylevel output from each of these filters will always follow the graylevel ordering of:

$$minimum[f(x, y)] \leq contra\text{-}harmonic\ mean[f(x, y), -K] \leq$$
$$power\ mean[f(x, y), -K] \leq harmonic\ mean[f(x, y)] \leq$$
$$geometric\ mean[f(x, y)] \leq standard\ mean[f(x, y)] \leq$$
$$contra\text{-}harmonic\ mean[f(x, y), K] \leq power\ mean[f(x, y), K] \leq$$
$$maximum[f(x, y)] \ .$$

$$(5.20)$$

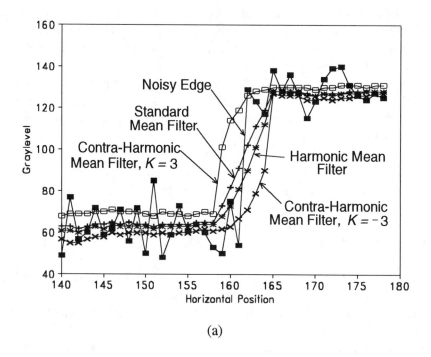

(a)

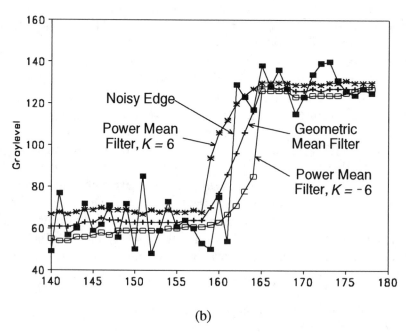

(b)

Figure 5.12: Examples of the edge preserving properties of: (a) the
standard mean, the harmonic, and the contra-harmonic
filters and (b) the geometric and the power mean
filters (adapted from Pitas and Venetsanopoulos, 1990).

Since the standard mean filter is an unbiased filter, Equation (5.20) shows that all of the nonlinear mean filters presented in this section as well as the minimum and maximum filters are biased filters, either darkening or lightening an image. The least in terms of bias are the harmonic and geometric mean filters. The reduction of noise from these nonlinear mean filters depends on the size of the filter mask, the variance of the noise present in the image, and the average graylevel of the pixel that is being filtered. This is unlike the standard mean filter, in which the amount of noise reduction depends solely on the filter mask size.

Figure 5.13(a) is a 256 grayscale image of a 1950s Zenith Corporation transistor radio. Figure 5.13(b) is a version of this radio image degraded with 20% salt noise containing graylevel 255. Figure 5.13(c) shows the result of applying a 3×3 power mean filter with a K value of -1. This filter has successfully removed the salt noise from the image, while maintaining approximately the same sharpness as the original image. Some noise is still present in the filtered image, giving this image a slight grainy appearance as compared to the original image given in Figure 5.13(a). The power mean operation did remove some of the "Zenith" name located just above the center of the radio. This is to be expected because of the minimum-filter-like characteristics of the power mean filter for negative values of K. The minimum filter reduces the size of light regions, while enlarging the size of dark regions within the image. Figure 5.13(d) is the 3×3 median filtered image, showing the difficulty the median filter had in removing all of the salt noise from the image. White spots can easily be seen within the body of the radio. The sharpness of both filtered images is about the same. The power mean filter definitely outperformed the median filter in the removal of salt noise.

Figure 5.13(e) is the 20% pepper noise degraded version of the radio image given in Figure 5.13(a). Note that the addition of the pepper noise has darkened this image in comparison to the salt noise degraded image given in Figure 5.13(b). This change in perceived brightness is because 20% of the pixels in the pepper degraded image had their graylevels reduced to zero, while in the salt degraded image 20% of the pixels had their graylevels increased to 255. Figure 5.13(f) is the 3×3 contra-harmonic mean filtered image obtained with a value of $K = 1$. Figure 5.13(f) shows the extreme effectiveness of this filter in removing pepper noise from an image. The contra-harmonic mean filter removed all of the pepper noise from the image while preserving the sharpness of the image. This filter performed better at removing the pepper noise than the power mean filter did in removing salt noise from the image. There is no residue noise left giving the contra-harmonic filtered image a grainy appearance. This filter even enhanced the white lettering of the Zenith label as a result of its maximum-filter-like characteristic for positive values of K.

A close inspection of the edge preserving graphs given in Figures 5.12(a) and (b) shows that the contra-harmonic and the power mean filters produce two

(a)

(b)

Figure 5.13: Examples of using the nonlinear mean filters to
          remove outlier noise from an image: (a) the original
          image and (b) the 20% salt degraded image.

(c)

(d)

Figure 5.13: Examples of using the nonlinear mean filters to remove outlier noise from an image: (c) the 3×3 power mean filtered image with $K = -1$ and (d) the 3×3 median filtered image.

(e)

(f)

Figure 5.13: Examples of using the nonlinear mean filters to
remove outlier noise from an image: (e) the pepper
degraded image and (f) the 3×3 contra-harmonic
mean filtered image with $K = 1$.

types of edge responses depending on whether the parameter K is positive or negative. In homogeneous regions, the output of these two filters is basically the same, independent of K being negative or positive. Combining the results of the contra-harmonic mean filter for both positive and negative values of K yields an edge detection filter that works quite well in the presence of nonoutlier noise:

$$contra\text{-}harmonic\ mean[f(x, y), K] - contra\text{-}harmonic\ mean[f(x, y), -K]. \quad (5.21)$$

In homogeneous regions of an image, both filter operations filter the noise the same. As a result, the output of Equation (5.21) reduces to zero even in the presence of nonoutlier noise. In edge regions, the positive K valued contra-harmonic mean filter will be larger than the negative K valued filter. The output from Equation (5.21) will always produce a positive, nonzero value within the edge regions. Hence, Equation (5.21) makes a robust edge detector in the presence of nonoutlier noise. Equation (5.21) reduces the noise present in the image while at the same time finding the edges that are present. The concepts used to develop Equation (5.21) can also be used to implement an edge detector based on the power mean filter:

$$power\ mean[f(x, y), K] - power\ mean[f(x, y), -K]. \quad (5.22)$$

The power mean edge filter has similar properties to the contra-harmonic edge filter.

Figure 5.14 shows the result of subtracting the $7 \times 7\ K = -3$ contra-harmonic filtered edge from the 7×7, $K = 3$ contra-harmonic filtered edge as given in Figure 5.12(a). Figure 5.14 also shows the 7×7 power mean edge filter output for the parameter K set equal to 6. Both filters performed equally well at removing the Gaussian noise (variance = 10) from the final edge detected outputs. Figure 5.15(a) is a Gaussian noise (variance = 400) corrupted version of the radio image given in Figure 5.13(a). Figure 5.15(b) is the corresponding contra-harmonic edge detected image generated using Equation (5.21) with K set to 3. Figure 5.15(c) is also an edge image of the radio image generated using the range filter. A comparison of Figures 5.15(b) and (c) shows that the contra-harmonic mean edge filter outperformed the range filter in the presence of Gaussian noise. There is still a considerable amount of noise present within the background regions of the range filtered image. It is typically expected that in the background regions of an image, an edge filter should produce an edge image with a zero graylevel. Yet, the background region of the radio as shown in Figure 5.15(b) is at a different graylevel than the background surrounding the entire radio. This effect is due to the bias property of the contra-harmonic mean filter. Positive K values increase the brightness of an image, while negative K values decrease the brightness. The difference between the positive and the negative valued contra-harmonic filters results in a graylevel background that depends on the background in the initial graylevel image.

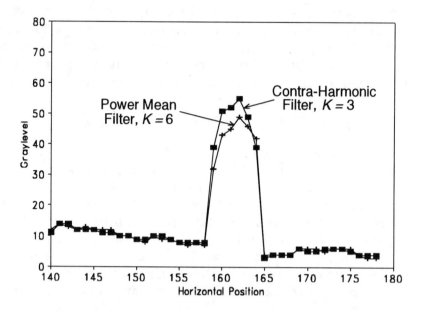

Figure 5.14: An example of using the contra-harmonic and the power
mean filters to detect edges within an image.

(a)

Figure 5.15: An example of using the contra-harmonic mean filter
to detect edges within an image: (a) the Gaussian noise
degraded image (variance = 400) of Figure 5.13(a).

(b)

(c)

Figure 5.15: An example of using the contra-harmonic mean filter
to detect edges within an image: (b) the contra-harmonic
edge filtered image and (c) the range filtered image.

5.3 Adaptive Filters

The difficulty with all of the filters that have been presented in this text up to now is that they perform the same filtering operation throughout the entire image. Images for the most part are nonhomogeneous and are better filtered with filters that change their filtering characteristics as they move throughout an image. The optimum goal of image filtering is to remove the noise from an image without causing any further degradation. Unfortunately, most filters blur an image in the process of removing noise. Consider an image that has been separated into edge regions and background regions. In the background regions, it is desired to perform as much filtering as possible to remove as much of the noise as possible. It is in these homogeneous regions that the noise present in an image is the most noticeable and objectionable. Within edge regions, it is desired to maintain the sharpness of the edges, hence maintaining the sharpness of the overall image. To maintain edge sharpness implies the lack of filtering within these edge regions. The goal of adaptive filtering is to detect edge and homogeneous regions within an image and to perform heavy filtering within the homogeneous regions while performing no or minimum filtering to preserve the image edges.

The first filter that will be presented is the *minimum mean square error* (*MMSE*) filter. This filter changes its filtering characteristics based upon the local variance about the pixel $f(x, y)$. The local variance gives a good measure in determining the presence of an edge within a local region. For large local variances, the MMSE filter performs no filtering, while for low local variances it applies the standard mean filter.

Consider an image $f(x, y)$ that has been corrupted with zero mean additive noise $n(x, y)$:

$$g(x, y) = f(x, y) + n(x, y) . \tag{5.23}$$

The goal is to estimate $f(x, y)$ given $g(x, y)$ and knowing some statistical information about the noise. Defining est(x, y) as the estimate of the noise free image, one method of estimating $f(x, y)$ is to use a first order linear equation:

$$\text{est}(x, y) = M \cdot g(x, y) + B , \tag{5.24}$$

where M and B are the parameters that must be determined. The difference between Equation (5.24) and the original noise free image gives a measure on how good the estimate image is:

$$err(x, y) = f(x, y) - \text{est}(x, y) = f(x, y) - M \cdot g(x, y) - B . \tag{5.25}$$

If Equation (5.25) is minimized, then the estimated image is a good representation of the noise free image.

The error function can be minimized in the mean square error (MSE) sense

$$P = E[err(x, y)^2] = E[\{f(x, y) - M \cdot g(x, y) - B\}^2] \ , \qquad (5.26)$$

where E is the average or expected value operator. Substituting Equation (5.23) for $g(x, y)$ and performing the squaring operation in Equation (5.26) yields

$$E[err(x, y)^2] = E[B^2 + (1 - M)^2 f(x, y)^2 - 2B(1 - M) \cdot n(x, y) \cdot f(x, y)$$
$$- 2(1 - M)B \cdot f(x, y) - 2MB \cdot n(x, y) + M^2 \cdot n(x, y)^2] \ . \qquad (5.27)$$

Assuming that the noise $n(x, y)$ is uncorrelated with the original image $f(x, y)$ and that the noise is zero mean, the expected value of Equation (5.27) reduces to

$$P = E[err(x, y)^2] = B^2 + (1 - M)^2 \cdot (\sigma_f^2 + m_f^2)$$
$$- 2B(1 - M) \cdot m_f + M^2 \cdot \sigma_n^2 \ , \qquad (5.28)$$

where m_f and σ_f^2 are the mean and variance of the noise free image $f(x, y)$ and σ_n^2 is the variance of the noise. Taking the derivative of Equation (5.28) with respect to M and B gives

$$\frac{\partial P}{\partial B} = 2B - 2(1 - M) \cdot m_f \qquad (5.29)$$

and

$$\frac{\partial P}{\partial M} = 2B \cdot m_f + 2M \cdot \sigma_n^2 - 2(1 - M) \cdot (\sigma_f^2 + m_f^2) \ . \qquad (5.30)$$

Setting Equations (5.29) and (5.30) to zero and then solving in terms of B and M produces

$$M = \frac{\sigma_f^2}{\sigma_f^2 + \sigma_n^2} \ . \qquad (5.31)$$

and

$$B = \left(\frac{\sigma_n^2}{\sigma_f^2 + \sigma_n^2} \right) \cdot m_f \ . \qquad (5.32)$$

Equations (5.31) and (5.32) give the best estimate in the least mean square sense of the noise free image as defined by Equation (5.24):

$$est(x, y) = \frac{\sigma_f^2}{\sigma_f^2 + \sigma_n^2} \cdot g(x, y) + \left(\frac{\sigma_n^2}{\sigma_f^2 + \sigma_n^2}\right) \cdot m_f \ . \tag{5.33}$$

But the sum $\sigma_f^2 + \sigma_n^2$ is just the variance of the noise degraded image $g(x, y)$, σ_g^2. Equation (5.33) can then be rewritten as

$$est(x, y) = \left(1 - \frac{\sigma_n^2}{\sigma_g^2}\right) \cdot g(x, y) + \left(\frac{\sigma_n^2}{\sigma_g^2}\right) \cdot m_f \ . \tag{5.34}$$

Equation (5.34) is known as the MMSE filter. The mean m_f of the noise free image varies throughout the image and must be estimated using a local neighborhood about the pixel $g(x, y)$. The mean can be estimated using the local average of the surrounding pixels of $g(x, y)$, which is obtained using the standard mean filter as defined by Equation (4.18):

$$m_f(x, y) = \frac{1}{N_t} \sum_{i, j \in H} \sum g(x - i, y - j) \ , \tag{5.35}$$

where $x - i \in N$, $y - j \in M$, and N_t is the number of elements that define the filter mask H. The variance of the degraded image σ_g^2 is found in a similar manner by using a local estimate of the variance:

$$\sigma_g^2(x, y) = \frac{1}{N_t} \sum_{i, j \in H} \sum [g(x - i, y - j) - m_f(x, y)]^2 \ . \tag{5.36}$$

The variance of the noise is the most difficult to determine if it is not known a priori. One method of determining the noise variance is to locate a homogeneous background region within the noise degraded image. The variation of graylevels within this region is predominantly due to the noise. Hence, the variance of this region is approximately equal to the variance of the noise.

The MMSE filter as described by Equation (5.34) changes its filter characteristics as the local variance of the noise degraded image changes. When the MMSE filter is located in homogeneous regions of the image, the local variance σ_g^2 is approximately equal to the noise variance σ_n^2 and the MMSE filter reduces to

$$est(x, y) = m_f(x, y) \ , \tag{5.37}$$

which is the output of the standard mean filter. Near edges, the variance σ_g^2 of the noise degraded image will be much greater than the noise variance σ_n^2, producing an output from the MMSE filter as

$$\text{est}(x, y) = g(x, y) \,. \tag{5.38}$$

Under this condition, the MMSE filter reduces to the original noise degraded pixel $g(x, y)$. The MMSE filter changes from a mean filter in homogeneous regions to performing no filtering near edges. The MMSE filter gives the best of both situations by removing the noise in homogeneous regions while preserving edges.

In summary, the first step in computing the MMSE filter is to obtain an estimate of the noise variance σ_n^2 either by a priori knowledge or from within a homogeneous region of the noise degraded image. Next, the standard mean filter is computed using Equation (5.35) to estimate the local mean, and then the local variance is calculated using Equation (5.36). Finally, Equation (5.34) is used to produce the MMSE filter output. This process repeats until all of the pixels within the image have been filtered. Figure 5.16(a) is an image of a boy that has been degraded with zero mean uncorrelated Gaussian noise with a variance of 800. Figure 5.16(b) is the corresponding 5×5 adaptive MMSE filtered image, while Figure 5.16(c) is the 5×5 standard mean filtered image. While both filters have successfully reduced the Gaussian noise, the adaptive MMSE filtered image has a slightly better sharpness than the mean filtered image, which can be seen in the glint of the boy's eyes.

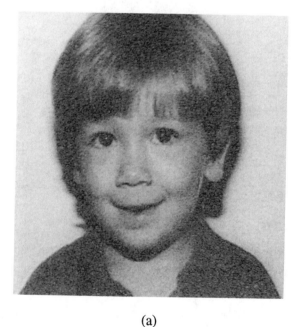

(a)

Figure 5.16: An example of using the MMSE adaptive filter:
(a) the Gaussian noise degraded image with
a variance = 800.

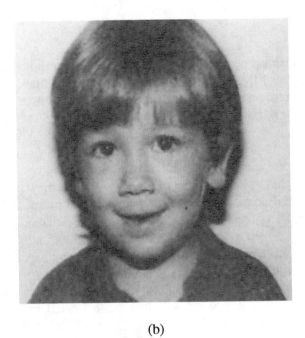

(b)

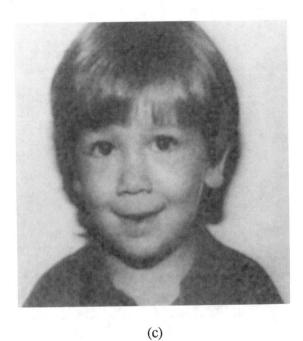

(c)

Figure 5.16: An example of using the MMSE adaptive filter:
            (b) the 5 × 5 adaptive MMSE filtered image and
            (c) the standard mean filtered image.

In the presence of outlier noise, the adaptive MMSE filter fails. The difficulty with the adaptive MMSE filter is that both edges and outlier noise will produce large local variances. Both outlier noise and edges are treated the same by this filter. Regions with large local variances are unfiltered by the adaptive MMSE filter as given by Equation (5.38). Hence, outlier noise is unfiltered and appears in the output filtered image. A different approach is to use a standard mean filter in homogeneous regions and to use a median filter near edge and outlier regions. Two adaptive filters of this type are the *adaptive double-window-modified-trimmed mean* (DW-MTM) filter and the *adaptive alpha-trimmed mean* filter.

The first of these is an adaptive version of the DW-MTM filter discussed earlier in this chapter. The first step in computing the adaptive DW-MTM filter is to compute the median over a small neighborhood of size P about the pixel $g(x, y)$. Next, the mean is computed using a larger neighborhood of size L ($P < L$) but not including any pixels with graylevels outside the range of

$$(\text{med} - K) \; to \; (\text{med} + K) \; . \tag{5.39}$$

The parameter K is chosen adaptively depending on the noise that is present within the image. One method of choosing the parameter K based upon the noise present within an image is:

$$K = C \cdot \sigma_n , \tag{5.40}$$

where the constant C is typically chosen between 2 and 3 and σ_n is the standard deviation of the Gaussian or short tailed noise. This sets the parameter K between 2 and 3 times the noise standard deviation. Pixels with graylevels outside this range are typically outliers and should be excluded in the mean calculation. A threshold value of 2 to 3 times the standard deviation of the noise comes from the fact that for Gaussian noise, the graylevel variations of the pixels vary 99.7% of the time between ±3 times the standard deviation about the mean.

In situations where an image is corrupted with outlier noise plus additive Gaussian noise, which varies from one point to another point within an image (such as signal dependent noise), another method of choosing the parameter K is to replace the noise standard deviation σ_n in Equation (5.40) by the local standard deviation σ_l so that K can vary depending on the noise present within a local region. A good estimator of the local standard deviation in the presence of outlier noise is the *median of the absolute deviations estimator* (MAD):

$$\sigma_l = \text{MAD} = 1.483 \cdot \text{med}\{f(x - i, y - j) - \text{med}[f(x - i, y - j)]\} \tag{5.41}$$

where $i, j \in H$ and $0 \le x - i < N$ and $0 \le y - j < M$. For $K = 0$, the adaptive DW-MTM reduces to a size P median filter and for K extremely large, it reduces to a size L mean filter. Still another method of choosing the parameter K is to set this parameter equal to $B \cdot \sigma_n^2 / \sigma_g^2$. Near edges or outlier regions $\sigma_g^2 > \sigma_n^2$ and the DW-MTM filter reduces to the median filter of size P. In homogeneous

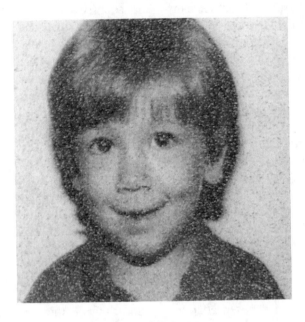

(a)

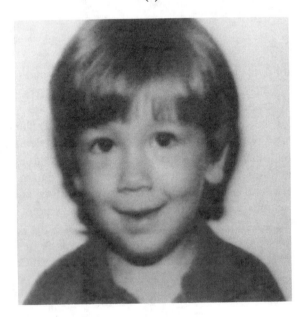

(b)

Figure 5.17: An example of using the adaptive DW-MTM filter:
(a) the 10% salt-and-pepper and Gaussian
(variance = 200) degraded image and (b) the
3×3 and 5×5 adaptive DW-MTM filtered image.

regions, $\sigma_g^2 = \sigma_n^2$ and the DW-MTM reduces to a size L mean filter. The constant B determines how fast the DW-MTM filter changes from a median to a mean filter. Typical applications of the adaptive DW-MTM use square filter sizes of $n \times n$ for the median filter and $m \times m$ for the mean filter, with the requirement that $n < m$. Figure 5.17(a) is an image that has been corrupted with Gaussian noise with a variance of 200 and 10% salt-and-pepper noise. Figure 5.17(b) is the adaptive DW-MTM filtered image generated using a 3×3 window and a 5×5 window. The parameter K was set to a constant for the entire filtering process, using Equation (5.40) with C chosen as 1.5 and σ_n set equal to 200. The adaptive DW-MTM filter was able to eliminate both the salt-and-pepper noise and the Gaussian noise but at the expense of blurring the filtered image, as seen in Figure 5.17(b).

Another adaptive filter similar to the adaptive DW-MTM filter is the adaptive alpha-trimmed mean filter. As described by Equation (5.8) the alpha-trimmed mean filter is

$$
alpha\text{-}mean\ [f(x, y)] = \frac{1}{N - 2p} \sum_{i = p}^{N-p-1} F_{(i)} \,, \tag{5.42}
$$

where $F_{(i)}$ is the ordered set of pixels included in the filtering operation as defined by the filter mask H and N is the total number of pixels included in the filtering operation. As p is varied from 0 to its maximum value of

$$
p_{max} = \begin{cases} \text{int}[N/2] & \text{for } N \text{ odd} \\ N/2 - 1 & \text{for } N \text{ even} \end{cases} , \tag{5.43}
$$

the alpha-trimmed mean filter changes its filtering characteristics from a median filter to a mean filter. The median filter performs quite well near edges and in the presence of outlier noise, while the mean filter performs better in homogeneous background regions and for Gaussian noise. The goal of the adaptive version of the alpha-trimmed mean filter is to adaptively select the parameter p. Like the adaptive MMSE filter, the local variance of the noise degraded image relative to the variance of the noise can be used as a means of determining the presence of edges or outliers.

One method of selecting the parameter p adaptively is

$$
p = p_{max} \cdot \left(1 - \frac{a \cdot \sigma_n^2}{\sigma_g^2} \right) \text{for } 0 \le a \le 1 \,. \tag{5.44}
$$

The parameter a controls the sensitivity of the adaptive alpha-trimmed mean filter, varying it from a median to a mean filter. For $a = 0$, the adaptive alpha-trimmed mean filter reduces to the median filter. In regions within the image containing outlier noise or edges, the local variance σ_g^2 of the degraded image will be much larger than noise variance σ_n^2. Under this condition, Equation (5.44) reduces to $p = p_{max}$ and the alpha-trimmed mean filter reduces to the median filter, filtering outlier noise and preserving edges. In homogeneous regions of the image, $\sigma_g^2 = \sigma_n^2$ reduces Equation (5.44) to $p = 0$, for $a = 1$. For $p = 0$, the alpha-trimmed mean filter becomes a mean filter, filtering the noise within the background regions of the image.

The Gaussian and 10% salt-and-pepper noise degraded image of Figure 5.17(a) was filtered using a 5×5 adaptive alpha-trimmed mean filter. Figure 5.18 shows the result of this filtering operation with the parameter p chosen from Equation (5.44). Table 5.2 gives the percentage of the pixels within the image that were filtered using the given parameter p. Even though the degraded image contained 10% salt-and-pepper noise, not one pixel was filtered using the equivalent of the median filter. About 28% of the pixels were filtered using the 5×5 mean, with the rest of the pixels within the image using p values that were about evenly distributed between 1 and 11. The output image of Figure 5.18 shows a reduction of noise but is blurrier than the original degraded image. The best the adaptive alpha-trimmed mean filter can do is to produce an image that is sharper than the output of the mean filter but blurrier than that of the median filter.

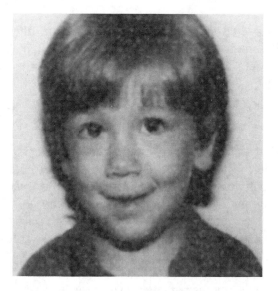

Figure 5.18: An example of using a 5×5 adaptive alpha-trimmed
             mean filter on the noise degraded image given in
             Figure 5.17(a).

An image can be separated into its low and high frequency components, as shown in Figure 5.19. As described previously in this text, the low frequency components of an image can be associated with background regions of an image. The high frequency components, on the other hand, control the sharpness of the edges within the image. The goal of the *two-component adaptive* filter is to adaptively select the number of high frequency components that are added back to the low frequency content of an image. Consider an image that has been degraded by Gaussian or uniform noise. Within the background regions of this image it is desired to lowpass filter these regions to reduce the noise that is present. At the location of edges, it is desired to perform very little if any filtering to prevent degradation of the edges, thus resulting in a blurry image.

Table 5.2: The percentage of pixels using a particular value of
p in the 5×5 adaptive alpha-trimmed mean
filter example given in Figure 5.18.

p	Percentage of pixels
0 mean filter	0.278
1	0.028
2	0.032
3	0.039
4	0.050
5	0.061
6	0.077
7	0.099
8	0.119
9	0.123
10	0.085
11	0.009
12 median filter	0.000

The output of the two-component filter given in Figure 5.19 can be written as

$$g(x, y) = f_L(x, y) + C \cdot f_H(x, y) \quad , \qquad (5.45)$$

where $f_L(x, y)$ and $f_H(x, y)$ are the lowpass and highpass filtered versions of the original image $f(x, y)$. With the parameter C set to 0, the output of the two-component filter reduces to the lowpass filtered image and if C is set equal to one, the output of this filter is the original unfiltered image. By adaptively varying C, the output of this filter can vary from a lowpass filter to a highpass enhancement filter. One way to adaptively choose C is to use the ratio of the local variance of the degraded image to the estimate of the noise variance, similar to the way the adaptive alpha-trimmed mean filter was adjusted:

$$C = \left(1 - \frac{\sigma_n^2}{\sigma_g^2}\right) \quad . \qquad (5.46)$$

In background regions, σ_g^2 is approximately equal to σ_n^2 and the output of the two-component filter reduces to the lowpass filtered image. Near edge regions, σ_g^2 is greater than σ_n^2, setting C to approximately one and adding the high frequency components back into the filtered image.

Typically, the lowpass filter is implemented using a mean filter, and the highpass filter is created from the original image and the lowpass filtered image:

$$f_H(x, y) = f(x, y) - f_L(x, y) \quad . \tag{5.47}$$

The median or any of the nonlinear mean filters can also be used to approximate the lowpass filter required by the adaptive two-component filter. If the mean filter is chosen as the lowpass filter and Equation (5.47) is used to compute the highpass filter, the characteristics of the adaptive two-component filter become very similar to the adaptive MMSE filter. A close inspection of these two filters shows that both filters vary from a mean filter to performing no filtering at all near edge pixels. As with the adaptive MMSE filter, the two-component filter as described here should not be used with images containing outlier noise. This filter cannot discriminate between edge pixels and outlier noise pixels.

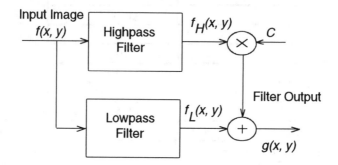

Figure 5.19: A block diagram showing a two-component adaptive filter.

All of the previously mentioned adaptive filters use a constant window size during the filtering operation. The size of the filter determines the amount of filtering that is performed and the total reduction of the noise that is present. The larger the filter size, the more filtering that is performed but the blurrier the resulting filtered image becomes. Even an edge preserving filter like the median filter will blur an image for large sized filter windows. The approach that the next two filters take is to adjust their window size depending on whether the filter is located near an edge within the image. In homogeneous regions, the filter window increases to increase the filtering of the noise, but near edges, the filter window size is decreased to the smallest size possible to preserve as much of an edge as possible.

The *signal adaptive mean* (SAM) filter is a two-component filter that changes its window size depending on whether an edge is detected. A block diagram of this filter is shown in Figure 5.20. Initially, the filter window is set to some maximum size $J \times J$. Typically, the maximum filter sizes might be 7×7 or 9×9. Next, an impulse detector is used to find and locate the impulses in the $J \times J$ filter window. The standard deviation of the graylevels can be used to detect the impulses for this step, but a better method is to use the spatial information within the filter window. Single pixel outliers can then be easily separated from edges or lines using the additional spatial information. Once the outliers have been removed, an edge detector is applied to the nonoutlier pixels within the filter window to determine if any edges are present. If so, the window size is decreased until there are no edges within the filter window or until a 3×3 filter window size is reached.

At this point, a median filter using the selected window size is applied. The output of the median filter represents a lowpass filtered version of the original pixel and its neighboring pixels. Next, the highpass filter components are generated using Equation (5.47). Like for the two-component filter, a portion of the highpass filtered components are added back to the lowpass filtered components to form the filtered image. If an edge is detected and the filter window size is equal to 3×3, the parameter C is set to one to prevent any degradation of the edge. In homogeneous regions containing noise, the value of C is set to zero, producing a filter output that depends only on the lowpass filtered components. The window size is now increased back to its maximum value and the filtering process is repeated for the next pixel until every pixel in the degraded image has been filtered. Since the median filter is used as the lowpass filter, the SAM filter is very robust in the presence of outliers and performs reasonably well in the presence of Gaussian or uniform noise. The main disadvantage of this filter is its computational complexity, since several window size iterations are used for each filtering operation.

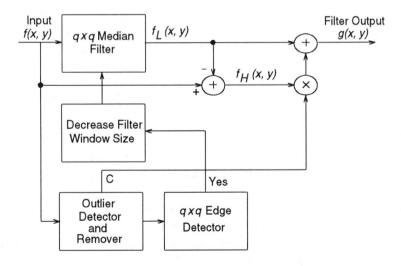

Figure 5.20: A block diagram showing the signal adaptive filter (adapted from Pitas and Venetsanopoulos, 1990).

The final adaptive filter discussed in this section is the *adaptive window edge detected* (AWED) filter. The block diagram of this filter is given in Figure 5.21. This filter also changes its window size whether an edge is detected. Initially, the filter window size is set to its maximum size of $n \times n$. Next, the local histogram is computed and the bins adjacent to the minimum and maximum graylevels are investigated for the presence of any outlier pixels. If outlier pixels are detected, they are tagged and not used in the detection of edges within the local window. If an edge is detected, the filter size is decreased by one until a 3×3 window is reached or no edges are found. At this point, either a 3×3 median filter is performed if the window size is at 3×3 or a mean filter is implemented for the selected size filter window. If the 3×3 window size is selected, then there is an edge present within the local window and the 3×3 median filter is used to prevent any degradation of edges. In homogeneous background regions, the filter window size increases and a large sized mean filter is applied to reduce any noise that is present.

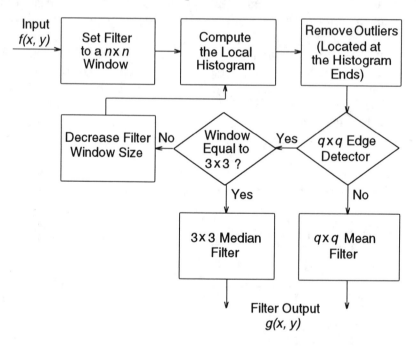

Figure 5.21: A block diagram showing the AWED Filter (adapted from Pitas and Venetsanopoulos, 1990).

The basis for most of the adaptive filters presented in this chapter is the detection of outlier noise and edges within an image. Once an edge has been located, the smallest sized edge preserving filter is applied to prevent any edge degradation. In homogeneous regions, the size of the filter window is increased to increase the filtering that is performed. There are many more adaptive filters that are available than have been discussed here. For further information and an excellent treatment of adaptive filters, the reader is referred to *Nonlinear Digital Filters: Principles and Applications* by I. Pitas and A. N. Venetsanopoulos given in the bibliography.

5.4 The Homomorphic Filter

Fourier frequency methods are commonly used to remove noise from an image, as shown in Chapter 4. The only requirement is that the noise degradation be linear. In other words, the noise must be additive and space invariant. Images that have been degraded by a multiplicative type of degradation cannot be enhanced directly using Fourier frequency methods. The approach taken by the *homomorphic* filter is to change the multiplicative degradation to an additive degradation that can be filtered using spatial Fourier frequency methods. The homomorphic filter is typically used to reduce multiplicative noise and to remove the shading from poorly illuminated images.

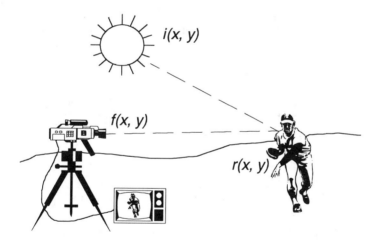

Figure 5.22: An illustration describing the formation of an image (images, © New Vision Technologies).

Consider the generation of an image on a camera as shown in Figure 5.22. The image on the camera as defined in Chapter 1 is

$$f(x, y) = i(x, y) \cdot r(x, y) \ , \tag{5.48}$$

where $r(x, y)$ is the reflectivity of the object and $i(x, y)$ is the illumination light source. The desire is to have $i(x, y)$ a constant, producing an image that contains only the image of the object. For poorly illuminated images, $i(x, y)$ varies spatially across the image $f(x, y)$, making parts of the image dark and parts light. Typical point methods of brightness and contrast enhancement are space invariant and will lighten or darken all parts of the image equally. What is needed is a filter operation that lightens the dark areas, while darkening the bright areas of an image. Equation (5.48) can also be used to represent the degradation of an image by multiplicative noise. In this case, the original undegraded image can be defined by $r(x, y)$ and the multiplicative noise by $i(x, y)$. The result is the same if the poorly illuminated or noisy image has been degraded by a multiplicative process.

The enhancement of the poorly illuminated image that is modeled by Equation (5.48) using Fourier frequency techniques requires the conversion of the multiplicative process to an additive process. The homomorphic filter shown in Figure 5.23 converts the multiplicative process to an additive process using the natural logarithmic function. The DFT of this logarithmic image yields spatial frequency components that can now be filtered by a spatial filter $H(n, m)$. The inverse DFT of this filtered image is obtained and the exponential function is applied to this image to remove the effect of the initial logarithmic operation. The final output is defined as the homomorphic filtered image.

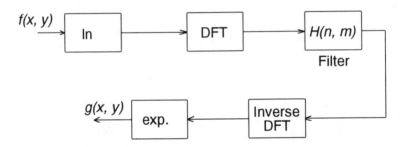

Figure 5.23: A block diagram of the homomorphic filter.

Starting with Equation (5.48) and applying the logarithmic function yields

$$q(x, y) = \ln[r(x, y) \cdot i(x, y)] = \ln[i(x, y)] + \ln[r(x, y)] \ . \tag{5.49}$$

Since Equation (5.49) is now an additive process, Fourier frequency methods can be used. Taking the DFT of each side of Equation (5.49) gives

$$Q(n, m) = I'(n, m) + R'(n, m) \ , \tag{5.50}$$

where n and m are the spatial frequencies in the x and y directions, respectively:

$$I'(n, m) = \mathcal{F}\{\ln[i(x, y)]\} \tag{5.51}$$

and

$$R'(n, m) = \mathcal{F}\{\ln[r(x, y)]\} \ . \tag{5.52}$$

The goal is to eliminate the frequency components of $i(x, y)$, which are either due to the multiplicative noise or from the poor illumination condition. Applying a linear spatial frequency filter $H(n, m)$ to Equation (5.50) produces a filtered output of

$$S(n, m) = Q(n, m) H(n, m) = I'(n, m) H(n, m) + R'(n, m) H(n, m) \ . \qquad (5.53)$$

If the spectral components of $I'(n, m)$ do not overlap the spectral components of $R'(n, m)$, then a filter can be chosen that completely eliminates $I'(n, m)$. This is the easiest of all of the filtering conditions. Typically, the spectral components of $I'(n, m)$ overlap those of $R'(n, m)$. Under this condition, the best that can be accomplished is to pick a filter that removes most of the frequency components of $I'(n, m)$, while minimizing the removal of the frequency components associated with $R'(n, m)$. A typical filter that is used in this situation is the highpass enhancement filter shown in Figure 5.24. If the spectral components due to $i(x, y)$ have been properly filtered, Equation (5.53) can then be approximated as

$$S(n, m) = Q(n, m) H(n, m) \approx R'(n, m) \ . \qquad (5.54)$$

The next step is to take the inverse DFT of the filtered spectral components as defined by Equation (5.54):

$$s(x, y) \approx \ln[r(x, y)] \ , \qquad (5.55)$$

and applying the exponential function to Equation (5.55) produces the enhanced and filtered image:

$$g(x, y) = e^{s(x, y)} \approx r(x, y) \ . \qquad (5.56)$$

The logarithm of the degraded image using Equation (5.49) requires that none of pixels within the image contain a graylevel of zero, since the logarithm function is not defined at zero. Adding one to each pixel's graylevel before taking the logarithm solves this problem:

$$q(x, y) = \ln[f(x, y) + 1] \ . \qquad (5.57)$$

This effectively increases the brightness of the degraded image by one pixel. Equation (5.56) is also modified accordingly by subtracting one from each pixel in the enhanced image:

$$g(x, y) = e^{s(x, y)} - 1 \approx r(x, y). \qquad (5.58)$$

Since the addition of the value one to each pixel within an image increases the image's brightness by an unnoticeable amount, Equation (5.56) is typically used instead of Equation (5.58) to avoid the additional subtraction calculation.

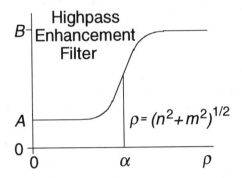

Figure 5.24: The highpass enhancement filter used to remove
the image shading from Figure 5.25(a).

The first example of using the homomorphic filter shows the enhancement of an image containing poor illumination, shown in Figure 5.25(a). The illumination function varies from bright in the center of the image and decreases radially from the center. This poor illumination function $i(x, y)$ varies slowly across the image, implying that it has predominantly low spatial frequency content $I(n, m)$. The enhancement of Figure 5.25(a) using only brightness and contrast adjustment is not able to lighten the outer regions of the image without saturating the center of the image, washing out the features of the girl's face. Proper enhancement of this image requires that the low spatial frequency components due to the poor illumination function be eliminated or reduced while preventing the elimination of the high spatial frequency components due to the reflectivity term $r(x, y)$.

A circularly symmetric highpass enhancement filter was used in the homomorphic filter to attenuate the low spatial frequency components of the poor illumination function:

$$H(\rho) = H(n, m) = \begin{cases} A + \dfrac{B - A}{1.0 + \dfrac{\alpha^2}{\rho^2}} & \text{for } \rho \neq 0 \\ 1 & \text{for } \rho = 0 \end{cases} \qquad (5.59)$$

where ρ is the radial spatial frequency, defined as $\rho^2 = n^2 + m^2$, A is the attenuation coefficient, B is the enhancement coefficient, and α is the cutoff frequency of the filter. Figure 5.24 shows a plot of $H(\rho)$ as a function of ρ. The coefficient A determines the amount of low frequency attenuation, while the coefficient B sets the desired level of the high frequency enhancement.

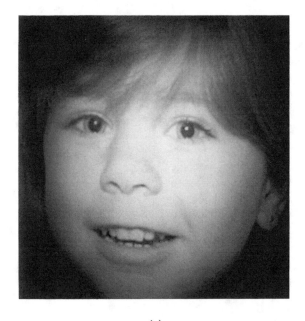

(a)

(b)

Figure 5.25: Examples of using the homomorphic filter to remove image shading: (a) the first shaded image and (b) its corresponding homomorphic filtered image.

(c)

(d)

Figure 5.25: Examples of using the homomorphic filter to
remove image shading: (c) the second shaded
image and (d) its corresponding homomorphic
filtered image.

Figure 5.25(b) shows the homomorphic filtered image obtained using the highpass enhancement filter given in Equation (5.59) and shown in Figure 5.24. The parameter values of $A = 0.5$, $B = 1.0$, and $\alpha = 4.0$ were chosen after several iterations of the homomorphic filter in which the best enhanced image was obtained. The choice of these parameters determines the quality of the enhanced image. Selecting too high a cutoff frequency washes out the image as a result of removing too many high spatial frequency components associated with the girl's face. A value of A too close to a value of one does not remove enough of the poor illumination function, resulting in very little observable illumination enhancement in the homomorphic filtered image. The outer edges of Figure 5.25(b) have been lightened, while the brightness of the center of the image has basically stayed the same.

Figure 5.25(c) is an example of an image that was acquired using a camera set to an incorrect exposure. This image was taken in Röthenburg, Germany, on a bright sunny day. The large dynamic range of intensities present in this image, ranging from the bright sky to the dark buildings on the left, makes the choice of a camera exposure difficult. Decreasing the exposure darkens the sky but further darkens the buildings. Increasing the exposure to lighten the buildings on the left side results in saturating the buildings on the right side of the image. This image is a good example of where the homomorphic filter can be used to improve the overall illumination of an image. Figure 5.25(d) is the enhanced homomorphic filtered image obtained using the highpass enhancement filter given in Equation (5.59) with $A = 0.6$, $B = 1.0$, and $\alpha = 3.0$. The homomorphic filter has darkened the bright sky and at the same time lightened the buildings located on the left side of the image. In the left-hand corner of the image, the three pedestrians, who are barely observable in the original image, have been lightened in the homomorphic filtered image. The cars and the window frames of the building in the center of the image have also been enhanced and are clearly more observable in the homomorphic filtered image.

CHAPTER 6

Color Image Processing

Color image processing is one of the newest and most exciting areas of electronic image processing. Until recently, the computing hardware required to store and manipulate color images was limited to a special few. Today, it is not uncommon to find a standard desktop computer system with a true-color 24-bit display, at least 8 million bytes of memory, and 2 gigabytes of hard disk storage. The use of desktop color scanners and color printers attached to a desktop computer system are also becoming commonplace. The demand has increased to integrate both color and black and white images into presentations and documents. The new operating systems that are being used with these desktop computer systems are integrating to make it easier to transfer color images from one application to another. Several completely electronic photography systems have appeared in recent years that totally eliminate the use of film and chemicals. An image is acquired by an electronic camera, digitized, and then stored digitally within the camera. These images are then transferred at a later time to a desktop computer for processing and inclusion in documents. Even though they presently do not have the resolution of film, they are becoming quite popular because of the speed with which images can be acquired and placed within a document or presentation.

This chapter first presents the fundamentals of color followed by a discussion of several commonly used color models. Next, several examples of using color image processing will be given, including the correction of the tint and saturation, the spatial filtering, the detection of edges, the histogram equalization, and the color white balancing of color images. The final section discusses a technique known as pseudocoloring used to highlight specific graylevels in a grayscale image in color to enhance important features so that they are clearly observable.

6.1 Color Fundamentals

One of the initial studies of color was done by Sir Isaac Newton in the eighteenth century and is contained in his treatise *Opticks*. Newton showed that white light was made from a combination of colors. Using an optical prism, he was able to separate sunlight into a rainbow of colors ranging from blue to red. Newton was also able to combine colors together to form other colors. For example, he combined green with red to produce yellow. From his research, Newton concluded that seven colors were needed to represent all the combinations of visual colors. It was Thomas Young, James Forbes, and James Clerk Maxwell in the nineteenth century who showed that only three primary colors are needed in different combinations to represent the visible spectrum of light. As discussed in Chapter 1, it was Maxwell who confirmed this when he showed that three primary colors were all that were needed to generate a color image in his now famous experiment performed during one of the Royal Society meetings in 1861. Using three separate black and white photographic plates, representing the filtered red, green, and blue color components of an image of a ribbon, Maxwell re-created the image by illuminating each photographic plate with the color light source corresponding to the color component of the plate. He then imaged the three color images from each photographic plate onto each other, producing a color image of the ribbon.

Figure 6.1 shows a diagram of the electromagnetic spectrum ranging from 0.001 nanometers to 1000 meters in wavelength. In the center of the spectrum is the visible spectrum, ranging in colors from violet to red and in wavelengths from 0.4 μm to 0.7 μm. Plate 1 (p. 243) shows the visible spectrum and the seven colors that will be of interest in this chapter, while Table 6.1 gives their approximate wavelengths. In the center of the visible spectrum is the color green, defined over the approximate wavelengths of 0.49 - 0.56 μm.

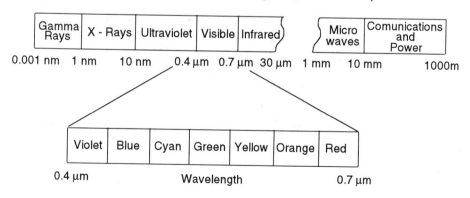

Figure 6.1: The electromagnetic spectrum.

Adjacent to the visible spectrum, are the ultraviolet and the infrared spectra. The ultraviolet spectrum is not used much in image processing because it is difficult to manufacture optical elements and detectors for use at these short wavelengths. As an example, to manufacture lenses for use at these short wavelengths, they must be extremely well polished to reduce the amount of light scattering off them, making them very expensive to build. The infrared spectrum is of interest to electronic image processing because of the enormous number of images that have been obtained with infrared camera systems. All objects at a temperature above absolute zero radiate electromagnetic energy, and if the object is at room temperature it radiates this energy in the infrared part of the electromagnetic spectrum. By the *Stefan-Boltzmann law*, a blackbody radiator radiates energy proportional to the fourth power of its temperature:

$$I = 5.67 \times 10^{-12} \, T^4 \quad \text{watts/cm}^2 \,, \tag{6.1}$$

where T is absolute temperature in degrees kelvin. Max Planck showed that this radiated energy is distributed over many wavelengths and at a given wavelength is predicted by

$$I(\lambda) = \frac{3.747 \times 10^4}{\lambda^5 \, [e^{\, 14,388 \, / \, \lambda T} - 1]} \quad \text{watts/cm}^2 \text{ per micrometer} \,, \tag{6.2}$$

where the wavelength λ is in micrometers. Integrating Equation (6.2) over all wavelengths $[-\infty, \infty]$ yields Equation (6.1).

Table 6.1: Wavelengths for the seven visible colors

Color	Wavelength
Violet	0.38 - 0.45 μm
Blue	0.45 - 0.48 μm
Cyan	0.48 - 0.49 μm
Green	0.49 - 0.56 μm
Yellow	0.56 - 0.58 μm
Orange	0.58 - 0.60 μm
Red	0.60 - 0.70 μm

The peak of this radiating energy is found by taking the derivative of Equation (6.2) and setting it to zero. The peak wavelength in microns occurs at

$$\lambda_{peak} = \frac{2898}{T} \,, \tag{6.3}$$

where T is the absolute temperature given in degrees Kelvin. For objects at a typical room temperature of 27°C, which corresponds to an absolute temperature of 27 + 273°K, the peak wavelength of the radiated energy from a room temperature object is 9.65 μm. Temperature differences are easily observable with camera systems that are sensitive to infrared wavelengths.

A common infrared camera system is the *forward looking infrared* (FLIR), which is sensitive to 10 μm. The FLIR camera system uses a linear array of detectors that scan a scene, producing a two-dimensional image that is converted to a standard video format for viewing. Other camera systems have been developed that use a two-dimensional array of sensors that are sensitive to wavelengths in either the 3 to 5 μm band or the 8 to 10 μm band. In the absence of any visible radiation (complete darkness to the human eye), these camera systems can detect and produce video image of objects that would otherwise be undetected. Initial applications of these cameras were for military purposes, but they have found a use in remote sensing of the earth, criminal surveillance, emergency rescue, and security.

Other electronic cameras have been developed that also produce images in the microwave part of the electromagnetic spectrum. The combining of images of the same scene collected from a set of imaging sensors that are each sensitive to different parts of the electromagnetic spectrum is known as *multispectral image processing*. Color image processing is multispectral image processing in which the multispectral images are limited to the visible spectrum. The three primary colors of light as used by Maxwell in his color imaging experiment were red, green, and blue. Figure 6.2 shows the relationship between these three primary colors and the colors produced by combinations of them. The new color produced by the combination of two primary colors is referred to as a *secondary color*. For example, the addition of the primary colors red and green produces the secondary color yellow. The other two secondary colors are cyan, created by the addition of green and blue, and magenta, created by combining equal amounts of red and blue. In the center of the three circles is the color white formed from the addition of equal amounts of the three primary colors.

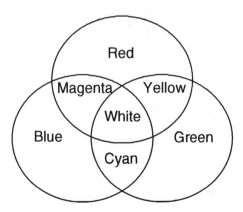

Figure 6.2: The primary and secondary colors of light (additive colors).

The mixture of the primary colors of light plays an important role in the generation of color images that are viewed on an electronic video monitor such as a television or a computer monitor. Each of these display devices uses a three gun cathode ray tube to generate three electron beams that are focused onto red, green, and blue phosphors, producing the primary colors of light. As the electron beams are scanned across the surface of the CRT, each beam is modulated, changing the amount of light that is generated proportional to the three primary colors contained within the displayed image.

The three primary colors of light are different from the primary colors of paint, or subtractive colors, as shown in Figure 6.3. The primary colors of paint are defined as the absorption of a primary color of light and the reflection of the other two primary colors of light. For example, consider a white light source incident on an object that absorbs the color green and reflects the colors red and blue. As a result, this object takes on a magenta color. The primary subtractive colors given in Figure 6.3 are created by absorption of a primary color of light, which is not included among the subtractive primary colors. For example, the subtractive primary color cyan, which is defined as the combination of green and blue, is created by an object that absorbs the color red. The secondary subtractive colors are created by the absorption of a secondary color of light by an object. An object illuminated with a white light source that absorbs the color yellow can only reflect the color blue. From Figure 6.3, the mixture of the subtractive primary colors magenta and cyan produces the color blue, while the mixture of magenta with yellow produces the color red.

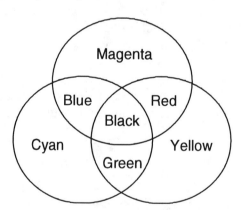

Figure 6.3: The primary and secondary colors of paint (subtractive colors).

The subtractive colors are primarily used to predict the color outputs from color photographic film and color printers. Both of these technologies use paper, a white light source, and the absorption and reflection of colors to produce a color image. Even though equal combinations of the three subtractive colors magenta, cyan, and yellow produce the color black, most high quality color printers add the color black as a fourth print color in order to produce a better perceived black in the final color print. When a black color is to be printed, this fourth print color is used instead of the equal mixtures of yellow, magenta, and cyan.

The use of three coefficients to describe a color is consistent with our understanding on how the human eye perceives color. As discussed in Chapter 1, the retina of the eye is composed of rods and cones. Located centrally about the fovea are the cones, which are sensitive to color. Figure 6.4 shows a typical response of the eye as a function of wavelength for both scotopic and photopic vision. For scotopic vision, the peak response occurs in the green region of the visible spectrum at a wavelength of approximately 0.51 μm. On the other hand, for photopic vision, the peak of the response shifts to the right to 0.56 μm, which corresponds to the yellow-green part of the visible spectrum. For both types of vision, the response of the eye decreases toward violet and red. This is why red and blue colors seem to have a perceived brightness which is lower than green colors. For many years, this was one of the reasons manufacturers chose green color display terminals for computers, believing that this color would produce less strain on the eyes under continuous hours of work.

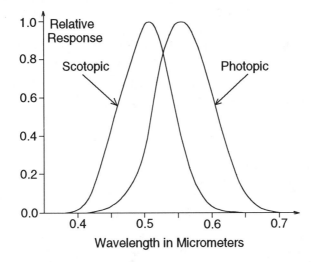

Figure 6.4: The visual response of the human eye as a function of wavelength (adapted from Carterette and Friedman, 1975).

In 1931, the Commission Internationale de L'Eclairage (CIE) devised a method of standardizing the definition of a color. A number of visual color experiments were performed in which observers were asked to match a composite color created from the combination of a set of known colors at a given wavelength. From these observations, it was determined that three color components were all that were needed to accurately match an observed color. Figure 6.5 shows, as defined by the CIE, the response of the eye to each color component as a function of wavelength. Essentially, this figure illustrates that the human visual system can be modeled using three color sensors each having a peak sensitivity corresponding to the colors red, blue, and green. Unlike the red and green sensors, which only have one spectral maximum, the red sensor has two maximums, at approximately 0.6 μm and 0.45 μm.

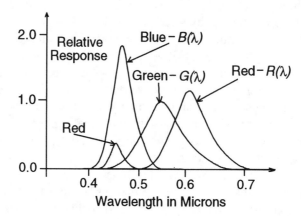

Figure 6.5: The three color components as defined by the CIE
(adapted from Martin, 1962).

Consider a light source with a color distribution $L(\lambda)$ that depends on the wavelength λ. The perceived color by the eye is then defined by its *tristimulus values X, Y,* and *Z*:

$$X = C \int_{-\infty}^{\infty} R(\lambda)\, L(\lambda)\, d\lambda \ , \tag{6.4}$$

$$Y = C \int_{-\infty}^{\infty} G(\lambda)\, L(\lambda)\, d\lambda \ , \tag{6.5}$$

and

$$Z = C \int_{-\infty}^{\infty} B(\lambda)\, L(\lambda)\, d\lambda \ , \tag{6.6}$$

where C is a constant that defines the total overall brightness response of the eye. Each of the three tristimulus values X, Y, and Z are directly related to the three primary colors of red, green, and blue. The use of Equations (6.4) through (6.6) includes the visual response of the human eye in determining the tristimulus values that defines the perceived color. Equations (6.4) through (6.6) give a better match of a perceived color than using the mixture of the three primary colors as proposed by Maxwell, Forbes, and Young. The sum of the tristimulus values gives the total luminance (brightness) perceived by the eye :

$$L = X + Y + Z \ . \tag{6.7}$$

Equation (6.7) produces a more accurate perceived brightness than adding the red, green, and blue components together equally by taking into account the different spectral responses of the eye for each of the three primary colors. Each of the three tristimulus values can be normalized by the luminance value L to give the percentage of each color:

$$x = \frac{X}{L} = \frac{X}{(X+Y+Z)} \ , \tag{6.8}$$

$$y = \frac{Y}{L} = \frac{Y}{(X+Y+Z)} \ , \tag{6.9}$$

and

$$z = \frac{Z}{L} = \frac{Z}{(X+Y+Z)} \ . \tag{6.10}$$

Equations (6.8) through (6.10) are known as the *trichromatic coefficients* of a color. The trichromatic coefficients do not depend on the total brightness of a color but only on the percentage of red, blue, and green color components.

A color can also be described by its luminance, hue, and saturation. The luminance defines the brightness of a color, for example, a dark blue sport suit versus a bright blue shirt. The hue (also referred to as tint) of a color defines its wavelength, and a color's saturation defines the percentage of white in it. Pastel colors, which contain large amounts of white, are low saturated colors, while deep colors such as those shown in the visible spectrum given in Plate 1 comprise fully saturated colors. A color's saturation value varies between 0 and 100%. The hue, saturation, and luminance color coefficients describe a color similar to the human visual system and are better than the description of a color using the tristimulus values X, Y, and Z or the red, blue, and green color coefficients. For example, consider a color containing 40% red, 30% green, and 30% blue. The reader must think about this combination of colors before realizing it contains 90% white and 10% red. But if this color was described as the color pink, the reader can immediately visualize it without thinking about the combination of primary colors.

The hue and saturation of a color define its *chromaticity*, while the luminance describes its overall brightness. The luminance component will be used later in this chapter to generate grayscale images from color images. Often the luminance image is referred to as the grayscale image of a color image. Equations (6.8) through (6.10) can also describe the chromaticity of a color by defining its hue and saturation in terms of x and y. The luminance dependence has been removed from the trichromatic coefficients by the normalization process. In particular, only the x and y trichromatic coefficients are needed to describe a color's chromaticity, since the third coefficient z can be written in terms of the other two coefficients:

$$z = 1 - x - y \ . \tag{6.11}$$

Plate 2 gives the distribution of color as a function of the x and y trichromatic coefficients. This chart was defined by the CIE in 1931 and is known as the CIE *chromaticity diagram*. Plotted along the horizontal axis is the x trichromatic coefficient which represents the amount of red in a color, and along the vertical axis is the y trichromatic coefficient, giving the percentage of green in a color. The amount of blue is defined using Equation (6.11). Located at the center of the chart is the color white, given by equal amounts of x, y, and z. Along the outer edge of the chart are the wavelengths of the spectral colors (in micrometers) describing the various color hues as defined by the CIE. The saturation of a color increases radially from the center of the chart to the outer edge, which describes 100% saturated colors. Along the bottom of the chart, the line connecting the wavelengths from 0.380 µm to 0.780 µm gives the nonspectral color of violet. The color violet is a perceived color due to both the blue and red color receptors of the eye having a sensitivity to the visible wavelength of approximately 0.43 µm. The color violet is created as the mixture of the perceived color blue with the perceived color red.

The three primary colors of red, green, and blue are also shown on the chart, but the CIE definitions show that they are not completely monochromatic in color as one would expect. The color red, for example, contains approximately 73% red, 27% green, and no blue. The color blue contains approximately 18% red, 18% green, and 64% blue. By the CIE definition, the primary color blue is only 46% saturated.

The CIE chromaticity diagram is useful in determining the wavelength (hue), the percentage of white (saturation), and the mixture of two colors. Given the location of two colors on the CIE chromaticity diagram, the line connecting them gives all of the colors that can be created by different amounts of the two colors. Also, a line drawn from any color to the color white, located at the point of equal amounts of the trichromatic coefficients, gives all possible saturations of that color. For example, drawing a line from the color white to the color red generates colors that change from pink at the center to red at the outer edge of the chart. Given a color on one side of the CIE diagram and a line drawn from this color through the color white, the point at which the line intersects the outer edge of the diagram on the other side of the color white is defined as the *complementary hue* of the selected color. For example, drawing a line from 0.7 µm through the color white gives a complementary hue of 0.493 µm.

The mixture of two colors can be extended to include the mixture of three colors. The three colors that are to be mixed together represent the vertices of a triangle. Enclosed within this triangle are all of the possible colors that can be created from different combinations of the three colors. Colors outside the triangle are not realizable by any mixture of the three colors. An interesting point is that the selection of no three colors within the CIE chromaticity diagram will produce a triangle that encloses all of the colors defined by the diagram. No matter what three colors are chosen, some colors will not be included within the triangle and hence are not realizable. A combination of more than three colors is needed to produce every color defined within the CIE chart. The realization of colors is an important part of color display and printer technology. Due to the limited color phosphors that are available, only some of the colors as defined in the CIE chromaticity diagram are realized by a color CRT in a color display. The color blue is the most difficult to accurately produce. In addition, since most color printers are also based upon a three color model, this limits the range of colors that are printable.

6.2 Color Models

Because of the limited understanding of the human visual system, several different color models have been proposed to model the characteristics of a color. All of these color models are based on using at least three components to describe a color, similar to the X, Y, Z components used to describe a color given by the CIE chromaticity diagram. Many of these color models start with the red, green, and blue color components and perform a transformation into a new color coordinate system. For example, the transformation from the red, green, and blue coordinate system to the hue, saturation, and intensity color system yields the *HSI color model*. Before any color models can be given, the definition of the red, green, and blue color components must be given, and there are two basic standards defining these. The first standard was given by the National Television Standards Committee (NTSC) for use in color television and the other standard was defined by the CIE.

The CIE red, green, and blue color components are related to the NTSC color components using the following linear transformation:

$$\begin{bmatrix} R_{\text{CIE}} \\ G_{\text{CIE}} \\ B_{\text{CIE}} \end{bmatrix} = \begin{bmatrix} 1.167 & -0.146 & -0.151 \\ 0.114 & 0.753 & 0.159 \\ -0.001 & 0.059 & 1.128 \end{bmatrix} \begin{bmatrix} R_n \\ G_n \\ B_n \end{bmatrix}, \tag{6.12}$$

where R_{CIE}, G_{CIE}, and B_{CIE} are the red, green, and blue CIE color components and R_n, G_n, and B_n are the corresponding NTSC color components. The inverse transformation is found by finding the inverse of the 3×3 transformation matrix given in Equation (6.12) and is defined as

$$\begin{bmatrix} R_n \\ G_n \\ B_n \end{bmatrix} = \begin{bmatrix} 0.842 & 0.156 & 0.091 \\ -0.129 & 1.319 & -0.203 \\ 0.006 & -0.069 & 0.897 \end{bmatrix} \begin{bmatrix} R_{\text{CIE}} \\ G_{\text{CIE}} \\ B_{\text{CIE}} \end{bmatrix}. \tag{6.13}$$

Equations (6.12) and (6.13) give the spectral relationships between the two color standards. A close inspection of Equation (6.12) shows that a single nonzero NTSC color component such as $R_n = 0$, $G_n = 1$, and $B_n = 0$ produces the nonrealizable CIE color $R_{CIE} = -0.146$, $G_{CIE} = 0.753$, and $B_{CIE} = 0.059$. The CIE color components must be nonnegative values, and this requires at least two nonzero NTSC color components. The CIE chromaticity diagram in Plate 2 shows that the CIE color components of red, green, and blue are not monochromatic in color but contain a composition of several colors. The requirement of at least two nonzero NTSC color components is consistent with this observation. Unless otherwise stated, the color models that will be presented in this section are based upon the NTSC RGB color components. The subscript n will be dropped to reduce the notation complexity of the equations in the rest of this chapter.

The simplest of the color models is the *RGB color model*. This model uses the three NTSC primary colors to describe a color within a color image. Each color component represents an orthogonal axis in a three-dimensional Euclidean space as shown in Figure 6.6. The RGB color model is a normalized color space and is defined by the color components

$$r_o = \frac{R}{R_{max}} , \tag{6.14}$$

$$g_o = \frac{G}{G_{max}} , \tag{6.15}$$

and

$$b_o = \frac{B}{B_{max}} , \tag{6.16}$$

where R_{max}, G_{max}, and B_{max} are the maximum color intensities for each of the corresponding color components and r_o, g_o, and b_o are the normalized RGB color components. For a 24-bit color system, $R_{max} = 255$, $G_{max} = 255$, and $B_{max} = 255$. The color components as defined by Equations (6.14) through (6.16) have been normalized between 0 and 1, guaranteeing that the lengths of the eight edges of the cube in Figure 6.6 will always be equal to one. All possible RGB colors are contained within this cube.

Six of the eight corners of the color cube in Figure 6.6 describe the three primary colors red, green, and blue and the three secondary colors yellow, magenta, and cyan. Figure 6.6 shows that the secondary color yellow is formed by equal amounts of red and green. The two additional corners describe the color white, created from equal amounts of r_o, g_o, and b_o, and the color black, defined as all three color components equal to zero. The dotted line between white and black corresponds to all of the possible combinations of equal values of all three color components. All the different graylevels of a grayscale image reside along this line.

The RGB color model treats a color image as a set of three independent grayscale images, each of which represents one of the red, green, and blue components of a color image. As described in Chapter 1, a properly digitized grayscale image must have at least 64 different grayscales. Typically, though, 256 graylevels are used to represent a grayscale image to meet the computer storage requirement of one byte per pixel of storage. RGB color images need one byte per pixel for each of the color planes, or three times the storage of a grayscale image of the same spatial dimensions. Each pixel in a color image requires 3 bytes or 24 bits to represent all of the possible colors. Color images of this type are now standard and are often referred to as 24-bit or *true-color* images. A 24-bit color image can contain up to 16.7 million different colors, 256^3.

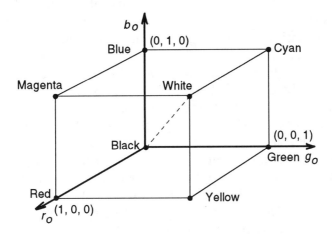

Figure 6.6: The RGB color model.

Because of the huge memory storage requirements of 24-bit color images and the limitation of many color computer display systems, a scheme called palette color images has been developed. The total number of colors within a palette color image is reduced from 16.7 million colors in a 24-bit color image to a smaller set of colors. The process of creating a palette color image from a 24-bit color image is similar to the concept of a painter's palette, in that only a small number of colors are displayable at any one time in a palette color image as compared to the total number of colors available in a 24-bit color image. A painter's palette can hold only a small subset of colors at any one time as compared to the total possible colors that are available by combining different color paints together. To create a palette image from a 24-bit color image, a *color lookup table* is created, with the number of entries equal to the number of colors desired in the palette image. The lookup table is often referred to as a *palette table* and has typical sizes of 256, 32768, and 65536. The size of the palette table determines the total number of distinct colors that can be contained in the final palette image. The N palette table entries are chosen from the N predominant colors in the original 24-bit color image. The value stored at each x and y location in the image is the palette table index value that accesses the color closet to the color that was given in the original 24-bit color image.

For a 256 palette color image, there is a factor of three reduction in the storage requirements over a 24-bit color image. Only one byte per pixel is needed to access one of the 256 palette table entries and 256×3 bytes of storage are needed to store the 24-bit palette table entries. Selecting the 256 colors that best represent the 24-bit color image is the most difficult part of the process and has caused a considerable debate among researchers. Typically, 256 palette images do not have enough colors to properly represent the total number of colors that are present in a 24-bit color image. Better quality palette color images use 32768 or 65536 palette colors, requiring at least two bytes per pixel. Palette color images solved two problems. First, they provided a means of reducing the storage size and computer memory requirements. Since memory has become inexpensive and several gigabyte hard drives are becoming common, the storage size requirement for color images is no longer as critical an issue as it has been in the past. Each of the color images shown in the color plate section of this chapter required only 921,600 bytes. This is small compared to the 1 gigabyte hard drive that was used to store these images.

Palette color images also solved the problem of displaying a color image on a computer system that could display only a small subset of colors, such as 256 or 32768, at any one time. During the time that palette images were being developed, most desktop computers also used a palette process for displaying a color image. The color display electronics of these systems used a palette table similar to that described above to display a small set of colors out of a large palette of colors. This was required to save memory, which was very expensive at the time. Initially, these systems displayed 256 colors out of a palette of 262,184 colors. As memory prices dropped and high speed video electronics became available, the palette table for these systems grew to 32768 and then to 65536.

Today, most color computer display systems come standard with 16.7 million color capability and are available at a fraction of the cost of the original palette display systems. Except for image transmissions such as via the internet or a modem, palette color images are becoming less popular and the use of 24-bit color images is gaining in popularity. The author recommends that 24-bit images should be used as the standard method of representing a color image and that palette images should be used only to maintain backward compatibility with older palette display systems that exist. The author also believes that if image storage requirements are a major issue, one of the advanced image compression schemes such as that of the Joint Photographic Experts Group (JPEG) should be used over a palette color image (see Chapter 9). These compression algorithms yield a better reduction in image storage size and produce a better quality image

COLOR PLATES

Plate 1

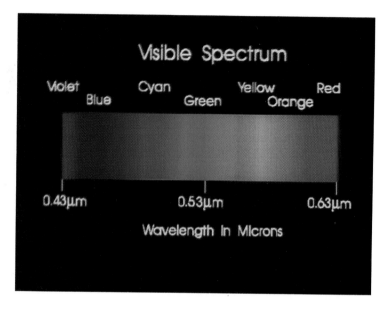

Plate 1: The visible part of the electromagnetic spectrum.

Plate 2

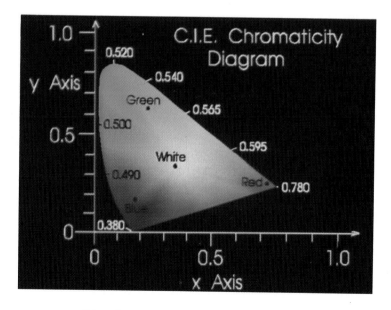

Plate 2: The CIE chromaticity diagram.

Plate 3

Plate 3: A color image showing several different color squares.

Plate 4

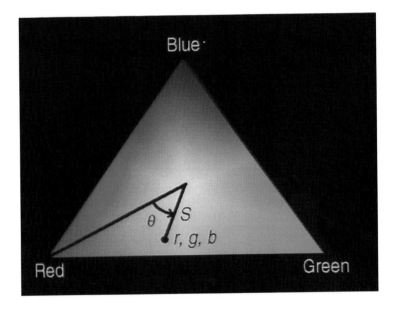

Plate 4: The HSI color space.

Plate 5

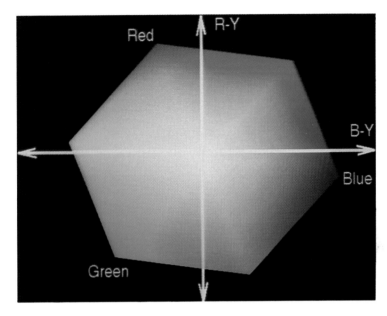

Plate 5: The C-Y color space.

Plate 6

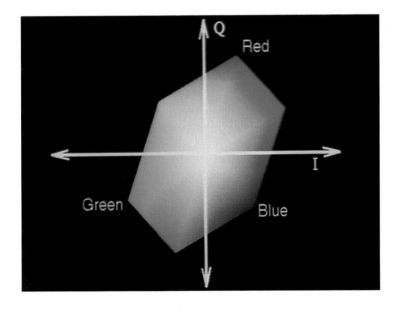

Plate 6: The YIQ color space.

Plate 7

Plate 7: The original low color saturation image.

Plate 8

Plate 8: The enhanced image of Plate 7 with the saturation increased by 1.8.

Plate 9

Plate 9: The original image containing the incorrect hue.

Plate 10

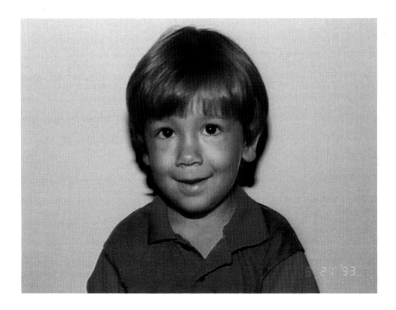

Plate 10: The enhanced image of Plate 9 with the hue rotated by 30°.

Plate 11

Plate 11: The original color image.

Plate 12

Plate 12: The selective color removal based upon the hue component of Plate 11.

Plate 13

Plate 13: The original Gaussian noise degraded image.

Plate 14

Plate 14: The enhanced image obtained using a 5×5 mean filter.

Plate 15

Plate 15: The original poorly white balanced image.

Plate 16

Plate 16: The enhanced white balanced image of Plate 15.

Plate 17

Plate 17: The original poorly colored image (from Weeks et al., 1995).

Plate 18

Plate 18: The C-Y color space of Plate 17 (from Weeks et al., 1995) .

Plate 19

Plate 19: The histogram equalized saturation image of Plate 17
(from Weeks et al., 1995).

Plate 20

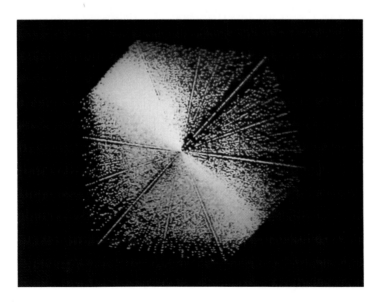

Plate 20: The C-Y color space of Plate 19 (from Weeks et al., 1995).

Plate 21

Plate 21: The histogram equalized saturation and luminance image
of Plate 17 (from Weeks et al., 1995).

Plate 22

Plate 22: The original low contrast and low saturation image
(from Weeks et al., 1995).

Plate 23

Plate 23: The histogram equalized saturation and luminance image of
Plate 22 (from Weeks et al., 1995).

Plate 24

Plate 24: The original color image used for color edge detection
(from Weeks et al., 1995).

Plate 25

Plate 25: The pseudocolored image of Figure 6.16 highlighting the keys in red.

Plate 26

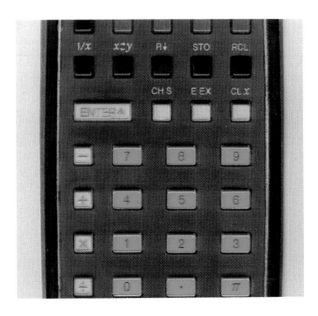

Plate 26: The pseudocolored image of Figure 6.16 highlighting the keys in green.

Plate 27

Plate 27: The pseudocolored image of Figure 6.17 using the mapping
function as defined by Equation (6.82).

Plate 28

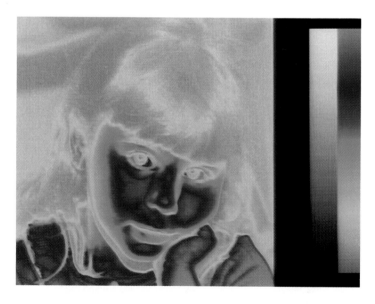

Plate 28: The pseudocolored image of Figure 6.17 using the inverse
mapping function of Plate 27.

than using a palette color image. For example, up to a 10 to 1 reduction in image size can be obtained with little observable image distortion.

Plate 3 is an example 24-bit color image showing a cube containing several different color squares. The colors of the squares were chosen to include the three primary colors as well as the secondary color yellow located in the center bottom square. The white squares were chosen to show that combining approximately equal amounts of the three primary color components produces the color white. The three grayscale images given in Figure 6.7 correspond to the red, green, and blue color images of Plate 3. The graylevel of each of these images represents the amount of red, green, or blue contained in the original color image, with white indicating a higher intensity. The squares in the three grayscale images that correspond to the green square in the original color image have a medium graylevel in the green image and are black in the other two grayscale images. This is as expected, since the color green should produce a bright graylevel in the green component image and approximately a black graylevel in the red and blue color images. Consider the two white squares in the upper left-hand corner of the original color image. Comparing the two corresponding squares in the three color component images shows that the graylevel in each is about the same. This shows that approximately equal amounts of red, green, and blue are needed to generate the color white.

The yellow and orange squares produce bright white squares in the red color component image, implying that large amounts of red are required to produce the colors yellow and orange. The blue color component image shows that both of these colors contain very little blue. The green component image shows that the difference in yellow and orange is the amount of green added to red. The yellow color square has more green than the orange color square. This is also confirmed by the CIE chromaticity diagram given in Plate 2. As stated earlier, drawing a line from the color red to the color green gives all of the possible combinations of colors that can be created by mixing these two colors. The CIE diagram shows that adding less green to the color red produces the color orange, while adding more green makes the color more yellow.

The difficulty with the RGB color model is that it produces color components that do not closely follow those of the human visual system. A better set of color models that produce color components that follow the understanding of color is that of hue, saturation, and intensity, or luminance. Of these three components, the hue is considered a key component in the human recognition process. For example, consider an image of a light blue body of water. Changing the luminance of the water would lighten or darken it, while changing its saturation would change its color from a pale blue to a deep blue. Neither of these changes in the saturation or luminance components would produce an image that seems greatly out of the ordinary. On the other hand, changing the hue of the water from blue to red produces an image of the body of water that seems unnatural and would be objectionable to most observers.

(a)

(b)

Figure 6.7: The three color component images of Plate 3: (a) the
red and (b) the green color images.

(c)

Figure 6.7: The three color component images of Plate 3: (c) the blue color image.

Hence, it is very important when applying electronic imaging techniques to a color image that the hue component be maintained and unaltered.

The requirement that the hue component be maintained during electronic image processing of a color image is one of the major difficulties associated with using the RGB color model to generate three color images that are then processed individually as grayscale images. Let $f_r(x, y)$, $f_g(x, y)$, and $f_b(x, y)$ be the three grayscale images representing the red, green, and blue color components of the color image $f(x, y)$. A linear image operator T operating individually on each component image produces three new color components:

$$o_r(x, y) = T[f_r(x, y)] \ , \tag{6.17}$$

$$o_g(x, y) = T[f_g(x, y)] \ , \tag{6.18}$$

and

$$o_b(x, y) = T[f_b(x, y)] \ , \tag{6.19}$$

where $o_r(x, y)$, $o_g(x, y)$, and $o_b(x, y)$ are the red, green, and blue components of the output image $o(x, y)$. A linear image processing operation applied equally to each of the RGB component images does not change the percentage of red, green, or blue, maintaining the hue present in the original image. A nonlinear filtering operator such as those discussed in Chapter 5 can change the percentage of each of the RGB components, changing the hue of a pixel. Using the HSI color model to represent a color image allows for its processing using only its luminance and saturation components, leaving the hue component unchanged.

There are several color models that are used to represent the hue and saturation components of a color image. The first of these models, the *HSI color model*, is based upon Maxwell's triangle, derived from the RGB color cube shown in Figure 6.6. Figure 6.8 shows the Maxwell or HSI triangle as a plane that intercepts the r_o, g_o, and b_o coordinates of (1, 0, 0),(0, 0, 1), and (0, 1, 0). The HSI model collapses the three-dimensional RGB cube into a two-dimensional triangle by separating the luminance component from the chromatic components of a color. Going through the center of this triangle is a line connecting the colors white and black. Since this line gives all the possible shades of gray of a color, the center of the HSI triangle corresponds to zero saturated colors. Plate 4 shows the color version of the HSI triangle, with the low saturated color at the center. Located at the vertices are the three primary colors. The color saturation S is measured as the length of the vector from a given r, g, b color to the center of the triangle. The outer edge of the triangle defines fully 100% saturated colors. The hue θ is defined as an angle between 0 and 360° and is measured from the reference line drawn from the center of the triangle to the red vertex.

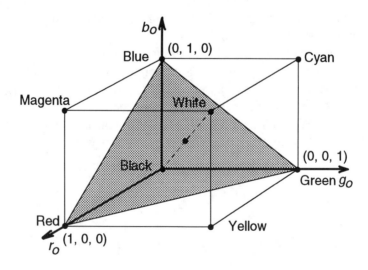

Figure 6.8: The HSI triangle shown in the RGB color cube.

To derive the equations that give the RGB to HSI transformation and its inverse, the derivation requires that the H component be divided into three regions, which essentially divides the triangle into three regions along each of its vertices. When the blue *b* component is the minimum of the three, the color is located in the bottom region of the triangle and the hue is in the range of 0 to 120°. Likewise when the red *r* component is the minimum, the *r*, *g*, *b* color is located on the right side of the triangle and the hue is in the range of 120° to 240°. Finally, when the green *g* component is the minimum, the *r*, *g*, *b* color is located on the left side of the triangle and the hue is in the range of 240° to 360°. Three normalized color components,

$$r = \frac{R}{R + G + B} \; , \tag{6.20}$$

$$g = \frac{G}{R + G + B} \; , \tag{6.21}$$

and

$$b = \frac{B}{R + G + B} \tag{6.22}$$

are used to define the normalized red, green and blue color components within the HSI color space. For equal values of *R*, *G*, and *B*, all three normalized color components given in the above three equations become equal to 1/3, which is the location intercept of the line connecting the colors of black and white with the HSI triangle in the RGB color space.

The intensity, saturation, and hue components in terms of the RGB color components are defined as

$$I = \frac{R + G + B}{3} \; , \tag{6.23}$$

$$S = 1 - 3 \cdot \min[r, g, b] \; , \tag{6.24}$$

and

$$\theta = \cos^{-1} \left[\frac{\frac{2}{3}\left(r - \frac{1}{3}\right) - \frac{1}{3}\left(b - \frac{1}{3}\right) - \frac{1}{3}\left(g - \frac{1}{3}\right)}{\sqrt{\left(\frac{2}{3}\right)\left[\left(r - \frac{1}{3}\right)^2 + \left(b - \frac{1}{3}\right)^2 + \left(g - \frac{1}{3}\right)^2\right]}} \right]. \tag{6.25}$$

Whenever $b > g$, the hue θ will be greater than 180°. For this case, since the inverse cosine is defined only over the range of 0 to 180°, θ is replaced by

$360° - \theta$. Three grayscale images that represent the hue, saturation and intensity components can be generated by scanning the RGB color image pixel by pixel and then using the values computed from Equations (6.23) through (6.25) as the graylevels for each of these grayscale images. Since most grayscale images are represented by 256 graylevels requiring one byte per pixel, the hue and saturation components are typically scaled between 0 and 255. If the original color image is a 24-bit color image, the intensity component will already be in the range of 0 to 255.

The inverse transformation from HSI color space to the RGB color space requires the use of three separate equations, depending on which one of the three regions the color is located, within the HSI triangle.

For $0° \le \theta < 120°$

$$b = \frac{1}{3}(1-S) \,, \tag{6.26}$$

$$r = \frac{1}{3}\left[1 + \frac{S\cos\theta}{\cos(60° - \theta)}\right] \,, \tag{6.27}$$

$$g = 1 - r - b \,. \tag{6.28}$$

For $120° \le \theta < 240°$

$$r = \frac{1}{3}(1-S) \,, \tag{6.29}$$

$$g = \frac{1}{3}\left[1 + \frac{S\cos(\theta - 120°)}{\cos(180° - \theta)}\right] \,, \tag{6.30}$$

$$b = 1 - r - g \,. \tag{6.31}$$

For $240° \le \theta < 360°$

$$g = \frac{1}{3}(1-S) \,, \tag{6.32}$$

$$b = \frac{1}{3}\left[1 + \frac{S\cos(\theta - 240°)}{\cos(300° - \theta)}\right] \,, \tag{6.33}$$

$$r = 1 - b - g \,. \tag{6.34}$$

The final unnormalized RGB color components are obtained by multiplying the normalized r, g, and b components by $3 \cdot I$.

Electronic image processing using the HSI color space usually entails the manipulation of the intensity and saturation components while leaving the hue component unchanged. Consider an image of a red square located against a blue background. This image can be divided into two different regions, containing the red square and the blue background. Applying a mean filter to the luminance component blurs the sharp edges associated with the red square but leaves the chromatic part unchanged. Applying a mean filter to the saturation component produces a gradual change in saturation between the two color regions. Since the human visual system is very insensitive to changes in color saturation, these changes are usually unnoticed in the processed color image.

The difficulty of processing the hue component can easily be explained with the mean filtering of it. In the hue space, the red and blue color regions generate two different hue regions with four edges located at the discontinuity between the two regions. Application of a mean filter changes the sharp discontinuities to gradual edges that linearly change from one hue value (blue) to the other hue value (red). This results in a rainbow of colors at the edges of the red square. The rainbow of colors that are displayed are the colors that the hue transitions through as it changes gradually from blue to red, which is easily observable and objectionable in the processed color image. Except for changing the overall hue of an image, similar to the tint control on a color television, the hue color component is normally ignored in the processing of color images. There has been some recent work in using the hue space in the detection of edges and some of that work will be presented later in this chapter.

Care must be taken when using any color model that transforms the RGB color components into intensity and chromatic components. Modification of a color's hue, saturation, and intensity can produce nonrealizable colors when transforming back from the HSI space to the RGB space. The RGB to HSI transformation implies that the hue and saturation components are independent of the intensity. This is true only if there are no limits on the maximum allowable R, G, B values. Dealing with digitized color images, there will always be a minimum and maximum limit on the allowable values for the R, G, B components. The RGB cube given in Figure 6.6 contains all the possible combinations of allowable colors for a given set of R_{max}, G_{max}, and B_{max}. A better way to look at the HSI color model is that for a given intensity there is a corresponding HSI triangle. As the intensity I of a color changes, the size and shape of the HSI triangle must change to guarantee that all colors as defined by the HSI color model are still contained within the RGB color cube.

For low intensity values the HSI triangle shape remains the same, but its size decreases as the intensity decreases. At the point of black, the size of the HSI triangle reduces to a single point. This makes sense in that the color black, which corresponds to all three RGB components being equal to zero, must also have a saturation value of zero. The size of the triangle is determined by the maximum saturation, with the outer edge of the triangle defining the maximum saturated colors allowed for a given intensity. Figure 6.9 shows the HSI color triangle for several different intensities. As the intensity increases, the size of the HSI triangle increases until the vertices of the triangle meets the outer edge of the RGB cube. At this point, the vertices of the triangle become clipped. For a 24-bit color image this corresponds to intensity values greater than 85. Eventually the triangle's shape becomes inverted and reduces to a single point at the coordinate R_{max}, G_{max}, B_{max} (white).

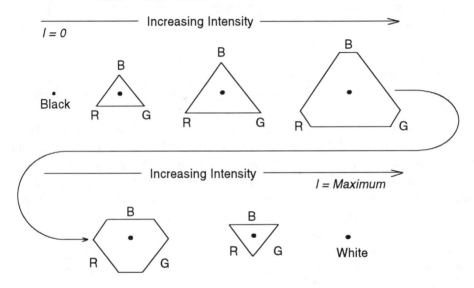

Figure 6.9: The shape of the HSI triangle as a function of intensity I.

The HSI color model has several limitations. The first is that it gives equal weighting to the RGB components when computing the intensity, or luminance, of an image. As shown previously in this chapter, the sensitivity of the eye varies for each of the RGB color components. A better mapping for the luminance or intensity is

$$Y = 0.299 \cdot R + 0.587 \cdot G + 0.114 \cdot B \ . \qquad (6.35)$$

Equation (6.35) gives a better correspondence with the perceived brightness of a color than using equal weightings of the colors of red, green, and blue. The second difficulty with the HSI color model is that the length of the maximum saturation vector varies depending on the hue of the color. As shown in Plate 4, the length of the saturation vector is longest at the vertices and is shortest halfway between the vertices, compressing the same saturation range into a smaller geometrical length. Since the saturation component many times is digitized to a finite dynamic range either for memory size restrictions or for real-time hardware processing of color images, there are fewer values of saturation available for hues that are halfway between the vertices of the HSI triangle as compared to colors with hues located near the vertices of the triangle. The third difficulty with the HSI color model is the complexity of the equations, in that different equations must be used depending on the location of a color within the HSI triangle.

The *C-Y color model* as proposed by the NTSC for use in color television solves these three limitations of the HSI model. The C-Y color model has three color components *B-Y*, *R-Y*, and *G-Y* and one luminance component *Y*. Only two of three color components are needed to define a color. The third component can be derived in terms of the other two. Typically, the *R-Y* and the *B-Y* color components are used and the *G-Y* component is derived from them. The luminance component *Y* in terms of the RGB components is given by Equation (6.35). The C-Y color components can be computed in two different ways. The first method is to subtract the luminance *Y* given in Equation (6.35) from each of the *R*, *G*, and *B* components to produce the *R-Y*, *G-Y*, and *B-Y* components. The other method is to use the following 3 × 3 transformation

$$
\begin{bmatrix} Y \\ R\text{-}Y \\ B\text{-}Y \end{bmatrix} = \begin{bmatrix} 0.299 & 0.587 & 0.114 \\ 0.701 & -0.587 & -0.114 \\ -0.299 & -0.587 & 0.886 \end{bmatrix} \begin{bmatrix} R \\ G \\ B \end{bmatrix} . \tag{6.36}
$$

The *G-Y* color component is then given in terms of the *R-Y* and *B-Y* components as

$$
G\text{-}Y = -0.509 \cdot R\text{-}Y - 0.194 \cdot B\text{-}Y . \tag{6.37}
$$

The *B-Y* and *R-Y* components form two orthogonal components that represent the chromaticity of a color. Plate 5 shows the C-Y color space plotted with the *B-Y* component along the horizontal axis and the *R-Y* component along the vertical axis. At the center of this space are the zero saturated colors. Radially from the center, the saturation increases until it reaches its maximum at the outer edges of the color space, shown as a hexagon shape. The angle relative to the *B-Y* axis gives the hue of the color. The hue θ and saturation *S* are derived from the *B-Y* and *R-Y* components as

$$
S = \sqrt{B\text{-}Y^2 + R\text{-}Y^2} \tag{6.38}
$$

and

$$
\theta = \begin{cases} \tan^{-1}\left[\dfrac{R\text{-}Y}{B\text{-}Y}\right] & \text{for } S \neq 0 \\ \text{undefined} & \text{for } S = 0 \end{cases} . \tag{6.39}
$$

Equation (6.39) shows that the hue is undefined for zero saturation. Colors with zero saturation contain no color information and are grayscales ranging from black to white. Also shown in Plate 5 are the three primary colors red, green, and blue each separated by 120°. A close observation of Plate 5 shows that the C-Y color space is a two-dimensional perspective image of the RGB color cube looking down the axis of the line from white to black in Figure 6.6. Even though the C-Y color space is not perfectly circular, this space is better at providing a more uniform maximum saturation than the HSI color model.

The inverse transformation to go from the C-Y color components to the RGB components again can be computed in two ways. The first is to add the Y component to the B-Y and R-Y components to yield the B and R components. The Y component is then added to Equation (6.37) to yield the G component. The other method is to use

$$\begin{bmatrix} R \\ G \\ B \end{bmatrix} = \begin{bmatrix} 1.0 & 1.0 & 0.0 \\ 1.0 & -0.509 & -0.194 \\ 1.0 & 0.0 & 1.0 \end{bmatrix} \begin{bmatrix} Y \\ R\text{-}Y \\ B\text{-}Y \end{bmatrix}. \tag{6.40}$$

The processing of color images using the C-Y color space is basically the same as was explained for the HSI color space. For every pixel in the color image, the Y, B-Y, and R-Y color components are computed. Next, Equations (6.38) and (6.39) are used to find the saturation and hue. The hue, saturation, and luminance components can then be processed to enhance the color image. The enhanced R-Y and B-Y components can then be computed using

$$R\text{-}Y = S \cdot \sin \theta \tag{6.41}$$

and

$$B\text{-}Y = S \cdot \cos \theta \quad . \tag{6.42}$$

The inverse transformation of C-Y to RGB yields the enhanced RGB color image. For example, multiplying the saturation component by a gain term for every pixel in the color image changes the saturation level of the processed color image. This is the process that is used to adjust the saturation level in a color television when the color level control is changed. The luminance component given in Equation (6.35) can also be used to convert a color image to a grayscale image. For each pixel in the color image, only the luminance component is computed to create the grayscale image of the original color image.

As with the HSI space, the shape and size of the maximum allowable saturation changes as a function of luminance, resulting in a change of the size and shape of the C-Y space. Figure 6.10 shows the allowable values for the R-Y and B-Y components of a 24-bit color image for several different luminance levels. The merging of each of these planes into one plane to form the total C-Y space as shown in Plate 5 defines the hexagon shape for the allowable colors. As with the HSI model, the area of the allowable colors decreases for very low and high luminance values. At the extreme luminance levels of white (255, 255, 255) and black (0, 0, 0), the allowable area for the *B-Y* and *R-Y* components reduces to a single point where the maximum possible saturation must be zero. The size and shape limitations of the C-Y color space are not due to the definition of the color model itself but to limiting the C-Y color space to within the finite volume of the RGB cube. As stated earlier, this is a practical limitation that has been imposed by using only a finite number of values to represent the original RGB color components.

For the color cube given in Plate 3, a set of 256 graylevel images corresponding to the luminance, saturation, and hue images can be generated using Equations (6.36), (6.38) and (6.39). As can be seen from Figure 6.11(a), the luminance image is simply the grayscale image of the original color image. The graylevels in the saturation image given in Figure 6.11(b) correspond to the color saturation level of a color. The black graylevels represent low saturated colors, while the high saturated colors are given in white. This saturation image was scaled so that the maximum saturation within the color image is scaled to graylevel 255. Inspection of the saturation image shows that seven out of the nine squares contain large saturation values. A comparison of Figure 6.11(b) with Plate 3 shows that the zero saturated squares located in the upper left-hand corner correspond to the white color squares in the color image. This is as expected since the color white has zero saturation.

The hue component as defined by Equation (6.39) ranges from 0 to 360°. So that the hue component could be displayed as a grayscale image, Equation (6.39) was also scaled between 0 and 255. Figure 6.11(c) gives the hue image, showing the angle of the hue for each of the seven color squares. Since the hue is undefined for pixels containing zero saturation, the hue of all pixels containing zero saturation was set to zero. Any other value could have been chosen, but setting the undefined hue to zero also sets the white background surrounding the nine color squares to zero, emphasizing the hues of the seven color squares. A comparison of Plates 3 and 5 with the hue image shows that the degree values do match the colors represented. For example, the hue value of 113° in the lower left-hand square is the angle this color makes with the *B-Y* axis in Plate 5. This places the square's color slightly to the left of the *R-Y* axis in the second quadrant. At this location, Plate 5 gives the color red, which matches the color of the square shown in the original color image in Plate 3. The hue of the lower right-hand corner square is 135°. This places the color at the center of quadrant 2, 135° from the *B-Y* axis. Inspection of Plate 5 yields the color orange, which is in agreement with the color of the square in the original image.

One of the major advantages of the C-Y color space is that its transformations are much simpler to compute than the HSI color model. Secondly, since the C-Y color space is the standard color model as proposed by the NTSC for use in color television, this makes the real-time hardware implementation of any of the color algorithms developed in the C-Y color space feasible due to the enormous amount of color video hardware that is available.

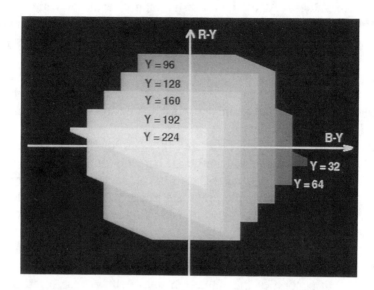

Figure 6.10: The shape of the C-Y space as a function of luminance
(From Weeks et. al., 1995).

(a)

Figure 6.11: The hue, saturation, and luminance images of Plate 3:
(a) the luminance image.

(b)

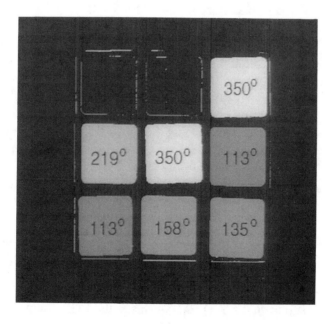

(c)

Figure 6.11: The hue, saturation, and luminance images of Plate 3:
(b) the saturation image and (c) the hue image.

Another color model that is closely related to the C-Y model and is also used in color television is the *YIQ color model*. The three components for this color model are the luminance as defined by Equation (6.35), the in-phase I, and quadrature-phase Q component. The I and Q components define the chromaticity of the color and form two orthogonal components. The transformation from the RGB color space to the YIQ color space is

$$
\begin{bmatrix} Y \\ I \\ Q \end{bmatrix} = \begin{bmatrix} 0.299 & 0.587 & 0.114 \\ 0.596 & -0.275 & -0.321 \\ 0.212 & -0.523 & 0.311 \end{bmatrix} \begin{bmatrix} R \\ G \\ B \end{bmatrix}.
\tag{6.43}
$$

Once in the YIQ space, the hue and saturation can be computed in a similar manner to the C-Y color space:

$$
S = \sqrt{I^2 + Q^2}
\tag{6.44}
$$

and

$$
\theta = \begin{cases} \tan^{-1}\left[\dfrac{Q}{I}\right] & \text{for } S \neq 0 \\ \text{undefined} & \text{for } S = 0 \end{cases},
\tag{6.45}
$$

where the hue is measured relative to the I axis and varies between 0 and 360° and is undefined if the saturation is equal to zero implying a grayscale color.

Plate 6 shows a color version of the YIQ space. Like the C-Y color space, the zero saturated colors are located at the origin, while the highly saturated colors are located at the outer edges. A comparison of the YIQ color space with the C-Y color space in Plate 5 shows that these two color spaces are very similar. Except for the compression of the space in the direction of the Q axis in comparison to the *B-Y* axis, they are the same but rotated by 33°. The I axis leads the *B-Y* axis by 33° and the same is true for the Q axis relative to the *R-Y* axis. Many books mention that the YIQ color space leads the C-Y color space but forget to mention that the scaling of the two color spaces is not equivalent. The processing of a color image using the YIQ color space is identical to the process used for the C-Y color space, except Equation (6.43) is used to convert from the RGB color space to the YIQ color space. Equations (6.44) and (6.45) are then used to compute the saturation and hue components and the inverse transformation to go from the YIQ components to the RGB components is used:

$$
\begin{bmatrix} R \\ G \\ B \end{bmatrix} = \begin{bmatrix} 1.0 & 0.956 & 0.62 \\ 1.0 & -0.272 & -0.647 \\ 1.0 & -1.108 & 1.705 \end{bmatrix} \begin{bmatrix} Y \\ I \\ Q \end{bmatrix}.
\tag{6.46}
$$

Another color space that is based upon the hue and saturation color components is the *HLS color space*. This color space is defined using the normalized set of RGB components as described by Equations (6.20) through (6.22). The HLS space produces the three color components hue θ, lightness L, and saturation S and by definition the RGB to HLS transformation is

$$\theta = H(r, g, b) \ , \tag{6.47}$$

$$L = \frac{\max(r, g, b) + \min(r, g, b)}{2} \ , \tag{6.48}$$

and

$$S = \begin{cases} \dfrac{\max(r, g, b) - \min(r, g, b)}{\max(r, g, b) + \min(r, g, b)} & \text{for } 0 < L \le 0.5 \\[4mm] \dfrac{\max(r, g, b) - \min(r, g, b)}{2.0 - \max(r, g, b) - \min(r, g, b)} & \text{for } 0.5 \le L < 1.0 \end{cases} \ , \tag{6.49}$$

where $H(r, g, b)$ is defined as

$$H(r, g, b) = \begin{cases} \dfrac{60 \cdot (g - b)}{\max(r, g, b) - \min(r, g, b)} & \text{for } r = \max(r, g, b) \\[4mm] 120 + \dfrac{60 \cdot (b - r)}{\max(r, g, b) - \min(r, g, b)} & \text{for } g = \max(r, g, b) \\[4mm] 240 + \dfrac{60 \cdot (r - g)}{\max(r, g, b) - \min(r, g, b)} & \text{for } b = \max(r, g, b) \end{cases} \ . \tag{6.50}$$

If $H(r, g, b) < 0$, then $H(r, b, g)$ is replaced with $H(r, g, b) = 360 + H(r, g, b)$. The values of L and S range between 0 and 1, while the θ varies between 0 and 360°. The L component given in Equation (6.48) is equivalent to the intensity or luminance component of the color image.

A color model similar to the HLS model is the *HSV color model* (sometimes referred to as the *HSB color model*). The three components of this model are hue θ, saturation S, and value V. This model uses the same equations for hue as the HLS color model and is defined by Equations (6.47) and (6.50). The value and saturation components are defined by

$$V = \max(r, g, b) \tag{6.51}$$

and

$$S = \begin{cases} \dfrac{\max(r,\, g,\, b) - \min(r,\, g,\, b)}{\max(r,\, g,\, b) + \min(r,\, g,\, b)} & \text{for } \max(r,\, g,\, b) \neq 0 \\[2ex] 0 & \text{for } \max(r,\, g,\, b) = 0 \end{cases} \qquad (6.52)$$

The major advantage of the HSL and the HSV color spaces over the HSI color space is that they produce color components that more closely follow that of what is perceived. Another advantage of these color models is that they do not require the computation of the square root and the inverse cosine as seen in comparing Equation (6.25) with Equation (6.50).

Two color spaces used for printing purposes are the subtractive color spaces *CMY* and *CMYK*. As with the HLS and HSV color spaces, both the CMY and the CMYK color spaces use the normalized RGB color components given in Equations (6.20) through (6.22). The subtractive color space CMY produces the primary colors cyan, magenta, and yellow. The RGB to CMY transformation is defined as

$$\begin{bmatrix} C \\ M \\ Y \end{bmatrix} = \begin{bmatrix} 1 \\ 1 \\ 1 \end{bmatrix} - \begin{bmatrix} r \\ g \\ b \end{bmatrix}. \qquad (6.53)$$

The inverse transformation is found by subtracting the C, M, Y components from one. As discussed earlier in this chapter, many printing processes use a fourth color to improve the quality of the black colors printed. The CMYK color space is the model used for the four color printing process and uses the four color components C_k, M_k, Y_k, K, with the fourth component K representing the additional color black. The RGB to CMYK transformation is first computed using Equation (6.53) to generate the CMY color components. The rest of the transformation into the CMYK space is accomplished using

$$K = \min\,(C,\, M,\, Y) \qquad (6.54)$$

and

$$\begin{bmatrix} C_k \\ M_k \\ Y_k \end{bmatrix} = \begin{bmatrix} C \\ M \\ Y \end{bmatrix} - \begin{bmatrix} K \\ K \\ K \end{bmatrix}. \qquad (6.55)$$

The final four color models that will be presented in this section are those that have been derived by the CIE to correct for the deficiency that exists in the CIE XYZ coordinate system. They have been summarized here for reference and the interested reader is referred to one of the many image processing texts listed in the bibliography for further discussion. Distance changes in either the x or y directions of the CIE chromaticity diagram of Plate 2 do not correspond to the perceived differences in color. In the green region of the CIE chart, larger linear distances have to be made to obtain the same perceived change in color as compared to the red or blue regions of the chart. Before these other color models are given, the equation that relates the CIE XYZ color space to the NTSC RGB color space will be given. By definition, the transformation from the RGB color space to CIE XYZ color space is

$$\begin{bmatrix} X \\ Y \\ Z \end{bmatrix} = \begin{bmatrix} 0.607 & 0.174 & 0.201 \\ 0.299 & 0.587 & 0.114 \\ 0.000 & 0.066 & 1.117 \end{bmatrix} \begin{bmatrix} R \\ G \\ B \end{bmatrix}, \qquad (6.56)$$

and its inverse is

$$\begin{bmatrix} R \\ G \\ B \end{bmatrix} = \begin{bmatrix} 1.91 & -0.534 & -0.289 \\ -0.984 & 1.998 & -0.027 \\ 0.058 & -0.118 & 0.897 \end{bmatrix} \begin{bmatrix} X \\ Y \\ Z \end{bmatrix}. \qquad (6.57)$$

To go from the CIE RGB color space to the XYZ color space, Equation(6.13) is used to compute the NTSC RGB components from the CIE RGB components. This result is then substituted into Equation (6.56).

The first of the color models that was developed to correct for the perceived color limitation of the original 1931 color space was the *UVW* system. The transformation from the XYZ coordinate system that derives the three components *U*, *V*, and *W* is

$$\begin{bmatrix} U \\ V \\ W \end{bmatrix} = \begin{bmatrix} 0.666 & 0.00 & 0.000 \\ 0.000 & 1.000 & 0.000 \\ -0.500 & 1.500 & 0.500 \end{bmatrix} \begin{bmatrix} X \\ Y \\ Z \end{bmatrix}. \qquad (6.58)$$

Its inverse transformation is defined as

$$\begin{bmatrix} X \\ Y \\ Z \end{bmatrix} = \begin{bmatrix} 1.500 & 0.000 & 0.000 \\ 0.000 & 1.000 & 0.000 \\ 1.500 & -3.000 & 2.000 \end{bmatrix} \begin{bmatrix} U \\ V \\ W \end{bmatrix}. \qquad (6.59)$$

The three trichromatic coefficients *u*, *v*, and *w* are defined as

$$u = \frac{U}{U + V + W}, \qquad (6.60)$$

$$v = \frac{V}{U + V + W}, \qquad (6.61)$$

and

$$w = \frac{W}{U + V + W} \quad . \tag{6.62}$$

Equations (6.60) through (6.62) form the uniform chromaticity scale *USC* system.

A further improvement on the UVW or USC color space is the CIE uniform perceptual color space $U^*V^*W^*$. This color space improves on the UVW color space by shifting the white reference point of the original CIE chart to a better perceived white, as defined by the 1960 CIE standard. The transformation from the XYZ color space to $U^*V^*W^*$ color space is

$$W^* = 25 \cdot (100 \cdot Y)^{1/3} - 17 \qquad \text{for } Y \geq 0.01 \; , \tag{6.63}$$

$$U^* = 13 \cdot W^* \cdot (u - u_o) \tag{6.64}$$

and

$$V^* = 13 \cdot W^* \cdot (v - v_o) \; , \tag{6.65}$$

where u and v are defined by Equation (6.60) and (6.61), and u_o and v_o are the coordinates of the 1960 CIE chromaticity reference white point.

The final two color models of interest are the $L^*a^*b^*$ and the $L^*u^*v^*$. For the $L^*a^*b^*$ color model, the L^* denotes the lightness of the color, the a^* component gives the red to green axis, and the b^* component gives the yellow to blue axis. The a^* and b^* color components completely define the chromatic component of a color. As with the USC system, the $L^*a^*b^*$ color model also provides a match between distances and changes in perceived color. The $L^*a^*b^*$ transformation from the CIE XYZ color space is

$$L^* = \begin{cases} 903.3 \cdot (Y / Y_o) & \text{for } 0 \leq Y < 0.01 \\ 25 \cdot (100 \cdot Y / Y_o)^{1/3} - 16 & \text{for } Y \geq 0.01 \end{cases} , \tag{6.66}$$

$$a^* = 500 \cdot [\, (X / X_o)^{1/3} - (Y / Y_o)^{1/3} \,] \; , \tag{6.67}$$

and

$$b^* = 500 \cdot [\, (X / X_o)^{1/3} - (Z / Z_o)^{1/3} \,] \; , \tag{6.68}$$

where X_o and Y_o are the reference white point.

In the $L^*u^*v^*$ color model, like in the $L^*a^*b^*$ color model, the u^* and v^* components completely describe the chromatic part of a color. The L^* component is defined by Equation (6.66) and the u^* and v^* components are defined as

$$u^* = 13 \cdot L^* \cdot (u - u_o) \qquad (6.69)$$

and

$$v^* = 13 \cdot L^* \cdot (v - v_o) \ , \qquad (6.70)$$

where u_o, v_o are the coordinates of the reference white point. The advantage of the $L^*a^*b^*$ and the $L^*u^*v^*$ color systems is that they provide a means of modeling the characteristics of different color displays, hard copy, and acquisition devices. Typically, each of these devices produces or uses a different reference white point. Both the $L^*a^*b^*$ and the $L^*u^*v^*$ color systems allow the selection of the reference white point.

Many of the color models mentioned in this section have a finite limited color space that varies with luminance or intensity. Typically the processing of color images involves transforming from the RGB color space to another color space. Since the RGB color space will be of finite volume, the allowable colors in the new space must be contained within the RGB color cube. The processing of color images in this new space could produce unrealizable colors when transforming back into the RGB color space that was used to represent the original color image. For example, consider the use of the HSI color model, in which the color red has been modified from $L = 85$, $S = 1.0$, $\theta = 0°$ to $L = 255$, $S = 1.0$, $\theta = 0°$ by enhancing just the luminance component of the color. The inverse transformation from HSI to RGB using Equations (6.23), (6.26)-(6.28), and (6.20)-(6.22) to obtain the unnormalized RGB components yields $R = 765$, $G = 0$, and $B = 0$. For a 24-bit color image, in which the range of values for the RGB components is 0 to 255, this color would not be realizable. In fact, for an intensity over 85 in a 24-bit color image the saturation value cannot be equal to 1.

Unrealizable colors must be brought back into the allowable color space. As previously mentioned, the eye is most sensitive to hue. The approach that is typically taken is to transform these unrealizable colors into new realizable colors in which their hues are left unchanged. This leaves changing either a color's luminance or saturation. Changing the luminance of a color is accomplished first by transforming the color back to the RGB color space. Next, if any one of the three color components exceeds the allowable range, then the value of each color component is decreased equally. For example, for a 24-bit color image each RGB component is multiplied by

$$\frac{255}{\max(R,G,B)} \cdot \qquad (6.71)$$

Equation (6.71) essentially decreases the luminance of the color while leaving the saturation and hue unchanged. The second approach, in which the saturation is changed, requires knowing the maximum allowable saturation for a given hue and luminance. Once the maximum saturation is known, the unrealizable color's saturation is scaled back to the maximum saturation. Of the two methods, scaling the saturation component is the most difficult in that it requires knowledge of the maximum saturation of a color. If the color space being used does not directly calculate a color's saturation, this adds additional computations to the processing of the color image. The author recommends the use of Equation (6.71) whenever possible.

6.3 Examples of Color Image Processing

In this section, several techniques for the processing of 24-bit color images will be presented. Because of the author's familiarity with the NTSC color television standard, the RGB and C-Y color spaces have been chosen in the processing of these 24-bit color images. Any of the other color spaces could have been used to demonstrate the techniques in this section. Color images were acquired either directly from a video camera into a computer using a 24-bit 640 × 480 image acquisition card or via photographic prints that were scanned using a 24-bit 300 × 300 dots per inch color desktop scanner. The color space chosen depended on the type of processing that was desired. If the RGB color model was used, each of the three color components was treated as a separate grayscale image and processed individually. For cases when the hue, saturation, and luminance components were required, the RGB color space was converted to the C-Y color space using Equation (6.36) and the saturation and hue components were then computed using Equations (6.38) and (6.39). After processing of the hue and saturation components, Equation (6.40) was used to transform from the C-Y back to the RGB color space. Finally, the RGB values computed from the inverse transformation were verified as valid colors. If one of the RGB triplet values exceeded the maximum allowable value of 255, Equation (6.71) was used to scale the luminance of the color back into the allowable range of color.

The first two color image processing examples show how the hue and saturation of a color image can be used to enhance the overall appearance of an image. Plate 7 shows an image of a field of tulips taken in central Holland. Except for the red row of tulips in the center of the image, this image is lacking in color saturation overall. The goal of this example is to enhance this image by adding more color saturation, similar to adjusting the color level control on a color television. For every pixel within this image, its luminance, hue, and saturation were computed so that its saturation value could be increased by a factor of 1.8. The enhanced saturation component along with the original hue and luminance components were then used to compute the RGB values of the enhanced image, as shown in Plate 8. A comparison of this image with the original shows the increase in the overall color saturation. In particular, the green colors of the tulip field in the lower part of the image are much more vivid as well as the buildings present in the image.

The enhancement of a color image by increasing the color saturation is one of the techniques that will produce colors outside the allowable range. As the luminance of a color approaches 0 or 255, the maximum saturation approaches zero, as shown in Figure 6.10. Multiplying every pixel's saturation value by a gain coefficient that is independent of luminance can produce hue, saturation, and luminance values that are outside the RGB range of colors when converted back to the RGB space. Equation (6.71) was used to limit the luminance of the color to prevent the generation of any unrealizable colors. This is why there is very little apparent increase in the saturation level of the blue sky in the upper part of Plate 8 as compared to the increase in the saturation of the tulip field. The blue colors of the sky were already at the maximum allowable saturation because of their high luminance values. If Equation (6.71) was not included in the saturation enhancement algorithm, eventually, as the saturation level of the enhanced image was increased, the color of the sky became distorted (red and green colors start to appear) due to the colors being pushed outside the allowable range of RGB colors.

In the observation of an image, the human visual system is most sensitive in detecting the presence of incorrect flesh tones. Slight changes in the hue can produce objectionable reddish or greenish flesh tone colors. Plate 9 shows an image of a boy with a reddish flesh tone. This image can be easily corrected by transforming the RGB components to their hue, saturation, and luminance components. The same process as explained for the saturation enhancement of a color image is used, except a constant value is either subtracted or added to the hue of each pixel. This rotates all the colors about the origin of the C-Y color space, changing the hue of each color within the image, which results in changing the hue of the boy's skin. The C-Y color space given in Plate 5 can be used to determine the angle in which the hue of the image in Plate 9 must be rotated to correct for the reddish flesh tone. Normally, flesh tones would produce colors around 135° relative to the *B-Y* axis. Since the flesh tone of the boy's skin in Plate 9 is on the reddish side, the hue angle of these colors is probably somewhere between 100° and 110°. Adding approximately 30° to the hue of each pixel will shift the overall hue of Plate 9 toward green. Plate 10 shows this result. Notice how correcting the hue of this image has removed the reddish flesh tone, giving the image a more natural appearance.

The two previous examples used the same hue and saturation correction throughout the entire image. The concept behind the next color image processing example is to process only pixels with luminance, saturation, or hue values within a predetermined range. Using this method, one can process only certain colors within a color image, leaving the other colors unchanged. This provides a means of enhancing objects within an image based upon their color. This type of color processing is known as *selective color processing*. Plate 11 shows an image of a farm plow with a blue frame and a red wheel against a stone wall. It is desired to highlight the blue frame of the plow by setting the entire image to a grayscale image except for the blue frame of the plow.

The C-Y color space given in Plate 5 shows that the blue frame of the plow will be in the blue region of the C-Y space and have hue values in the range of 260° to 315°. To convert a color image into a grayscale image is easily accomplished by converting the RGB color components to the C-Y to obtain the saturation and hue components as given by Equations (6.38) and (6.39) and then setting the saturation to zero for all pixels within the image. The goal of this example is to set the saturation of each pixel within the image to 0 except those corresponding to the blue frame of the plow:

$$S_{new} = \begin{cases} S_{old} & \text{for } \theta_1 \leq \theta \leq \theta_2 \\ 0 & \text{otherwise} \end{cases} , \qquad (6.72)$$

where θ_1 and θ_2 specify the range of hues that define the blue color of the plow, S_{new} is the new saturation value, and S_{old} is the saturation value of the original color. After several iterations, the best values for θ_1 and θ_2 to set the image to black and white, except for the frame of the plow, were found to be $\theta_1 = 290°$ and $\theta_2 = 315°$. Plate 12 shows that the entire image is now black and white, except for the blue frame of the plow.

Decreasing the range of hues smaller than the range given begins to set part of the frame to black and white. Determining the range of luminance, saturation, or hue values to implement selective color image processing is easily accomplished when the desired colors are well separated from the other colors within the image. For example, if it was desired to highlight the red wheel instead of the blue frame, the range of hue for Equation (6.72) would be difficult to obtain. A hue range set too small would set the color of the pixels defining the red wheel to black and white, while a hue range too large would leave the parts of the wall behind the plow in color. This would not accomplish the desired goal of setting the color image to a grayscale image, except for the red wheel of the plow.

Even though the range of color component values may be difficult to choose in some images, selective image processing is key to adding overlays and special effects to movies (film or video). The process known as *chroma keying* gives a movie special effects artist the capability of mixing two images together automatically. The concept is to use a particular color to define areas of an image to be overlaid with another image. In most natural scenes, a color blue similar to that given for the plow's frame in Plate 12 occurs rarely. For this reason, the color blue is chosen as the standard overlay color. Consider the special effect of having an adult woman appear as if she is two inches tall and standing in the palm of a child's hand. The first step is to acquire an image of the woman against a blue background, posing as if she is standing on a hand. A second image of the child's hand as if it is holding an object is obtained. The placement and orientation of the woman and the child's hand when creating the two images are such that when they are overlaid they give the desired special effect.

Before combining the images, the image of the woman is reduced in size to give the illusion that she is only two inches tall. The way in which selective processing comes into play is in the combining of the two images. A new combined image is created using the color blue to determine how the two images are to be combined. For every region in the image of the woman in which the color is blue, the output image is set to the image of the child's hand. For regions within the image of the woman that are not blue, the output image is set to the woman's image. When completed, the final image appears to have a two inch woman standing in the hand of a child. This process is particularly useful when creating special effects for movies. For obvious reasons, it is also known as *blue screening*.

The next example shows how to perform linear filtering on color images using the RGB color space. The use of linear spatial filtering methods enables the RGB color image to be treated as three separate grayscale images, representing the red, green, and blue components. The same linear filter is applied to each of the grayscale images, producing three new linearly filtered images that are then recombined to form the new filtered color image. This approach of separating a color image into three separate color images is valid only for linear operations. Linear filters modify the RGB components equally, which maintains the hue. Nonlinear filters could change the percentage of RGB values, which effectively changes the hue of the RGB triplet.

Plate 13 shows a 24-bit color image of a daffodil flower that has been corrupted by Gaussian noise. Not only has the Gaussian noise added a grainy appearance to the image, it has changed the hue of the colors. This is particularly noticeable in the different colors of red and green that appear in the yellow petals of the daffodil. The hue noise that is present in this image shows that the Gaussian noise is uncorrelated between the three RGB components. Additive noise that has a high degree of correlation between each of the RGB components can add only luminance noise, since the noise changes the value of the RGB components equally. Plate 14 shows the 5×5 mean filtered image of Plate 13. Notice how the noise present in the image has been reduced, but at the expense of blurring the image. The same characteristics of the mean filter for grayscale images apply for the filtered image of Plate 14. The improvement of the noise in Plate 14 is not as great as it appears. The use of the mean filter has reduced the noise present in each of the three RGB components. In effect, this has also reduced the hue noise that is present in the image. Since the eye is most sensitive to changes in hue, reducing the noise present within the image reduces the changes in hue, resulting in a large perceived improvement in the filtered color image.

Because the eye is very sensitive to the slightest change in hue, the incorrect tint of low saturated or pastel colors is easily detected. Colors in an image that should be perceived as white are easily detected by the eye and are typically found to be objectionable if these colors have a slight colored tint. An image in which the white colors of the image have obtained a slight colored tint is known as an improper *white balanced image*. An incorrect white balanced image can be attributed to several different factors. The most common being images acquired under improper lighting conditions. Different light sources emit light energy with different spectral responses. For example, an incandescent light appears slightly yellow when compared to a fluorescent light. When the same image is acquired using different light sources, dramatically different results can occur. Color photographs taken with standard indoor/outdoor film under fluorescent lighting conditions appear to have an overall greenish tint, while images acquired using sunlight appear natural. Chromatic filters exist that compensate for different lighting conditions, but in many instances, these white balance errors can be removed during the printing process by a laboratory technician manually adjusting the color filters on the printer until the best color photograph is obtained.

An additional problem with photographs is that the overall white balance changes as the photograph ages. Over time, the emulsion of the photographic paper changes color and alters the white balance of the image from a neutral white to a reddish or yellow colored image. Most of these old photographs can be corrected by taking new photographs of them and modifying the white balance. Correct white balance of an image is also important for color video cameras, which must be able to correctly reproduce the colors present in a scene under a variety of lighting conditions. Hence, white balance correction must be performed to compensate for the different lighting conditions to reduce the white balance errors in the images. Typically, many color video cameras offer manual white balance correction via an external user control. This control is varied until the image appears natural for the given lighting condition.

Manual white balance correction typically consists of transforming an RGB color image into a color space, which generates two orthogonal chromatic components and a luminance component. Typically, for NTSC color video either the YIQ or the C-Y color spaces are used. This is easily accomplished using either the I and Q components in the YIQ color space or the R-Y and B-Y components in the C-Y color space. Manual white balance correction of a color image ignores the luminance component of the image while translating the origin of the chromatic space, comprising either the I and Q or the B-Y and R-Y components. The translated chromatic components and the unmodified luminance component are then transformed back to the original RGB color space. The resulting RGB image is the white balance adjusted color image. Translation of the chromatic components continues interactively until the white balance corrected image appears natural to an observer.

Plate 15 is an image of a family portrait taken sometime during the early 1960s that has discolored with age. The reddish tint is very evident in the sky and the rocks on the left side of the photograph. This image was transformed from the RGB color space to the C-Y color space using Equation (6.36). Next, the B-Y and R-Y locations of each pixel were translated by a fixed amount. The unmodified luminance Y and the shifted B-Y and R-Y components were then used in Equation (6.40) to produce the RGB components of the white balanced modified image. After several iterations, the best white balanced image was obtained by shifting the *B-Y* components by -29.5 and the *R-Y* components by 40.5. Plate 16 shows the corrected white balanced image. Note the overall improvement in the colors of this image. The whites of the children's clothes appear to be white and the sky now appears to be blue. In comparing Plate 16 to Plate 15, the overall reddish color of the original image has been reduced. A more advanced method is to divide the image into two or three luminance regions. Next, the *R-Y* and *B-Y* chromatic spaces are computed for each luminance region. White balance correction now involves the translation of each of these *R-Y* and *B-Y* color spaces. For three luminance regions this involves the adjustment of six coefficients, one to translate the *B-Y* component and one to translate the *R-Y* component in each luminance region.

The next example is the enhancement of both the luminance and chromatic information in a color image by histogram equalization. In Chapter 3, grayscale histogram equalization was presented and used to enhance the contrast and brightness of an image. Histogram equalization redistributes pixels across all graylevels so as to yield a near-uniform distribution, improving the overall contrast and brightness in the equalized image and therefore enhancing its features. Features of an image that are barely visible in the original grayscale image become clearly visible after histogram equalization. There is no direct extension from grayscale histogram equalization to color histogram equalization. Grayscale histogram equalization is a one-dimensional process and is not easily converted into the three dimensions needed for color image processing. Treating the color image as three separate color images and applying histogram equalization separately to each of the three RGB grayscale images can change the relative percentage of red, green, and blue for each pixel, resulting in changes in hue. Consequently, color equalization must process each of the RGB components equally to prevent any changes in the relative percentage between the RGB values. The implementation of one RGB color histogram covering the entire 16.7 million colors can require as much as 64 megabytes of storage to compute the color histogram for a 640×480 24-bit color image (4 bytes for each bin and 16.7 million bins).

The most common method of histogram equalization of a color image is the transformation of the RGB color space into an HSI color space and then equalization of the luminance component only. This improves only the luminance of the image while ignoring the chromatic information present. The color histogram equalization example presented here equalizes both the luminance and saturation components, while using the hue component in the equalization process. The equalization of a color image must be based upon the hue of a color, while leaving the hue value unchanged. Consider the equalization of the luminance and saturation components of an image of a small orange square occupying 10% of the image against a blue background. The equalization of the luminance and saturation components, without any concern for the hue of a color, would be biased toward the blue background pixels due to the large number of them. Because histogram equalization is over the entire set of pixels, colors with more pixels will be equalized more than colors with fewer pixels.

To provide an unbiased equalized color image, the C-Y color space will be divided into N equally angularly spaced hue regions occupying $360/M$ degrees. Next, each region is then separated into K equally spaced luminance regions. Figure 6.12 illustrates one of these hue regions, showing the dependence of the maximum saturation on the luminance component. The equalization process begins with the N hue by K luminance segmented C-Y color space. Next, the values for the maximum realizable saturation $S_{max}(n, k)$ ($n = 1, 2, 3 \cdots N$, $k = 1$, $2, 3 \cdots K$) for each segment are computed by searching all the possible RGB combinations and selecting the corresponding maximum saturation value for the hue-luminance combination. This computation is performed once in each region, and then the results are stored in a table where they can be used repeatedly by the color histogram equalization algorithm.

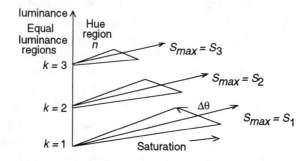

Figure 6.12: A single hue region showing the dependence on the maximum saturation as a function of luminance in the C-Y color space, as shown in Plate 5.

The histogram equalization performed on the saturation component in each $n = 1, 2, 3, \cdots N$ hue region and $k = 1, 2, 3, \cdots K$ luminance region yields a saturation histogram equalization of

$$S'_{nk}(s) = S_{max}(n, k) \sum_{i = 0}^{s} \frac{s_{ink}}{s_{tnk}} , \qquad (6.73)$$

where s_{ink} is the number of pixels in the ith bin of the nth hue and kth luminance region and s_{tnk} is the total number pixels in the nth hue and the kth luminance region. Given an input saturation level s, Equation (6.73) gives the new output saturation level $S'_{nk}(s)$ for that region.

Complete color equalization requires an equalization of both the saturation and luminance components. Once the saturation equalization is complete, equalization is performed on the luminance component using standard graylevel equalization. This process equalizes the entire luminance image and results in a single equalization of

$$Y(y) = Y_{max} \sum_{i=0}^{y} \frac{n_i}{n_t} , \qquad (6.74)$$

where Y_{max} is the maximum possible luminance, n_i is the number of pixels with luminance i (the *ith* bin), and n_t is the total number of pixels in the image. For a 24-bit color image, Y_{max} is 255. Equation (6.74) gives the new luminance values $Y(y)$ as a function of input luminance y. Even though the realizable luminance values depend on the saturation component, ignoring this dependence in the equalization of the luminance is not a major disadvantage. Some colors will be pushed outside the realizable RGB color space but can be brought back by normalizing each component using Equation (6.71).

Plate 17 is a photograph of the Swiss Alps taken in the presence of early morning fog, which partially obscures the mountain range in the background. It was chosen to illustrate the ability of color histogram equalization to enhance geographical features. Color histogram equalization should be able to enhance both the colors that are present in the mountain regions as well as the foliage present in the foreground. Plate 18 is an image of its corresponding C-Y color space distribution. This image was obtained by plotting the RGB colors at their appropriate *B-Y* (horizontal) and *R-Y* (vertical) coordinates. Plate 18 shows that most of the colors in the image lie near the origin (center of the image), indicating a low-saturation color image. It also shows the distribution of hues present in the image. Although all hues are present, only red and blue occur with any degree of saturation. This can be confirmed from the original image shown in Plate 17.

Color equalization was performed using 96 hue regions and 64 luminance regions. The 96 hue regions provided enough separate hue regions to independently equalize the different colors in the image, and the 64 luminance regions provided enough regions to adequately include the effect of luminance on the maximum saturation value. Plates 19 and 20 are the enhanced color image generated by equalizing the saturation component and its corresponding C-Y color space image. In comparing the original and the enhanced C-Y color space image, it is easily observed that colors within the original image are more uniformly distributed about the C-Y color space. Inspection of Plate 19 shows that the histogram equalization of the saturation component has enhanced the features on the mountain face while preserving the white sky. Plate 21 is the final histogram equalized image obtained by saturation equalization followed by luminance equalization. Histogram equalization of the luminance component enhanced the details in the trees as well as the clouds in the sky. Plate 22 shows an image of the ceiling inside an Austrian church. This image was chosen for its high spatial detail and its abundance of low saturation colors. Plate 23 is the final saturation followed by the luminance equalized image. Notice how both the luminance and color saturation have been increased in this image.

The final color image processing example presented in this section shows how to implement edge detection of color images by performing a separate edge detection of each of the RGB grayscale images and then combining them to form one combined edge image. But many times it is desired to find the edges within a color image based on the perception of color, using the hue, saturation, and luminance components. For example, consider the detection of the boundaries of a lake using the HSI color space. The use of the hue component alone provides the information necessary to locate the lake boundaries. Edge detection obtained using the RGB color space yields an edge detected image that contains not only edges due to the changes in hue but also edges that are the result of the changes in saturation and luminance.

The edge detection of the luminance and saturation components is the same as the edge detection of any grayscale image. For example, the nonlinear range filter discussed in Chapter 5 could be used as an edge detector, or several of the spatial edge detectors that will be discussed in Chapter 8 can also be used, such as the Sobel edge filter. Care must be used in the edge detection of the hue space. Application of an edge detector directly to the hue space can produce edges that are not present in the original color image. The problem stems from the fact that the hue space is modulus 360°. In fact, applying any grayscale edge detector to the hue image produces edges not present in the original image.

Consider the 3 × 3 sub-image in Figure 6.13. Applying a grayscale edge detector to this sub-image produces a vertical edge, even though hues of 0° and 360° are the same color. The sub-image is simply a constant color, containing no vertical edges. To eliminate the modulus 360° effect, two orthogonal components from the hue are computed:

$$R = \cos \theta \tag{6.75}$$

and

0°	360°	360°
0°	360°	360°
0°	360°	360°

Figure 6.13: A 3 × 3 sub-image of the hue space.

$$T = \sin \theta \ , \tag{6.76}$$

where θ is the hue value between $0°$ and $360°$. Scanning the hue image pixel by pixel and using Equations (6.75) and (6.76) generates two new images $R(x, y)$ and $T(x, y)$, which represent the two orthogonal components of the hue image. Edge detection is performed on each of these images independently to produce two new edge images $R_e(x, y)$ and $T_e(x, y)$. The final hue edge image is found by absolute summing both edge images:

$$H_e(x, y) = \mid R_e(x, y) \mid + \mid T_e(x, y) \mid \ . \tag{6.77}$$

Both orthogonal components are required to properly apply edge detection to the hue component. For example, consider if only the R component were used in the detection of edges. Given two colors with hues of $\theta = 60°$ and $\theta = 300°$, computation of R yields $\sqrt{3}/2$ for both hues. Using a grayscale edge detector on the R component alone would miss edges that are present in the hue image. Because of the orthogonal nature of Equations (6.76) and (6.77), two different hue values will always produce different values for R and T. Yet two colors with the same hue value produce exactly the same R and T values, eliminating the modulus $360°$ problem associated with the hue color component.

Plate 24 is a 24-bit color image of Thomas A. Edison's cars at the Edison Museum in Fort Myers, Florida. This image was chosen for its high spatial detail and its changes in color saturation and hue throughout the image. A 3×3 Sobel edge detector from Chapter 8 was chosen as the edge filter to demonstrate the detection of edges within this color image. Figure 6.14(a) gives a Sobel edge image of the luminance component, while Figure 6.14(b) gives the corresponding Sobel edge image of the saturation component. Notice how well the edges of the car and the wood frame of the garage are defined in Figure 6.14(a). Most of the edges in Figure 6.14(b) are dark in comparison to the luminance edge image because most of the colors throughout the image have a low saturation, except for the body of the left car. Figure 6.14(b) shows the advantage of using an HSI color model to highlight key edge features. In this figure, the edges associated with the highly saturated regions of the image are easily observable. This image does a good job of isolating the left car's features from the frame of the garage.

Figure 6.14(c) is the Sobel edge image of the hue component using Equations (6.75) through (6.77) to eliminate the modulus $360°$ effect. Figure 6.14(d) is the corresponding Sobel edge detected hue image obtained by applying the 3×3 Sobel filter directly to the hue space, without any regard for the modulus $360°$ nature of the hue component. Note the amount of noise present and how poorly the edges are defined throughout this image. The edges in Figure 6.14(c) are much better defined. In particular, the edge features of the

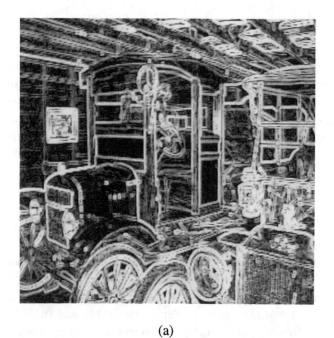

(a)

(b)

Figure 6.14: Examples of color edge detection: (a) the luminance
                edge image and (b) the saturation edge image (from
                Weeks et al., 1995).

(c)

(d)

Figure 6.14: Examples of color edge detection: (c) the hue edge
image with 360° correction and (d) the hue edge
image without 360° correction (from Weeks et al., 1995).

car on the right are much more observable. Also notice how the features in front of the left car are now clearly observable.

6.4 Pseudocoloring and Color Displays

An important use of color is the enhancement of grayscale images to highlight key features in color. Since the eye can observe color more readily than differences in graylevels, color can be used to emphasize a selected range of graylevels. The process of adding color to a grayscale image is called *pseudocolor* or *falsecolor*. The process of pseudocoloring should not be confused with the process of coloring a grayscale image. The process of coloring a grayscale image is to convert the various objects within it to colors that present what the image should look like in color. For example, the process of colorizing a grayscale image of a red automobile would require that the color image of the automobile contain the correct color red. The process of pseudocoloring, on the other hand, is to highlight a grayscale image based upon the graylevels present in the image, without any regard to the colors of the original objects.

Pseudocoloring of a grayscale image is typically performed using the RGB color model. Figure 6.15 gives a block diagram describing the process of pseudocoloring a grayscale image. The input image $f(x, y)$ is assumed to be an N graylevel image that is mapped into a red $r(x, y)$, a green $g(x, y)$, and a blue $b(x, y)$ image that represent the RGB color components of a color image. Usually, the grayscale image contains 256 graylevels that are mapped to a 24-bit color image using 256 intensity values for each one of the RGB component images. The type of function chosen for $R[f(x, y)]$, $G[f(x, y)]$, and $B[f(x, y)]$ determines the type of pseudocoloring enhancement possible. For example, if

$$c(x,y) = \begin{cases} R[f(x, y)] = f(x, y) \\ G[f(x, y)] = f(x, y) \\ B[f(x, y)] = f(x, y) \end{cases} \tag{6.78}$$

the output image is simply the original grayscale image.

Individual graylevels in a grayscale image can be highlighted in color by selectively modifying the mapping function to emphasize individual graylevels. For example, consider the mapping of a particular graylevel T in a grayscale image to the color blue. The mapping function for this pseudocoloring operation is

$$c(x,y) = \begin{cases} R[f(x, y)] = 0;\ G[f(x, y)] = 0;\ B[f(x, y)] = b_{max} \text{ for } f(x, y) = T \\ R[f(x, y)] = G[f(x, y)] = B[f(x, y)] = f(x, y) \qquad \text{otherwise} \end{cases},$$

$$\tag{6.79}$$

where b_{max} is the maximum graylevel allowed for the blue color image. The mapping function in Equation (6.79) sets the red $r(x, y)$ and green $g(x, y)$ component images to zero when the input grayscale image $f(x, y)$ is at graylevel T but sets the blue $b(x, y)$ component image to its maximum allowable graylevel. Otherwise this mapping function produces the original grayscale image. For $f(x, y) = T$, the output image displays a bright blue color.

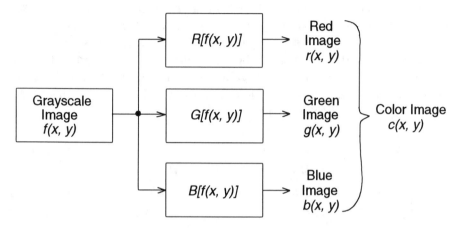

Figure 6.15: A block diagram showing the process of pseudocoloring.

Figure 6.16 is a 256 graylevel image of a Hewlett Packard model 35 calculator. The symbols on the keys in the leftmost column are barely visible in the grayscale image. The goal of pseudocoloring this image to form a 24-bit RGB color image is to highlight this column of keys in color to emphasize the symbols located on them. After several iterations, the best pseudocolored image that highlighted these keys was obtained by setting graylevels between 105 and 185 to red and leaving all other graylevels as is:

$$c(x,y) = \begin{cases} R[f(x, y)] = 255; \ G[f(x, y)] = B[f(x, y)] = 0 & 105 \leq f(x, y) \leq 185 \\ R[f(x, y)] = G[f(x, y)] = B[f(x, y)] = f(x, y) & \text{otherwise} \end{cases}.$$

$$(6.80)$$

Plate 25 shows the result of this pseudocoloring operation. Note that the symbols are clearly defined. It becomes immediately obvious by looking at this image which pixels are in the graylevel range of 105 to 185. Plate 26 is another example of pseudocoloring of Figure 6.16, with the graylevels between 210 and 255 set to green and all other pixels unchanged

$$c(x,y) = \begin{cases} R[f(x, y)] = B[f(x, y)] = 0; \ G[f(x, y)] = 255 & 210 \leq f(x, y) \leq 255 \\ R[f(x, y)] = G[f(x, y)] = B[f(x, y)] = f(x, y) & \text{otherwise} \end{cases}.$$

$$(6.81)$$

The 12 keys in the bottom center and right of the image have been highlighted in green. In addition, the symbols in the leftmost column have also been highlighted in green.

Another pseudocoloring mapping function is to map the entire set of graylevels to the rainbow of colors that represent the visible spectrum, shown in Plate 1. This pseudocoloring mapping function is typically used in the display of infrared imagery to highlight the changes in grayscale, which represent changes in temperature. Since the perception of color is that blue is cooler than red, blues map to the lower set of graylevels that represent the lower thermal temperatures. The red colors map to the higher graylevels that represent the hot temperatures in a thermal image. The following mapping function will produce a rainbow of colors, ranging from blue to red, for a 256 graylevel image that is mapped to a 24-bit RGB color image:

$$c(x,y) = \begin{cases} R = 0, & G = 254 - 4 \cdot f, & B = 255 & 0 \le f \le 63 \\ R = 0, & G = 4 \cdot f - 254, & B = 510 - 4 \cdot f & 64 \le f \le 127 \\ R = 4 \cdot f - 510, & G = 255, & B = 0 & 128 \le f \le 191 \\ R = 255, & G = 1022 - 4 \cdot f, & B = 0 & 192 \le f \le 255 \end{cases}$$

$$, \tag{6.82}$$

where $R = R[f(x, y)]$, $G = G[f(x, y)]$, $B = B[f(x, y)]$, and $f = f(x, y)$. Figure 6.17 is a grayscale image of a young girl. Plate 27 gives the pseudocolored image using the mapping function described in Equation (6.82). On the right side of Plate 27, the mapping function described by Equation (6.82) is given, showing the transformation of the graylevels to color. Notice how the girl's eyes have been highlighted by the pseudocoloring process. Plate 28 is the inverse mapping of Plate 27, with black mapping to red and white mapping to blue. Comparing Plate 27 to Plate 28 shows how changing the mapping of the graylevels to color changes the perception of the pseudocolored image. The appearance of Plate 27 seems more natural than Plate 28. Plate 28 appears more like a negative image.

Figure 6.18(a) shows an example of a simple 640×480 NTSC compatible grayscale display system that is capable of displaying 256 graylevels. The 256 graylevel image is stored into 307,200 bytes of memory. At the rate of 30 times a second, each byte of storage is fed to the digital-to-analog converter and converted from a digital number to an analog voltage that is used to modulate the electron beam of the black and white CRT, producing a grayscale image on the monitor. Usually a graylevel of 0 produces a black intensity, while a graylevel of 255 produces a white intensity. To display a pseudocolored image, either a palette display system or a 24-bit color display system is needed. Figure 6.18(b) shows an example of a palette display system that is capable of displaying 256 colors out of a palette of 16.7 million colors. The palette display system works

Figure 6.16: A grayscale image of a Hewlett Packard model 35 calculator that will be pseudocolored to enhance the keys.

Figure 6.17: A grayscale image that is pseudocolored using the mapping function given in Equation (6.82).

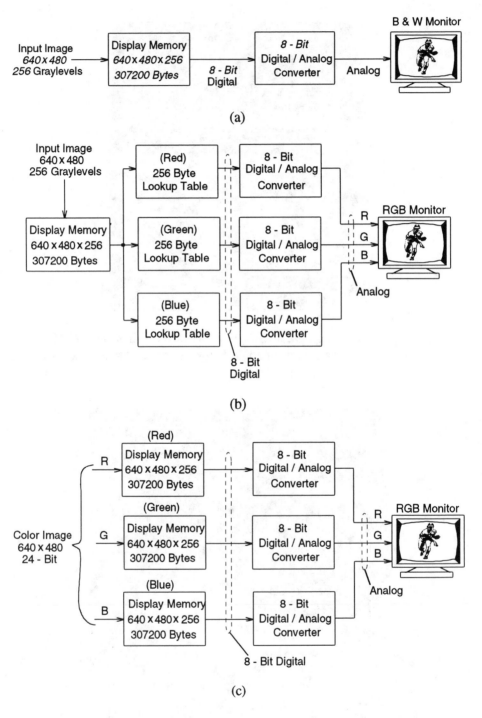

Figure 6.18: An example of (a) a grayscale display, (b) a palette
display and (c) a 24-bit color display (images,
© New Vision Technologies).

on the same principle as a palletized color image. The palette display system shown in Figure 6.18(b) can display only a 640 × 480 by 256 color palette image. For pseudocoloring, the original 640 × 480 grayscale image is stored in the 307,200 bytes of display memory. At the rate of approximately 30 to 75 times a second (depending on the system) each byte from this memory is fed to the address lines of three 256 byte lookup tables. The outputs of each of the RGB lookup tables are then converted from a digital number to an analog voltage using three digital-to-analog converters. Each of these RGB analog signals are then used to modulate the red, green, and blue electron beams in an RGB color monitor, producing the desired color image.

The lookup tables provide the means of implementing the three color mapping functions $R[f(x, y)]$, $G[f(x, y)]$, and $B[f(x, y)]$. The values stored in each of the 256 bytes of memory that compose each lookup table determine the final mapping function. For example, if every location in the three lookup tables is set to zero except the 255th location of the red lookup table, which is set to 255, the output image will display, in red, only those pixels within the grayscale image that are at graylevel 255. Another method of displaying a pseudocolored image is to use a 24-bit color display system directly. The grayscale image is converted to a 24-bit color image using the pseudocoloring block diagram given in Figure 6.15 with the desired mapping functions. Then each of the RGB color component images that comprised the 24-bit color image is loaded into its corresponding display memory. At a rate of approximately 30 to 75 times a second, the bytes from the three display memories are fed to the three input digital-to-analog converters. The output voltages from the three digital-to-analog converters are again used to modulate the red, green, and blue electron beams in the RGB color monitor.

In this chapter, the reader was introduced to the concepts of color image processing. Several different color models were given along with several color image processing examples using the RGB and C-Y color models. Pseudocoloring of a graylevel image was given, showing the advantages of highlighting important features within a graylevel image in color. Not discussed in this chapter, which is beyond the scope of an introductory discussion of color image processing, is a detailed comparison between each of the color models, the limitations of each model, and the effects of finite numerical precision on each model. Also not discussed, is the processing of color images using multichannel techniques. Since most of these topics are relatively new, the interested reader is referred to the open literature for a detailed discussion of these topics.

CHAPTER 7

Image Geometry and Morphological Filters

Both image geometry and morphological filtering are important aspects of electronic image processing. These topics have been combined into one chapter because they both deal with the geometry of an image and of objects within it. Through the use of image geometry, the size and orientation of an image can be changed. For example, an inverted and upside down image can be reoriented so that it can be viewed properly. Morphological filters on the other hand can change the form and structure of objects within an image. For example, the sharp corners of a rectangular object can be rounded, and small circular regions can be removed from an image while preserving the larger circular regions. This chapter begins with the discussion of the spatial interpolation of pixels within an image that is required to implement several of the image geometry operations presented in Section 7.2. The concepts behind binary morphology are then introduced and several binary example images are given, illustrating the benefits of binary morphological filtering. Finally, the concepts of binary morphological filters are expanded to cover grayscale morphological filters.

7.1 Spatial Interpolation

Many of the image geometrical operations require knowledge about the graylevel of points on an image located somewhere between a set of pixels, as illustrated in Figure 7.1. For example, consider doubling the size of an $N \times N$ image to a $2N \times 2N$ image. The original coordinates x, y map to $2x$, $2y$ in the new image resulting in every other pixel in the new image having an odd coordinate value, $2x + 1$ or $2y + 1$, which yields pixels with undefined graylevels. For example, the pixel at 3, 2 in the newly scaled image corresponds to the coordinate 1.5, 1 in the original image. Because the image has been discretely sampled, the graylevel at this point is unknown. From the graylevels of the surrounding pixels, the graylevel for this point must be interpolated.

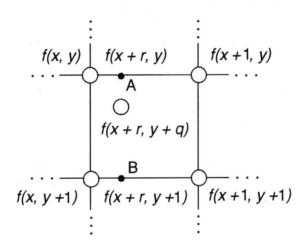

Figure 7.1: The interpolation of a point by its surrounding four neighbors.

There are several different types of interpolation methods that are available for equally spaced sampled images. The simplest of the interpolation methods is *pixel replication*. The closest defined pixel is used as the graylevel for the interpolated pixel. Figure 7.1 shows four neighboring pixels surrounding the pixel to be interpolated. Defining $0 \le r \le 1$ and $0 \le q \le 1$ so that the interpolated pixel can be placed anywhere within the four neighboring pixels, pixel replication can then be defined as

$$f(x + r, y + q) = f(\text{int}[x + r + 0.5], \text{int}[y + q + 0.5]) . \qquad (7.1)$$

For r or $q < 0.5$, the interpolated pixel becomes $f(x + r, y + q) = f(x, y)$, while for both $r \ge 0.5$ and $q \ge 0.5$, the interpolated pixel becomes $f(x + r, y + q) = f(x + 1, y + 1)$. With pixel replication there is no interpolation between the graylevels of the four neighboring pixels. Pixel replication was used in Chapter 2 in the presentation of the rotation transformation required by the Hotelling transform.

The next widely used interpolator is the *nearest neighbor interpolator*, most often referred to as *linear* or *zero order interpolation*. This interpolator uses the first order derivative of the graylevel change between the four neighboring pixels to estimate the graylevel of the interpolated pixel. Figure 7.1 shows the point A, given by $f(x + r, y)$, that is somewhere between the pixels $f(x, y)$ and $f(x + 1, y)$.

Linear interpolation between $f(x, y)$ and $f(x + 1, y)$ uses the equation for a straight line, with the slope given by the graylevel change between the two pixels:

$$y = m \cdot r + \textit{offset} , \tag{7.2}$$

where $m = f(x + 1, y) - f(x, y)$ and the offset $= f(x, y)$. Substituting for the slope and offset in terms of $f(x, y)$ and $f(x + 1, y)$ yields

$$f(x + r, y) = [f(x + 1, y) - f(x, y)] \cdot r + f(x, y) . \tag{7.3}$$

Equation (7.3) reduces to $f(x, y)$ for $r = 0$ and to $f(x + 1, y)$ for $r = 1$. Likewise, the interpolated value of point B is

$$f(x + r, y + 1) = [f(x + 1, y + 1) - f(x, y + 1)] \cdot r + f(x, y + 1) . \tag{7.4}$$

To find $f(x + r, y + q)$ requires using the same linear interpolation as in Equations (7.3) and (7.4), but the slope is now the difference between the graylevels of points A and B, and the offset is the graylevel of point A given by Equation (7.3):

$$f(x + r, y + q) = [f(x + r, y + 1) - f(x + r, y)] \cdot q + f(x + r, y) , \tag{7.5}$$

where $0 \le q \le 1$.

Substituting Equations (7.3) and (7.4) into Equation (7.5) gives the nearest neighbor interpolation as

$$f(x + r, y + q) = (1 - r) \cdot (1 - q) \cdot f(x, y) + r \cdot (1 - q) \cdot f(x + 1, y) + \\ q \cdot (1 - r) \cdot f(x, y + 1) + r \cdot q \cdot f(x + 1, y + 1) . \tag{7.6}$$

Equation (7.6) can be easily verified with $r = 0$ and $q = 0$, $r = 1$ and $q = 0$, $r = 0$ and $q = 1$, and $r = 1$ and $q = 1$. These four combinations of values for r and q yield the four surrounding neighbors $f(x, y)$, $f(x + 1, y)$, $f(x, y + 1)$, $f(x + 1, y + 1)$. The interpolated graylevel $r = q = 0.5$ for the pixel in the center of the four neighboring pixels reduces the average of the four neighboring pixels:

$$f(x + r, y + q) = \frac{f(x, y) + f(x + 1, y) + f(x, y + 1) + f(x + 1, y + 1)}{4} . \tag{7.7}$$

The disadvantage of the nearest neighbor interpolator is that it does not take into account the curvature of the graylevel change between the four neighboring pixels. The nearest neighbor method assumes that the graylevel variation between the four pixels can be modeled as a linear plane. A better interpolation method is to also include an estimate of the second derivative of the graylevel changes between the four neighboring pixels, since the second derivative is a measure of the curvature of a function. The *second order* or *six-point interpolator* is defined as

$$
\begin{aligned}
f(x+r, y+q) = \quad & (1 + rq - r^2 - q^2) \cdot f(x, y) + r \cdot q \cdot f(x+1, y+1) + \\
& 0.5 \cdot r \cdot (r - 2q + 1) \cdot f(x+1, y) + \\
& 0.5 \cdot r \cdot (r - 1) \cdot f(x-1, y) + \\
& 0.5 \cdot q \cdot (q - 2r + 1) \cdot f(x, y+1) + \\
& 0.5 \cdot q \cdot (q - 1) \cdot f(x, y-1) \; .
\end{aligned}
\tag{7.8}
$$

The second order interpolation function given in Equation (7.8) adds the two additional pixels $f(x - 1, y)$ and $f(x, y - 1)$, which are required to estimate the second derivative of the graylevel changes of the four neighboring pixels.

Another interpolation approach that is based upon the concept of curve fitting a set of points to a smooth continuous line is that of the *cubic B-spline interpolation*. Cubic B-spline interpolation is equivalent to using a French curve to fit a continuous smooth line to a discrete set of points. It has found use in computer graphics for generating polygons, in statistics in the curve fitting of probability density functions, and in signal processing to estimate the power spectral density function of a signal. Figure 7.2 shows a set of ten discrete points and a smooth continuous curve that interpolates between the points. Interpolation using a continuous curve takes into account the curvature of the sampled points, like the six-point interpolator. The nearest neighbor interpolator on the other hand assumes that the curve is a sequence of straight lines between each of the discrete points. The linear interpolator is known to produce discontinuous slopes at each of the discrete points.

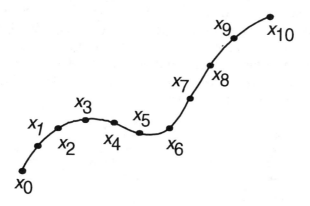

Figure 7.2: An example of a smooth curve interpolation to ten discrete points.

The two-dimensional cubic B-spline interpolator is implemented using two one-dimensional cubic B-spline interpolations. The interpolation first is in the x direction over the variable r and then in the y direction over the variable q as defined by Figure 7.1. Four separate interpolations are required in the x direction to produce the functions $\alpha(x+r, y-1)$, $\alpha(x+r, y)$, $\alpha(x+r, y+1)$, and $\alpha(x+r, y+2)$, where each is defined by

$$\alpha(x+r, y+k) = \frac{1}{6} f(x-1, y+k) \cdot [(3+r)^3 - 4(2+r)^3 + 6(1+r)^3 - 4r^3] +$$

$$\frac{1}{6} f(x, y+k) \cdot [(2+r)^3 - 4(1+r)^3 + 6r^3] +$$

$$\frac{1}{6} f(x+1, y+k) \cdot [(1+r)^3 - 4r^3] + \frac{1}{6} f(x+2, y+k) \cdot r^3 \quad,$$

(7.9)

with $r, k = -1, 0, 1$, and 2. Interpolation in the y direction over the variable q is

$$f(x+r, y+q) = \frac{1}{6} \alpha(x-1, y-1) \cdot [(3+q)^3 - 4(2+q)^3 + 6(1+q)^3 - 4q^3] +$$

$$\frac{1}{6} \alpha(x, y) \cdot [(2+q)^3 - 4(1+q)^3 + 6q^3] +$$

$$\frac{1}{6} \alpha(x+1, y+1) \cdot [(1+q)^3 - 4q^3] + \frac{1}{6} \alpha(x+2, y+2) \cdot q^3 \quad.$$

(7.10)

Implementation of the cubic B-spline interpolator requires using Equation (7.9) four times followed by using Equation (7.10) once. The only disadvantage of using cubic B-spline interpolation is the computational complexity required in computing the cube powers.

The final commonly used interpolation method presented here is the *cardinal spline, sampling,* or *sinc-function interpolator.* This interpolator is based on the fact that a continuous function can be properly reconstructed from its sampled version if the original function was sampled at least at the Nyquist rate. For an $N \times M$ image and using the same configuration as defined in Figure 7.1, the cardinal spline interpolator is defined as

$$f(x+r, y+q) = \sum_{m=0}^{M-1} \frac{\sin \pi(y+q-m)}{\pi(y+q-m)} \sum_{n=0}^{N-1} f(n, m) \frac{\sin \pi(x+r-n)}{\pi(x+r-n)} . \quad (7.11)$$

Of all the interpolators that have been presented in this section, the cardinal spline interpolator is the most difficult to use because of its computational complexity. For every pixel in an image that must be interpolated, Equation (7.11) requires $N \times M$ calculations. Hence, the interpolation of an $N \times M$ image would require a total of $N \times M \times N \times M$ calculations. Truncation of Equation (7.11) to a smaller size produces ringing in the interpolated image as a result of the Gibbs phenomenon (see Chapter 2). The above spatial interpolation methods will be used to interpolate pixels in the various image geometry techniques that will be presented in the next section.

7.2 Image Geometry

Since image geometry can change the shape of an image, the coordinates used to define the original image may have to change. For example, increasing the size of an $N \times M$ image by a factor of two produces a new $2N \times 2M$ image. In many of the image geometry operations, implementation is much easier if the output image is scanned pixel by pixel and the inverse mapping from new to old is performed. Pixels in the new image that do not directly coincide with the location of pixels in the original image can then be interpolated using one of the many interpolation routines given in the last section.

The first of the image geometry operations is *translation*. Translation is used to move regions of an image to other locations within the image. If the translation operation moves a region outside the area defined by the original image, then a new sized image must be created that encompasses the original image plus the translated region. Image translation is defined as

$$\begin{bmatrix} x_{new} \\ y_{new} \end{bmatrix} = \begin{bmatrix} x_{old} \\ y_{old} \end{bmatrix} + \begin{bmatrix} x_o \\ y_o \end{bmatrix} , \qquad (7.12)$$

where x_{old}, y_{old} are the pixel coordinates of the region to be translated, x_{new}, y_{new} are the coordinate locations of the translated region, and x_o and y_o define the amount of translation in the x and y directions, respectively. For each pixel within a region to be translated, Equation (7.12) is applied to produce a new set of translated coordinates. Even though it does not matter for image translation, the author chooses to map from the new coordinate system to the old coordinate system. In translating a region, the original image is first copied to the output image and then the region to be translated is moved to its new position within the image using Equation (7.12). If the pixels within the original region to be translated are left unchanged, the translation process becomes equivalent to an image copy. If, on the other hand, the original region to be translated is filled with a constant graylevel (erased), the translation operation becomes equivalent to a move operation.

Figure 7.3 shows a graphical sketch of a rectangle that is rotated at an angle θ about the coordinate 0, 0. The transformation that describes *image rotation* is given by

$$\begin{bmatrix} x_{new} \\ y_{new} \end{bmatrix} = \begin{bmatrix} \cos \theta & -\sin \theta \\ \sin \theta & \cos \theta \end{bmatrix} \begin{bmatrix} x_{old} \\ y_{old} \end{bmatrix} . \qquad (7.13)$$

Image rotation, in particular, is one of the image geometry operations that is best implemented by using the inverse of (7.13). Each pixel in the new rotated image is scanned pixel by pixel and the inverse mapping of Equation (7.13) is used to locate the coordinate position of each pixel in the original image. These coordinate positions are then used to determine the graylevel of each pixel within the rotated image using one of the interpolation methods given in the previous section. If the pixel maps outside the coordinates of the original image, the image can be truncated to the size of the original image or the coordinates of the rotated image can be expanded to include the entire newly rotated image. A special case of Equation (7.13) is for $\theta = 90°$, $-90°$, and $180°$. Under these conditions, the sine and cosine functions in Equation (7.13) reduce to either 0 or ± 1. The rotation of the entire $N \times N$ image by these angles results in a simple change of indices in accessing the image. For $\theta = 90°$

$$x_{new} = (N-1) - y_{old}$$
$$y_{new} = x_{old} \quad , \tag{7.14}$$

$\theta = -90°$

$$x_{new} = y_{old}$$
$$y_{new} = (N-1) - x_{old} \quad , \tag{7.15}$$

and $\theta = 180°$

$$x_{new} = (N-1) - x_{old}$$
$$y_{new} = (N-1) - y_{old} \quad . \tag{7.16}$$

Because Equations (7.14) and (7.15) require no multiplications or computations of trigonometric functions, just a swap of indices, 90° rotations are easily implemented in hardware by swapping the values in the address buffers used to access the memory frame buffer of the stored image.

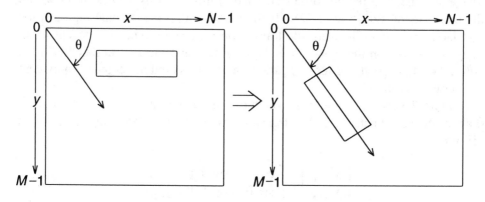

Figure 7.3: An example of rotating a rectangle by an angle θ.

Combining both geometrical translation and rotation allows for the rotation of a region within an image about its geometrical center M_x, M_y.

$$\begin{bmatrix} x_{new} \\ y_{new} \end{bmatrix} = \begin{bmatrix} \cos\theta & -\sin\theta \\ \sin\theta & \cos\theta \end{bmatrix} \left\{ \begin{bmatrix} x_{old} \\ y_{old} \end{bmatrix} - \begin{bmatrix} M_x \\ M_y \end{bmatrix} \right\}. \tag{7.17}$$

The geometrical center (centroid) of the region is given by

$$M_x = \frac{1}{b} \sum_{i=1}^{b} x_i \tag{7.18}$$

and

$$M_y = \frac{1}{b} \sum_{i=1}^{b} y_i \;, \tag{7.19}$$

where x_i and y_i are the coordinates for each pixel in the region to be translated and the parameter b is defined as the number of pixels within the region being translated. Equation (7.17) can also be used to rotate an entire image about the particular point x_o, y_o by setting $M_x = x_o$ and $M_y = y_o$. Once the rotation is completed, the image is then translated back to its original position x_o, y_o.

Figure 7.4 (a) is a 512×480 by 256 grayscale image of the color cube given in Plate 3 with the center of the cube translated to the coordinate location 165, 175 using Equation (7.12). Figure 7.4(b) is the rotated and translated image of Figure 7.4(a). Several different translation and rotation steps were involved to create Figure 7.4(b). First Equation (7.12) was used to translate the center of the cube from coordinate location 160, 175 to the origin 0, 0. Next, using Equation (7.13) the image was rotated by +45° as defined in Figure 7.3. Finally, the center of the cube was then translated to the center of the image, to the coordinate 256, 240. As stated earlier, there are two ways of handling the boundaries of a rotated image. One way is to increase the size of the new image to encompass the rotated image and the other way is to clip the rotated image to the size of the original image. In the generation of Figure 7.14(b), the rotated image was clipped to the size of the original image so that the size of the two images remained the same. The six-point interpolation method as defined by Equation (7.8) was used to interpolate pixels during the rotation operation. Cubic B-spline interpolation could have been used, but the six-point interpolator yields a sharper image and was easier to compute.

Another common type of geometrical operation is that of *scaling*. Scaling provides a means of reducing or enlarging the size of an image. Desired

(a)

(b)

Figure 7.4: An example of image translation and rotation: (a) the
original image and (b) the translated and rotated image.

regions within an image can be magnified to spatially enlarge features that would otherwise be difficult to observe. Geometrical image scaling is defined as

$$\begin{bmatrix} x_{new} \\ y_{new} \end{bmatrix} = \begin{bmatrix} S_x & 0 \\ 0 & S_y \end{bmatrix} \begin{bmatrix} x_{old} \\ y_{old} \end{bmatrix}. \tag{7.20}$$

To scale a total image, x_{old}, y_{old} are defined over the coordinates of the entire image, and for region scaling x_{old}, y_{old} are defined by the pixels within the region to be scaled. For S_x and $S_y > 1$, the output image will be an enlarged version of the input image, while for S_x and $S_y < 1$ the scaled output image is a reduced version of the input image. For either S_x or S_y negative, the image is rotated about the axis of the negative scaling parameter. For example if $S_x = -3$, $S_y = 1$, the image is increased by three and is flipped about the x axis.

Geometrical scaling in particular requires the use of interpolation prior to scaling an image. For image magnification, interpolation will produce an estimate of the graylevels of the pixels in the scaled output image that exist between the known graylevels of the pixels in the original image. For image reduction, interpolation guarantees that image features are not completely removed during the reduction process. Consider the reduction of an image by a factor of two by removing every even column and row within the original image. One pixel wide features located within even rows or columns would be removed by this image reduction process. If a new cubic B-spline interpolated image of the same size as the original image were used for this image reduction method instead of the original image, these features would not be completely removed from the reduced image. Since the cubic B-spline interpolation method averages over the adjacent four neighboring pixels, this effectively redistributes the features in the even rows and columns to the odd rows and columns. Hence, odd rows and columns now contain features that were once located only within the even rows and columns in the original image. This guarantees that image features are not completely lost during image reduction transformation. For example, consider a vertical one pixel wide line located in column 32 of the original image. Using no interpolation, this line will be completely removed from the image during image reduction of even columns and rows. During the averaging process of the cubic B-spline interpolator, this vertical line will be averaged into columns 31 and 33. Hence, the line will still be present in the reduced image. The cubic B-spline interpolator is not the only way to prevent image features from being completely removed during image reduction. Any lowpass filter operation prior to image reduction will produce a similar effect. For example, a 3×3 mean filter could have been used in the above image reduction example.

In the upper left-hand corner of the 480×480 image shown in Figure 7.5(a) is the original 120×120 grayscale image of a young girl. The 480×480 image was created by magnifying the x and y directions of this 120×120 image by a factor of 4 using simple pixel replication. Note the jagged edges present throughout this image. A close examination reveals the square pixel structure

(a)

(b)

Figure 7.5: Examples of image scaling: (a) the 4x image using
             pixel replication and (b) the 4x image using cubic
             B-spline interpolation.

(c)

Figure 7.5: Examples of image scaling: (c) increasing the sharpness
of the cubic B-spline interpolated image given in (b).

as a result of using pixel replication. Figure 7.5(b) is the 480×480 magnified
version of the 120×120 image, except this time the image was scaled using the
cubic B-spline interpolator given in Equations (7.9) and (7.10). Even though
this image is blurred, the square pixel structure that was present in Figure 7.5(a)
is no longer visible. Figure 7.5(c) is the highpass filtered version of the cubic B-
spline interpolated image used to increase the overall sharpness of the image.

The process of lowpass filtering followed by image reduction produces a
lower resolution image of the original image. Repeating the process one
additional time on the reduced image produces a new reduced image with an
even smaller spatial resolution. Image reduction can continue until the desired
resolution is obtained. The set of reduced resolution images form what is known
as an *image pyramid*. Image pyramids are very important in image classification
and recognition in that they use the principle of "finding the forest before
locating the trees." It is much easier to locate a particular region of interest using
a classification algorithm on a lower resolution image and then increasing the
resolution to find fine detail features that are located within this region. Image
pyramids have been successfully used in the location and tracing of edges in the
reconstruction of an image from its power spectral density, and in the
compression of an image.

Figure 7.6 shows a pyramid as a collection of lower resolution images. At
its base level l_0 is the original $N \times M$ image. The next level l_1 up is the lowpass

filtered and reduced image of size $N/2 \times M/2$. The level above this, l_2, is an $N/4 \times M/4$ image generated from the previous level by lowpass filtering and applying image reduction one step further. This process repeats until no further reduction in either the x or y direction is possible. At this point, the top of the pyramid is reached. Typically, image pyramids are used with square images with sizes that are a power of 2. If M and N are a power of 2 and if a linear lowpass filter is used as the lowpass filter, the top of the pyramid reduces to a single point, yielding the average of the original image.

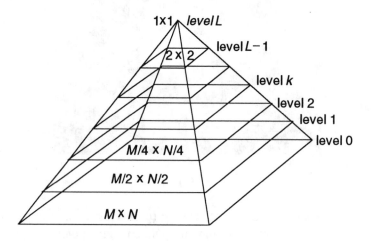

Figure 7.6: An example of an image pyramid.

Assuming a square image of size $N \times N$, which is a power of 2, the images comprising the image pyramid are constructed by lowpass filtering followed by reducing the size of the image by using either every even or odd column and row. Let $f_k(x, y)$ be the filtered and reduced image at level k. Then the next level image in the pyramid $f_{k+1}(x_{k+1}, y_{k+1})$ is given by

$$f_{k+1}(x_{k+1}, y_{k+1}) = lowpass[f_k(x_k, y_k)] , \qquad (7.21)$$

where

$$x_{k+1} = \frac{x_k}{2} , \quad y_{k+1} = \frac{y_k}{2}$$

or

$$x_{k+1} = \frac{x}{2^{k+1}} , \quad y_{k+1} = \frac{y}{2^{k+1}} , \qquad (7.22)$$

and $x_0 = x$ and $y_0 = y$. One of the most common pyramids is the *Gaussian pyramid,* defined by the type of filter mask that is used to perform the lowpass filter operation. Figure 7.7 shows the 5×5 filter mask $h(i, j)$ used in conjunction with the spatial convolution equation

$$g(x, y) \quad = \quad \sum_{i, j \in H} \sum f(x - i, y - j) \cdot h(i, j) \qquad (7.23)$$

to perform the lowpass filtering operation for the Gaussian pyramid. The parameter a is typically chosen in the range between 0.3 to 0.6.

$\left[\dfrac{1}{4} - \dfrac{a}{4}\right]^2$	$\dfrac{1}{16} - \dfrac{a}{16}$	$\dfrac{1}{4} - \dfrac{a}{4}$	$\dfrac{1}{16} - \dfrac{a}{16}$	$\left[\dfrac{1}{4} - \dfrac{a}{4}\right]^2$
$\dfrac{1}{16} - \dfrac{a}{16}$	$\dfrac{1}{16}$	$\dfrac{a}{4}$	$\dfrac{1}{16}$	$\dfrac{1}{16} - \dfrac{a}{16}$
$\dfrac{1}{4} - \dfrac{a}{4}$	$\dfrac{a}{4}$	a^2	$\dfrac{a}{4}$	$\dfrac{1}{4} - \dfrac{a}{4}$
$\dfrac{1}{16} - \dfrac{a}{16}$	$\dfrac{1}{16}$	$\dfrac{a}{4}$	$\dfrac{1}{16}$	$\dfrac{1}{16} - \dfrac{a}{16}$
$\left[\dfrac{1}{4} - \dfrac{a}{4}\right]^2$	$\dfrac{1}{16} - \dfrac{a}{16}$	$\dfrac{1}{4} - \dfrac{a}{4}$	$\dfrac{1}{16} - \dfrac{a}{16}$	$\left[\dfrac{1}{4} - \dfrac{a}{4}\right]^2$

Figure 7.7: The 5×5 filter mask used in computing the Gaussian pyramid.

The large and small grayscale images of Figure 7.8(a) are the level 0 and level 2 images of an image pyramid created using a 3×3 mean filter as the lowpass filter operator, where the level 0 image is a 480×480 image. Note, the lack of high spatial frequencies in the level 2 image as the result of reducing the spatial resolution. Figure 7.8(b) are the Sobel edge images of the two multiresolution images given in Figure 7.8(a). The Sobel edge image of the level 0 image contains many fine edges due to the girl's hair and many unwanted edges due to noise present in the image. The Sobel edge image of level 2 contains only the edges from the original image that are predominant. The low gradient edges due to noise in the original image have been removed. This is why the multiresolution pyramid is used in edge detection and tracing. The lower resolution image has made it much easier to find predominant edges by removing small nonpredominant edges that were present in the original image. This is a good application of "looking for the forest before looking for the trees".

(a)

(b)

Figure 7.8: An example of two levels in an image pyramid: (a) the
level 0 and level 2 grayscale images and (b) the
corresponding Sobel edge images.

The next image geometry operation is that of *skewing*. Figure 7.9 shows an image of a rectangle that has been skewed to the right in the *x* direction by an angle θ. The skewing geometrical transformation in the horizontal direction is defined by

$$\begin{bmatrix} x_{new} \\ y_{new} \end{bmatrix} = \begin{bmatrix} 1 & \tan \theta \\ 0 & 1 \end{bmatrix} \begin{bmatrix} x_{old} \\ y_{old} \end{bmatrix} . \tag{7.24}$$

As with image rotation, this geometrical transformation requires the use of image interpolation to determine graylevels of pixels in the skewed image. Rotating the skewing operation by 90° to produce a skew in the positive *y* direction by an angle of θ yields the transformation

$$\begin{bmatrix} x_{new} \\ y_{new} \end{bmatrix} = \begin{bmatrix} 1 & 0 \\ \tan \theta & 1 \end{bmatrix} \begin{bmatrix} x_{old} \\ y_{old} \end{bmatrix} . \tag{7.25}$$

Figure 7.10 shows an image of the cube given in Figure 7.4(a) that has been skewed to the right by 45° using Equation (7.24). Typically, the skewing operation will produce a new image that is bigger than the original image. As with image geometry, the size of the skewed image can be increased to encompass the entire original image or cropped to the size of the original image. Figure 7.10 has been cropped to maintain the same size image as that of Figure 7.4(a). In the black region to left of the cube are pixels that are completely undefined in the original image so interpolation is no longer valid. Hence, this region was filled with a 0 graylevel. Finally, it should also be mentioned that Figure 7.10 was generated using six-point interpolation.

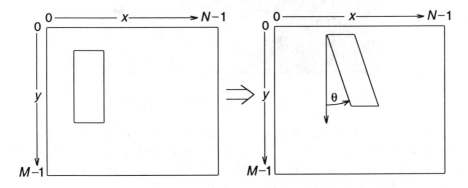

Figure 7.9: An example of skewing an image by an angle θ.

The final image geometry operation discussed in this section is that of *warping*. In many instances, it is desired to change the overall geometrical shape of an image, for example, to be able to change curved lines within an image to straight lines. Aerial maps typically have distortions due to the curvature of the earth. Lines that should be straight appear in the image as curved. Additionally, images of the earth acquired using a camera that is not looking straight down but is slightly tilted relative to the horizon introduces a trapezoidal effect due to the three-dimensional perspective mapping of the earth's surface to a two-dimensional image. In other words, objects that are farther away appear smaller in the image. This makes it difficult to compare a set of images of the exact same surface but taken with different camera positions. A solution to this problem is to use image warping to geometrically transform the distorted image to the desired or standard shape. Another common use for warping is in the generation of a large surface map by combining a set of smaller map images to form a mosaic image. Warping each of the smaller images so that the overlapping areas from each are aligned guarantees that the combined map appears continuous, with no artificial boundaries. Warping has also been successively used to remove geometrical "barrel" distortion that is on some electronic camera systems due to the lens that is used.

Figure 7.10: An example of image skewing.

Figure 7.11 shows an original grid on the left that has undergone a geometrical distortion, making it appear trapezoidal in shape, as shown on the right. Located at the center of the distorted grid are four points that can be related to their corresponding points in the original grid:

$$
\begin{aligned}
g_0 &\to f_0 \\
g_1 &\to f_1 \\
g_2 &\to f_2 \\
g_3 &\to f_3
\end{aligned}
\qquad (7.26)
$$

The geometrical relationship between these eight points describes the geometrical distortion between the original and distorted grids. In fact, the four points in the original grid are tied to the four points of the distorted grid by the geometrical distortion. Because of this geometrical relationship, these eight points are commonly referred to as *tiepoints*. Let $g(r, s)$ describe the original grid and $f(x, y)$ describe the distorted grid. The coordinates of these tiepoints can be related through a set of bilinear equations:

$$x = a_o \cdot r + a_1 \cdot s + a_2 \cdot rs + a_3 \tag{7.27}$$

and

$$y = b_o \cdot r + b_1 \cdot s + b_2 \cdot rs + b_3 \quad , \tag{7.28}$$

where the unknown coefficients determine the actual geometrical relationship between the original and distorted grids.

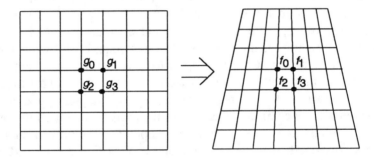

Figure 7.11: An example of trapezoidal distortion of a rectangular grid.

There are a total of eight tiepoints and eight unknown coefficients. Assuming that the g_0 tiepoint is at r_0, s_0, the g_1 tiepoint is at r_1, s_1, the f_0 tiepoint is at x_0, y_0 and so forth, these eight tiepoints can then be substituted into Equations (7.27) and (7.28),

$$
\begin{aligned}
x_0 &= a_o \cdot r_0 + a_1 \cdot s_0 + a_2 \cdot r_0 s_0 + a_3 \\
y_0 &= b_o \cdot r_0 + b_1 \cdot s_0 + b_2 \cdot r_0 s_0 + b_3 \\
x_1 &= a_o \cdot r_1 + a_1 \cdot s_1 + a_2 \cdot r_1 s_1 + a_3 \\
y_1 &= b_o \cdot r_1 + b_1 \cdot s_1 + b_2 \cdot r_1 s_1 + b_3 \\
x_2 &= a_o \cdot r_2 + a_1 \cdot s_2 + a_2 \cdot r_2 s_2 + a_3 \\
y_2 &= b_o \cdot r_2 + b_1 \cdot s_2 + b_2 \cdot r_2 s_2 + b_3 \\
x_3 &= a_o \cdot r_3 + a_1 \cdot s_3 + a_2 \cdot r_3 s_3 + a_3 \\
y_3 &= b_o \cdot r_3 + b_1 \cdot s_3 + b_2 \cdot r_3 s_3 + b_3
\end{aligned} \tag{7.29}
$$

to yield eight equations and eight unknown coefficients. The solution of Equation (7.29) produces the eight unknown coefficients that relate the original grid to the geometrically distorted grid. The bilinear equations given in

Equations (7.27) and (7.28) can produce only linear geometrical transformations. To correct for arbitrary curvatures, higher order terms are needed:

$$x = a_0 \cdot r + a_1 \cdot s + a_2 \cdot rs + a_3 + a_4 \cdot x^2 + a_6 \cdot y^2 + \cdots \qquad (7.30)$$

and

$$y = b_0 \cdot r + b_1 \cdot s + b_2 \cdot rs + b_3 + a_4 \cdot x^2 + a_6 \cdot y^2 + \cdots \ . \qquad (7.31)$$

The higher the order of the polynomials used to model the transformation, the better the model is for complex geometrical distortions. Similar to curve fitting, the higher the polynomial, the better the curve can be fitted through a complex distribution of points. Typically, Equations (7.30) and (7.31) are expanded to include up to the second order terms.

The concepts in the derivation of removing the geometrical distortions from a grid are easily transferred to that of an image. In fact, the tiepoints used in the grid example above could be considered as relating four pixels in the undistorted image to four pixels in the geometrical distorted image. For very complex geometrical distortions, there are two different approaches that can be taken. The first is to use the smallest set of tiepoints required to solve the set of unknown coefficients for a very complicated geometrical mapping function, obtained by expanding Equations (7.27) and (7.28) to contain higher order terms. The other approach is to use many closely spaced tiepoints that model the geometrical distortion as linear. Each set of tiepoints is then used to correct the pixels just within the region enclosed by the tiepoints. The only disadvantage of using the second method is the amount of tiepoints that are needed for very complex geometrical distortions. Implementation of image warping using the bilinear equations given in Equation (7.27) and (7.28) is as follows. For every pixel in the restored image $g(r, s)$, Equations (7.27) and (7.28) are used to obtain the mapping coordinates x, y in the geometrical distorted image $f(x, y)$ using the tiepoints associated with that pixel. Next, one of the image interpolation methods is used to determine the graylevel of the pixel at the coordinate x, y. This interpolated value is then stored at the coordinate location r, s in the restored image. The mapping of the geometrical distorted image's pixels to the restored image continues until all pixels have been mapped and interpolated.

Figure 7.12(a) is an image of a grid that was acquired with an 8 mm camera placed 6 inches in front of the grid. A macro lens was used to image the grid onto the camera, but unfortunately, this lens produced some barrel distortion, as can be seen by the curvature of the grid lines near the edges of the image. Figure 7.12(b) is the restored image, obtained using one of the many commercial warping programs that are available. Two linear rectangular grids that represent the tiepoints in the distorted and restored images were used to perform the warping process. The horizontal and vertical spacing of these tiepoint grids were then adjusted so that the two tiepoint grids overlaid the vertical and horizontal lines within the center of the distorted and restored images. This placed the

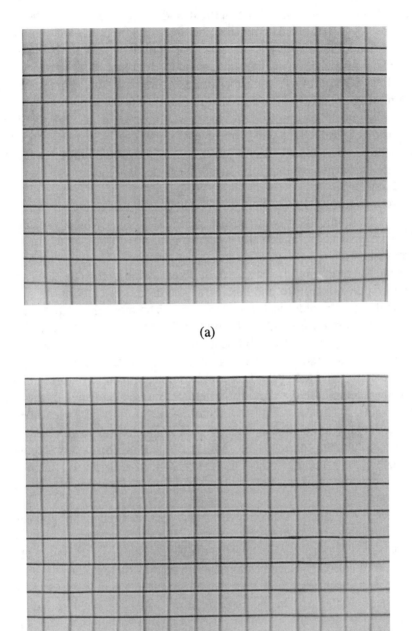

(a)

(b)

Figure 7.12: An example of image warping: (a) the original
geometrically distorted image and (b) the
restored image.

tiepoints at the intersection of the vertical and horizontal lines. The restored tiepoint grid was left unchanged and the intersection points of the distorted tiepoint grid were then adjusted one at a time until each tiepoint lay completely on top of the intersection of the vertical and horizontal lines in the distorted image. Image warping was then performed to remove the geometrical barrel distortion. Figure 7.12(b) shows that the barrel distortion has been removed. The slight "wiggles" of the lines are due to slight errors in placing the tiepoints for the distorted grid. This error can be easily reduced by using a finer spaced set of tiepoints. The most difficult job in image warping is not the geometrical transformation but determining the location of the tiepoints in the restored and distorted images.

A process similar to warping that has received recent attention is that of *morphing*. Morphing is a controlled warping process in which, typically, two reference images are geometrically mixed together using a set of lines within each image to describe the geometrical transformation. Unlike warping, which uses tiepoints, morphing is based upon lines called *fields* to determine the mapping of one function to another. The length and orientation of each field determine the final geometrical transformation. The more fields that are used to define the two reference images, the higher the complexity of the geometrical transformation can be. Using bilinear interpolation, this provides a smooth geometrical transformation from one image to the other. Figure 7.13(a) is an image of a quarter and a nickel. The morphed image given in Figure 7.13(b) was obtained by placing a field around the outer circumference of each of the coins in Figure 7.13(a) to create the first reference image. These fields were then converted into a square in the other reference image. As with warping, the difficulty of using morphing is the ability of picking the proper number and location of fields to map one object to the other. If more control points had been used in the generation of Figure 7.13(b), the outline of the coins would approach an exact square.

Morphing can also be used with two separate images to geometrically transform one image into the other image. The primary use of morphing has been to add special effects to movies and television commercials by transforming one object into another. Several morphing examples include the transformation of an animal to an automobile, a man into a robot, and a boy to a girl. Morphing has also found application in the identification of missing children. If a young child has been missing for several years, it is difficult to predict what the child may look like. Using an image of a older relative and an image of the child before abduction, morphing can be used to estimate the looks of the child as he/she ages. Morphing has also been used successfully in the alignment of satellite imagery. Since the earth is anything but flat, satellite imagery of it does not conform to the standard flat perspective views given by most maps. To solve this problem, the satellite imagery is morphed onto the map. This is one of the ways state boundaries are applied to a satellite weather image to give the composite image that is often presented in weather reports.

(a)

(b)

Figure 7.13: An example of morphing: (a) the original image of
a quarter and a nickel and (b) its morphed image,
producing square coins.

7.3 Binary Morphology - Dilation and Erosion

Morphological image processing is an area of nonlinear image processing that deals with the geometrical structure of objects within an image. Based upon Minkowski set addition and subtraction, it was Jean Serra and George Matheron who initially applied the concepts of morphology to binary images. Recently, the concepts of binary morphology have been expanded to include graylevel images. Morphological filters have been successfully used to remove noise from images, as edge detectors, in image compression, and for feature extraction. Binary morphology will be introduced in this section and then expanded into graylevel morphology later in this chapter.

Binary morphological image processing is based on treating an object within a binary image as a set of pixels. Each element in the set defines the set of two-dimensional vectors that give the coordinates of each pixel within the object. Figure 7.14 shows a 5×5 binary sub-image with a "U" shaped object in its center. The coordinate locations of each of the pixels that define the "U" shaped object can be grouped together to form a set A that completely describes the object:

$$A = \{(1,1),\ (1,2),\ (1,3)\ (2,3),\ (3,3),\ (3,2)\ (3,1)\}\ . \tag{7.32}$$

Equation (7.32) and Figure 7.14 show how pixels can be grouped together to form sets that define objects in a binary image. The many properties of set theory hold and can be used to modify objects within an image. For example, the union of two objects yields a new larger sized object that contains both of the original two objects. The intersection of two objects defines a new object that is composed of pixels given by the overlap between the two objects. Treating objects within binary images as sets is an important concept used in binary morphological image processing.

Figure 7.14: An example 5×5 binary image of a "U" shaped object.

The two fundamental filter operations of binary morphological image processing are *binary erosion* and *binary dilation*. Consider two objects A and B in a binary image. Based upon Minkowski set addition, the dilation of object A by B is defined as all possible vector additions of the elements in A with the elements in B. Letting I^2 define the two-dimensional space of an image, the dilation of object A by object B is defined as

$$A \oplus B = \{t \in I^2 : t = a + b, a \in A, b \in B\} \,. \tag{7.33}$$

Equation (7.33) states that the dilation of object A is the collection of all the vector additions of elements of A with elements of B. The contour of the new dilated object can be found using only the elements of objects A and B that form the contour in Equation (7.33).

Dilation of object A with object B can also be interpreted as the union of all the possible translations of object A by the elements of object B. Defining the translation of object A by the vector b as

$$A_t = \{t \in I^2 : t = a + b, a \in A\} \tag{7.34}$$

yields the definition of dilation as the union of all of these translates

$$A \oplus B = \bigcup_{b \in B} A_t \,, \tag{7.35}$$

or

$$A \oplus B = \bigcup_{b \in B} \{t \in I^2 : t = a + b, a \in A\} \,, \tag{7.36}$$

where $\cup$ means set union. Figure 7.15 shows a rectangular object A that has been translated by a second object B containing one point located at $a, 0$, using Equation (7.35). Dilation for this example is simply the translation of the rectangle A in the plus x direction by a, resulting in A_1.

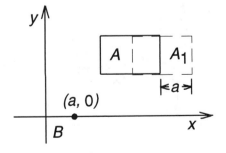

Figure 7.15: An example of translating object A by object B.

Figure 7.16 demonstrates the use of Equations (7.35) and (7.36) to dilate an object A by an object B. Located about the x axis are two points that compose the elements of object B that are offset by a in the plus and minus x directions. There are two possible translations of object A as a result of the two elements in object B. Object A_1 defines the translation of object A in the negative x direction (shown in dotted lines) as a result of the element in object B at $-a$, 0, while object A_2 is the result of translating object A in the plus a direction for the element at a, 0. As defined by Equation (7.35), the dilated object is defined as the union of object A_1 with A_2. The union of the two objects has produced a new object that has increased in size. This is why this binary morphological operation is referred to as binary dilation.

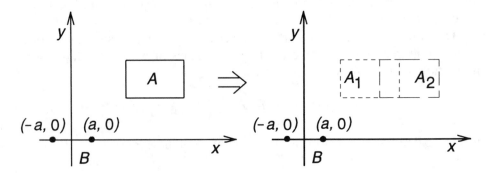

Figure 7.16: An example of binary dilation of object A by object B.

Still another way of interpreting binary dilation is that of rolling one object around the contour of another object, similar to rolling a wheel around the outside contour of an object. Figure 7.17 shows the dilation of the rectangular object A with object B using this concept. The wheel shaped object B is rolled around the outside of the contour of object A. The contour traced by the center of the wheel becomes the contour of the dilated object, as shown in Figure 7.17. Notice how the corners have been rounded by the dilation operation. The rolling wheel concept holds for any shaped object, but is much harder to visualize with more complex shaped structuring functions.

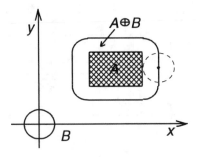

Figure 7.17: An example of binary dilation using the rolling wheel approach.

The rolling wheel concept can also be written mathematically as

$$A \oplus B = \bigcup_{x \in I^2} \{B_x \cap A \neq \emptyset\}, \qquad (7.37)$$

where

$$B_x = \{t \in I^2 : t = b + x, b \in -B\}, \qquad (7.38)$$

where x is defined over the entire image I^2, $\emptyset$ equals the null set and $\cap$ means set intersection. Equations (7.37) and (7.38) can be explained as follows. First object B is symmetrically flipped about the origin to form the new object $-B$. Next, this new object is translated by x over the entire image. The union of all these translations in which these translated objects B_x overlapped with object A becomes the new dilated object. In other words, any translation of object B in which it "hits" object A becomes part of the dilated object.

Figure 7.18 shows the use of the rolling wheel concept for dilation using a square object B. Note that this time the resulting dilated object has square corners, as compared to Figure 7.17, which had rounded corners. Changing the shape of object B determines the final shape of the dilated object. Object B acts like a filter function that changes the geometrical structure of object A by the dilation operation. For this reason, object B is referred to as a *structuring element/function*, and the dilation of object A with object B is called the dilation of object A with structuring element B.

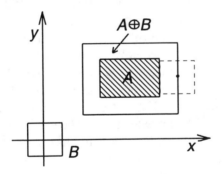

Figure 7.18: An example of binary dilation using a square structuring element.

Figure 7.19 shows an example of using a circular structuring element to smooth the contour of an object. The concept of smoothing contours using dilation is similar to an automobile traveling over a street with crevices. As the size of the automobile tires increase, the less the tire can follow the contour of the road and the smoother the automobile ride becomes. Hence, by increasing the size of the structuring element, the smoother the contour of the object becomes and the larger the size of the dilated object. Figure 7.19 shows that the inward bumps on the contour of object *A* have been reduced, while the outward bumps have been increased in width. Further increasing the size of the structuring element *B* would further smooth the contour of object *A*, but eventually its original shape would be lost, resulting in the dilated object taking on the shape of a circle. As the size of a structuring function increases, the more the dilated object takes on the shape of the structuring function.

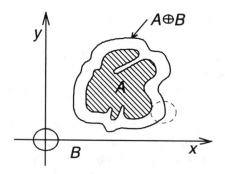

Figure 7.19: An example of binary dilation showing the enlarging of object *A* and the smoothing of its contour.

Figure 7.20(a) shows a three element structuring function that will be used to dilate the 5 × 5 binary sub-image of Figure 7.14. Any one of the methods of visualizing dilation can be used to create the dilated sub-image. Using the translation interpretation yields three "U" shaped objects that have been translated to the right by one pixel, down by one pixel, and down in the diagonal direction by one pixel. The union of these three translated objects yields the dilated object given in Figure 7.20(b). Close inspection of this figure shows that the center of the dilated object has moved down and to the right and is no longer in the center of the 5 × 5 sub-image, as was the case with the original "U" shaped object. If the structuring element is not symmetric about the origin, the resulting dilated object is both dilated and translated. Hence, the definition of a discrete structuring element requires knowledge of where the origin is located. The convention that will be used in this book is to place a dot in the pixel location of the structuring element that defines its origin (also referred to as the *hot spot*).

The second fundamental operation of binary morphology is that of binary erosion, which is based upon Minkowski subtraction. In contrast to binary dilation, which increases the size of an object, binary erosion decreases the size of an object. As dilation was the union of the translations of object *A* by a structuring element *B*, erosion is the intersection of these translations, with one additional modification. The structuring function is first rotated 180° about the origin before object *A* is translated. Binary erosion is defined as

$$A \ominus B = \bigcap_{b \in -B} A_t \, , \qquad (7.39)$$

or

$$A \ominus B = \bigcap_{b \in -B} \{t \in I^2 : t = a + b,\, a \in A\}, \tag{7.40}$$

where A_t is defined by Equation (7.34).

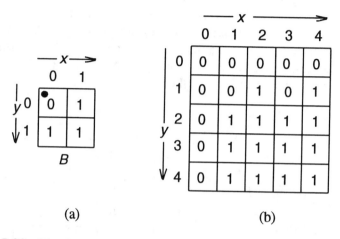

(a)   (b)

Figure 7.20: The dilation of the 5×5 binary image given in Figure 7.14 with a three element structuring function: (a) the structuring element and (b) the dilated image.

Figure 7.21 shows an example of binary erosion using a two point structuring element with elements located at $-a$, 0 and a, 0. The first translation of object A, given by A_1, is the result of the structuring element at a, 0. Object A_1 has been shifted in the direction of the negative x axis by $-a$ as a result of rotating the structuring element by 180° about the origin. The other translation of object A, defined by A_2, has been shifted to the right by a, due to the other structuring element member at $-a$, 0. The intersection of the two translated objects is the erosion of object A by structuring element B. As shown in Figure 7.21, the eroded object has been reduced in size in comparison to the original object A.

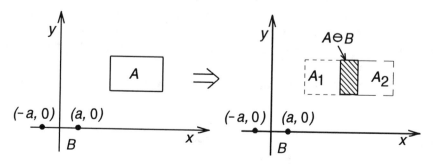

Figure 7.21: An example of binary erosion of object A by structuring element B.

Figure 7.22 is the erosion of the "U" shaped object given in the 5×5 sub-image of Figure 7.14 using the structuring element shown. Two translations are required for the erosion. The first translation moves the "U" shaped object to the left by one pixel and the other translation moves the object to the left and up, also by one pixel. The intersection of these two translated objects is the four pixels shown in Figure 7.22(b). The eroded object has been reduced in size in comparison to the original "U" shaped object given in Figure 7.14. Like dilation, if the structuring element is not centered at the origin, binary erosion results in a reduced sized object that is also translated, but in the opposite direction of dilation. This is due to rotating the structuring element by $180°$ prior to performing the translations of the object.

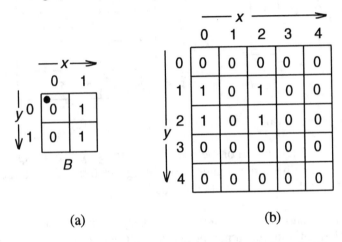

| | (a) | | (b) |

Figure 7.22: The erosion of the 5×5 binary image given in Figure 7.14 with a two element structuring function: (a) the structuring element and (b) the dilated image.

Like binary dilation, binary erosion can also be defined in terms of the rolling wheel approach. Instead of rolling the structuring element around the outside contour of the object to be eroded, it is rolled around the inside of the object. Figure 7.23(a) shows an example of binary erosion using the rolling wheel concept. The structuring element B is rolled along the inside of object A, with the center of the circle tracing the new eroded object. Unlike dilation, the erosion process has preserved the four sharp corners of the rectangle. Since the structuring element is symmetric about the origin, the eroded object is centered within the original object A.

The rolling wheel concept for binary erosion can also be written mathematically as

$$A \ominus B = \bigcup_{x \in I^2} \{B_x \cup A \neq \emptyset\}, \tag{7.41}$$

where

$$B_x = \{t \in I^2 : t = b + x, b \in B\}.\tag{7.42}$$

Equation (7.41) describes erosion as the union of all the translations of the structuring element B by x in which the object A completely encloses B. Figure 7.23(b) gives an example of using the rolling wheel concept to perform binary erosion on a complex shaped object. As the wheel is rolled around the inside of object A, the center of the circle traces the new contour of the eroded object. Erosion reduces the size of outward bumps, while enlarging the size of inward bumps. Notice how the contour has been smoothed by the process of erosion. Like dilation, the size of the structuring function determines the final size of the eroded object and the amount of contour smoothing of the original object. As the size of the structuring function increases, the size of the final eroded object becomes smaller. Eventually, when the structuring element has increased in size so that it no longer fits inside the original object, the eroded object becomes the null set; the object has been completely removed by the process of erosion and has been completely eliminated.

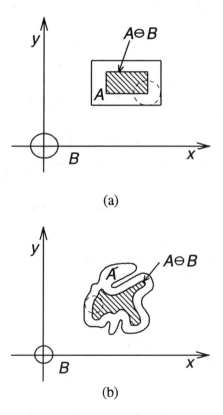

(a)

(b)

Figure 7.23: Examples of binary erosion: (a) the erosion of a rectangle by a circular structuring element and (b) the erosion of a complex shaped object, showing the reduction in size of object A and the smoothing of its contour.

Binary erosion of object A by structuring element B can also be written in terms of dilation using the complement A^C of object A,

$$A \ominus B = (A^C \oplus -B)^C \ . \tag{7.43}$$

Figure 7.24 shows the steps involved in performing binary erosion using Equation (7.43). At the left of the figure is the structuring element B, defined as a circle. Adjacent to the structuring element is the object A to be eroded. The first step in the erosion process as defined by Equation (7.43) is to obtain the complement of object A. For a binary image, this is the collection of all pixels within the image that are not included in object A. Next, the complementary image A^C is dilated using a structuring element created by rotating the structuring element B $180°$ about the origin. Dilation is accomplished using the rolling wheel concept, by placing the structuring element in the region that is not defined by A^C and tracing the new contour. This is shown in Figure 7.24 as step 3. The last step is to take the complement of the dilated result in step 3 to obtain the eroded image as shown in step 4.

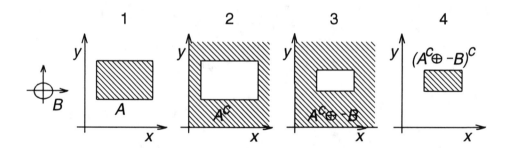

Figure 7.24: The steps to implement binary erosion using Equation (7.43).

There are several properties of erosion and dilation that are of importance. Several of these properties come directly from properties of set theory. The first property is that dilation is *commutative*:

$$A \oplus B = B \oplus A \ . \tag{7.44}$$

It does not make a difference if object A is dilated by structuring element B or object B is dilated by structuring element A. Dilation also follows the *associative* property:

$$(A \oplus B) \oplus C = A \oplus (B \oplus C) \ . \tag{7.45}$$

Both erosion and dilation are *translation invariant*. It does not make a difference if the object is translated first and then eroded or dilated, or if it is eroded or dilated and then translated. Let A_x define the translation of an object A by x. Then the translation invariant property for erosion and dilation states

$$A_x \ominus B = (A \ominus B)_x$$
$$A_x \oplus B = (A \oplus B)_x \ . \tag{7.46}$$

The translation invariance applies only to the image, not to the structuring element. The translation of the structuring element yields

$$A \ominus B_x = (A \ominus B)_{-x}$$
$$A \oplus B_x = (A \oplus B)_x \ . \tag{7.47}$$

For erosion, the translation of the structuring element prior to performing the erosion is equivalent to first performing the erosion followed by translating the object in the direction opposite of x.

If the sizes of both the object and the structuring function are scaled equally before performing erosion or dilation, the eroded or dilated object is the same as if the erosion or dilation has been performed first without any scaling, followed by scaling the eroded or dilated object. Defining scaling as

$$A^s = \{t \in R^{2:} \ t = sa, a \in A, s \in R\} \ , \tag{7.48}$$

where R is the set of real numbers. Both erosion and dilation follow the *scaling* property of

$$(A \ominus B)^s = (A^s \ominus B^s)$$
$$(A \oplus B)^s = (A^s \oplus B^s) \ . \tag{7.49}$$

For $s > 1$ the size of the object is increased, while for $s < 1$ the size of the object is decreased.

A very important property in morphology is the *increasing* property. This property holds for several morphological operations and states that if an object A is a subset of C, then the erosion or dilation of A and C will produce two new objects, one of which is still a subset. If $A \subseteq C$ then

$$(A \ominus B) \subseteq (C \ominus B)$$
$$(A \oplus B) \subseteq (C \oplus B) \ . \tag{7.50}$$

Dilation follows the *distributive* property of set theory for unions and intersections. The union or intersection of two objects A and C followed by dilation is equivalent to the dilation of each object followed by the union or intersection of the two dilated objects:

$$(A \cup C) \oplus B = (A \oplus B) \cup (C \oplus B)$$
$$(A \cap C) \oplus B = (A \oplus B) \cap (C \oplus B) \ . \tag{7.51}$$

On the other hand, the intersection of two objects A and C followed by erosion is equivalent to eroding each object separately and then finding the intersection of the two eroded objects:

$$(A \cap C) \ominus B = (A \ominus B) \cap (C \ominus B) \ . \tag{7.52}$$

But, the erosion of an object A by the union of two objects B and D is equivalent to obtaining the erosion of object A with object B and with object D and then finding the intersection of the two eroded objects:

$$A \ominus (B \cup D) = (A \ominus B) \cap (A \ominus D) \ . \tag{7.53}$$

Another important property is the *distributive* property between erosion and dilation:

$$A \ominus (B \oplus D) = (A \ominus B) \ominus D \ . \tag{7.54}$$

The erosion of object A by the dilation of two objects B and D is the same as the erosion of object A twice, by B and by D. This is a very important property in that it enables structuring elements to be decomposed into smaller structuring elements. Let BT be the overall structuring element composed of a smaller set of structuring elements:

$$BT = (\cdots(B1 \oplus B2) \oplus B3) \oplus \ \cdots \ \oplus BN) \ , \tag{7.55}$$

then for erosion

$$A \ominus BT = (\cdots(A \ominus B1) \ominus B2) \ominus B3) \ominus \cdots \ominus BN) \ . \tag{7.56}$$

The same holds true for dilation from the associate property given in Equation (7.45):

$$A \oplus BT = (\cdots(A \oplus B1) \oplus B2) \oplus B3) \oplus \cdots \oplus BN) \ . \tag{7.57}$$

Figure 7.25 shows the composition of a 3×3 square structuring function as the dilation of a vertical 1×3 structuring element with a horizontal 3×1 structuring element. A special case of Equations (7.56) and (7.57) is the erosion or dilation of an object by the same structuring function n times:

$$A \ominus Bn = (\cdots(A \ominus B) \ominus B) \ominus B) \ominus \cdots \ominus B) \tag{7.58}$$

and

$$A \oplus Bn = (\cdots(A \oplus B) \oplus B) \oplus B) \oplus \cdots \oplus B) \ , \qquad (7.59)$$

where

$$Bn = (\cdots(B \oplus B) \oplus B) \oplus \cdots \oplus B) \ . \qquad (7.60)$$

The structuring element Bn defines n dilations of structuring element B. Equations (5.58) and (5.59) state that erosion or dilation of an object with the same structuring element n times is equivalent to erosion or dilation of the object one time using the *nth* dilation of the structuring element.

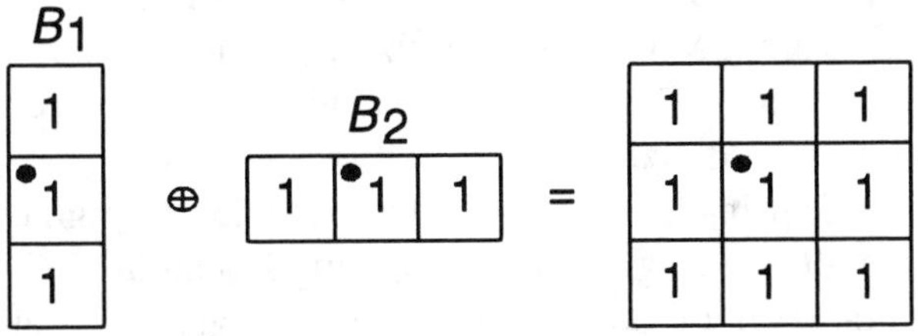

Figure 7.25: The composition of a 3×3 square structuring element from a 1×3 vertical structuring element and a 3×1 horizontal structuring element.

The implementation of erosion or dilation on binary images can be accomplished in many ways. The nonlinear minimum and maximum filters implement erosion and dilation, respectively. The disadvantage of using these two filters is the time required to sort the pixels, which is not necessary to perform the erosion and dilation operations. Another approach is to use the logical operations OR and AND in conjunction with a mask used to represent the structuring function. Then erosion and dilation can be defined using a modified implementation of spatial convolution. Since a binary image contains two graylevels 0 and 1, the goal of erosion is to reduce the size of objects within a binary image at a graylevel of one. Erosion can be defined as

$$g(x, y) = \begin{cases} 0 & \text{for} \quad \underset{i, j \in H}{\text{AND}} \{f(x - i, y - j)\} = 0 \\ 1 & \text{for} \quad \underset{i, j \in H}{\text{AND}} \{f(x - i, y - j)\} = 1 \end{cases}, \qquad (7.61)$$

where H defines the structuring element, $g(x, y)$ is the eroded image, and AND{ } implies the logical AND of all pixels defined by the structuring element

H, which is used instead of its rotated version −H to be consistent with the definition of erosion using translations. A binary image is scanned pixel by pixel and Equation (7.61) is used to perform the erosion operation. If any of the pixels in the AND operation as defined by the structuring function H are zero, the output from the equation will also be zero. The only time the output from Equation (7.61) will be 1 is when all of the pixels that are included in the erosion operation as defined by the structuring element H are nonzero.

Figure 7.26 shows a 3 × 3 structuring element H and a 3 × 3 square defined within a 5 × 5 binary sub-image f(x, y). Shown in f(x, y) are two pixels that have been highlighted with a circle and a triangle. Applying Equation (7.61) to the circle highlighted pixel can be interpreted as placing the structuring element H over that pixel with the origin element aligned on it. Next, only pixels in which the overlapping structuring member is one are included in the erosion calculation of Equation (7.61). For this example, this includes the four neighboring vertical and horizontal pixels along with the center pixel, producing the five graylevels of (0, 0, 1, 1, 1). Applying the logical AND operation to these five pixels yields a value of zero for the eroded pixel.

Moving the structuring element over the triangle highlighted pixel yields five pixels that are all one in graylevel. Applying Equation (7.61) to these five pixels yields a graylevel of one in the eroded image. Repeating this same process for every pixel in the 5 × 5 sub-image yields the eroded image g(x ,y) given in Figure 7.26. The process of computing the erosion of the boundary pixels for this 5 × 5 sub-image required that the image be expanded to a 7 × 7 image by replicating the boundary pixels. This eroded image is the same image that would have been obtained if the entire 5 × 5 sub-image had been translated 5 times according to the structuring element −H and the intersection of these five translated images had been obtained.

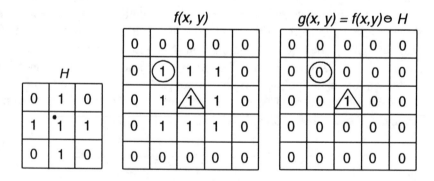

Figure 7.26: An example of implementing binary erosion.

The implementation of binary dilation follows the same steps as for erosion, except Equation (7.61) is changed to include the logical OR operation:

$$g(x, y) = \begin{cases} 0 & \text{for} \quad \underset{i,\, j\, \in\, -H}{\text{OR}} \{f(x-i, y-j)\} = 0 \\ 1 & \text{for} \quad \underset{i,\, j\, \in\, -H}{\text{OR}} \{f(x-i, y-j)\} = 1 \end{cases}, \qquad (7.62)$$

where $-H$ is the version of H rotated by $180°$ about its origin, $f(x, y)$ is the input image to be dilated, $g(x, y)$ is the dilated image, and OR{ } implies the logical OR of all pixels defined by the structuring element $-H$. The $180°$ rotated version of the structuring element H is required to be consistent with the original definition of dilation using translation. As with erosion, a binary image is scanned pixel by pixel and Equation (7.62) is used to perform the dilation operation. The output from Equation (7.62) will be 1 if any of the pixels in the dilation operation as defined by the structuring element $-H$ are 1.

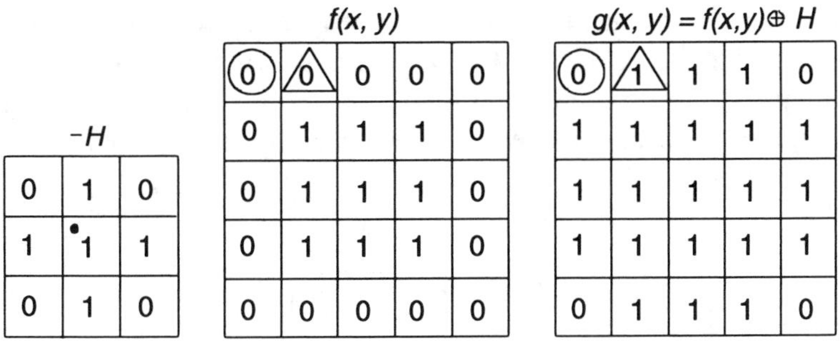

Figure 7.27: An example of implementing binary dilation.

Figure 7.27 shows the same structuring element and 5×5 sub-image that were used to demonstrate the implementation of erosion. The first step in dilating the image with its origin aligned with the circle highlighted pixel is to place the structuring element over the pixels in the image with the circle highlighted pixel. As with erosion, the 5×5 sub-image was expanded to a 7×7 image by replicating the boundary pixels. Next, only pixels in which the overlapping structuring element $-H$ members are one are included in the logical OR operation. For the circle highlighted pixel this yields five pixels that all contain a zero graylevel. The logical OR operation over these five pixels yields a dilated pixel with a zero graylevel.

Repeating the same process for the triangle highlighted pixel gives graylevels of the set of five pixels as (0, 0, 0, 1, 0). The logical OR of these five pixels produces the dilated pixel with a graylevel of 1. Figure 7.27 shows that the dilation operation has expanded the 3×3 square to a 5×5 square with rounded corners. As shown earlier in this section, the dilation of a rectangle with a circular structuring element produces a larger rectangle with rounded corners. Actually, the 3×3 structuring element used in this example is the best approximation of a circle using only a 3×3 neighborhood. The dilation of $f(x, y)$ in Figure 7.27 could also have been obtained by translating the 5×5 sub-image 5 times using H and then obtaining the union of these five translated images.

Figure 7.29(a) is an image of randomly spaced circles that are of two predominant sizes. The large circles range from approximately 30 to 50 pixels in diameter, while the smaller circles range from about 2 to 10 pixels in diameter. This image presents an excellent opportunity to use binary erosion to remove the smaller circles while leaving the larger circles. Figure 7.28 shows a 5×5 approximation to a circular structuring element, which was chosen since all of the objects within Figure 7.29(a) are circular in nature. Since the largest diameter of interest of the circles to be removed is 10 pixels, three passes of erosion using the 5×5 structuring element are necessary to completely remove the smaller circles. At most, each erosion of the image by the structuring element given in Figure 7.28 reduces the diameter of the circles by 4 pixels. Three passes of erosion using the 5×5 circular structuring function will reduce the diameter of the larger circles by 12. All circles with diameters less than 12 will be removed from the image. Figure 7.29(b) shows the results. All of the smaller circles have been removed and the larger circles preserved, but with their diameters reduced by 12 pixels.

$$H$$

0	1	1	1	0
1	1	1	1	1
1	1	•1	1	1
1	1	1	1	1
0	1	1	1	0

Figure 7.28: An example of a 5×5 circular structuring element.

Figure 7.30(a) is an example of a handwritten version of the word *"Morphology."* The individual who wrote this word tends to not close the letter "o" completely, as can be seen in the three versions of this letter in the word. Binary dilation will be used to expand the widths of the letters and thus connect the tops of the three "o" letters. Figure 7.30(b) is the dilated image of Figure 7.30(a) after one pass using the 5×5 circular structuring element given in Figure 7.28. Inspection of this figure shows that the tops of the three letters have been connected, and the width of the line used to write the word has been thickened by the dilation operation. Another interesting feature is that the inner loop of the letter "l" has been completely filled in by the dilation operation.

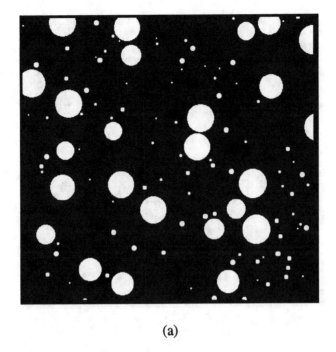

(a)

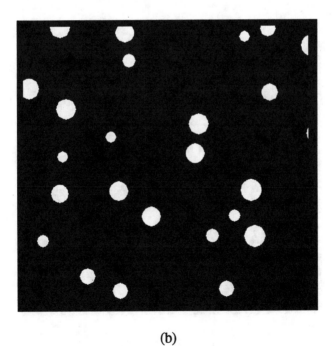

(b)

Figure 7.29: An example of binary erosion: (a) the original image and (b) the eroded image (adapted from Dougherty, 1992).

(a)

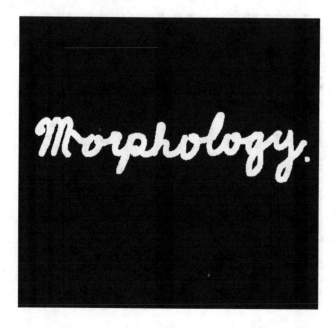

(b)

Figure 7.30: An example of binary dilation: (a) the original image and
(b) the dilated image.

7.4 Binary Morphology - Opening, Closing, Edge Detection, and Skeletonization

The examples given in Figures 7.29 and 7.30 show the usefulness of erosion and dilation in changing the geometrical shapes of objects within a binary image. Unfortunately, these two morphological filtering operations change the size of objects within an image. As Figure 7.30 showed, dilation increases the size of an object, while Figure 7.29 showed that erosion decreases the size of an object. The combination of these two binary morphological operations provides a means of maintaining the relative sizes of objects, while providing a method of performing geometrical filtering on an object. The morphological filtering operation *binary opening* is defined as erosion of an object by a structuring element, followed by dilation using the same structuring function:

$$A_B = (A \ominus B) \oplus B .$$
(7.63)

Various other symbols such as $O(A, B)$, OPEN[A, B], and $A \bigcirc B$ have also been used to define binary opening. The notation presented here follows that given by Pitas and Venetsanopoulos given in the bibliography.

Equation (7.63) does not produce the original object because erosion and dilation are not inverse operations. Figure 7.23(a) showed the erosion of a square by a circular structuring element, producing a smaller square with square corners. Yet, the dilation of this rectangle using the same structuring element produces a larger sized rectangle with rounded corners. Binary opening is commonly used to remove small geometrical objects from an image and to smooth an object's contour by reducing outward bumps and rounding off square corners. The binary opening operation will also increase the size of inward bumps.

Figure 7.31 shows the steps involved in performing an opening operation using a circular structuring element and the rolling wheel concept to perform the erosion and dilation operations. The object to be opened is a rectangular object containing both inward and outward bumps. The first step in opening is to erode the object. It is during this process that outward bumps are reduced and inward bumps are enlarged. This effect is shown in the eroded image of Figure 7.31. The next step is to dilate this eroded object to produce the opened object as shown in the opened image A_B of Figure 7.31. The square corners of the rectangle have taken on rounded corners with radii equal to the radius of the structuring element B. In addition, the outward bump has been reduced, while the inward bump has been enlarged by the rounding of its corners. Unlike the erosion and dilation operations, the opening operation produces an object of relatively the same size as the original object, effectively changing only the object's contour.

A similar operation to binary opening is *binary closing*, defined as the dilation of an object followed by the erosion of the dilated object using the same structuring element:

$$A^B = (A \oplus -B) \ominus -B \ . \tag{7.64}$$

Again, various other symbols such as $C(A, B)$, CLOSE$[A, B]$, and $A \bullet B$ have been used to denote binary closing. The operation of closing is not equal to that of opening. This can easily be explained by considering the opening of an object with a structuring element that is bigger than the object. During the erosion step of opening, the object is eroded completely away, leaving a null image containing no object. Dilation of this null image to produce the open image also yields the null image. In other words, once an object has been eliminated from an image, dilation does not bring it back. Closing, on the other hand, first dilates an object, making it bigger, followed by erosion. Objects are not eliminated by the closing operation. Closing is used to remove holes in an object and to smooth the contour of an object by reducing inward bumps. Square corners remain square after the closing process.

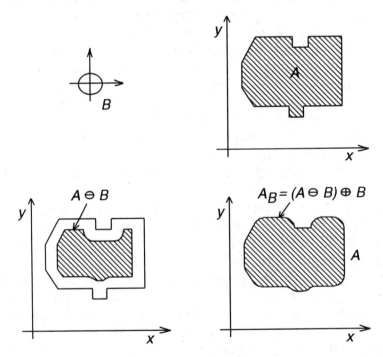

Figure 7.31: An example of the binary opening of a rectangular object with inward and outward bumps.

Figure 7.32 shows the steps involved in performing binary closing using the same rectangular object and structuring element given in Figure 7.31. The first step is to rotate the structuring function B 180° about the origin to produce $-B$. For this example, though, the rotation of the structuring element is not required since it is symmetric about the origin. The next step, dilation, has enlarged the rectangular object, increased the size of the outward bump, and decreased the size of the inward bump as shown in Figure 7.32. The final step, erosion, has reduced the size of the dilated object, yielding a rectangular object very near the same size as the original object. The corners have remained square, the inward bump has been reduced, and the outward bump has been increased.

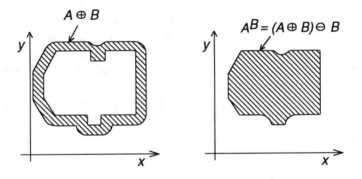

Figure 7.32: An example of binary closing of the rectangular object A given in Figure 7.31.

The operations of closing and opening can also be written in terms of the complement of the other:

$$A^B = [(A^C)_B]^C \qquad (7.65)$$

and

$$A_B = [(A^C)^B]^C . \qquad (7.66)$$

As with the operations of binary erosion and dilation, both closing and opening operations are translation invariant:

$$(A^B)_x = [(A_x)^B] \qquad (7.67)$$

and

$$(A_B)_x = [(A_x)_B] . \qquad (7.68)$$

It does not make a difference if the object is translated first, followed by opening or closing, or if opening or closing is performed first and then the object is translated.

Both closing and opening follow the increasing property, in that if object A is less than object C then opening and closing by the same structuring element B yields

$$A^B \subseteq C^B \qquad (7.69)$$

and

$$A_B \subseteq C_B \; . \qquad (7.70)$$

Equations (7.69) and (7.70) state that if object A is a subset of object C, then the opening or closing of A by structuring element B will be a subset of C opened or closed by B. A property that does not apply to erosion and dilation but applies to closing and opening is that of *idempotent*. A morphological operation is idempotent if additional use of the operation yields the same unchanged object. In other words, once an object is opened or closed, successive use of the same operator produces the same opened or closed image:

$$A^B = (A^B)^B \qquad (7.71)$$

and

$$A_B = (A_B)_B \; . \qquad (7.72)$$

The size of an object after closing will always be larger or equal to the original object:

$$A^B \supseteq A \; . \qquad (7.73)$$

Equation (7.73) follows the property of *extensive*. Opening on the other hand produces an object that will always be smaller or equal to the original object:

$$A_B \subseteq A \; . \qquad (7.74)$$

Both closing and opening operations perform well in removal of binary salt or pepper noise (0 or 1 graylevel) from a binary image, with closing removing the pepper noise and opening removing the salt noise. Unfortunately, closing is unable to remove salt noise, while opening is unable to remove pepper noise. If both salt-and-pepper noise is present in an image, the sequential use of closing and opening is used. This combination of closing and opening forms two additional filters:

$$\mathrm{OPENCLOSE}(A, B) = (A_B)^B \qquad (7.75)$$

and

$$\mathrm{CLOSEOPEN}(A, B) = (A^B)_B \; . \qquad (7.76)$$

Figure 7.33 shows an example of using binary opening to separate two different sized objects. Figure 7.33(a) shows a binary image of circles with two predominant diameters. The diameters of the smaller circles are approximately 20 pixels, while the diameters of the larger circles are about 50 pixels. Figure 7.33(b) is the opened image using a circular structuring element with a diameter of 21 pixels. Binary opening has removed the smaller circles and has left the larger circles intact. During the erosion operation filter, the smaller diameter circles were eroded completely away so that during the dilation operation only the larger diameter circles were left to be dilated back to their original diameters. This example illustrates why opening and closing are not equal operations. Once an object is eliminated from an image it cannot be regenerated by the dilation process that follows the erosion process in the opening operation. Closing of Figure 7.33(a) would not have eliminated the smaller diameter circles but would have combined the circles to form new combined regions as a result of dilating the image first.

Figure 7.34(a) is an image of ten circular rings that represent a binary image of ten washers. It is desired to fill in the center of these rings to yield a new image that contains ten filled circular regions of the same size diameter as the original ten rings. A dilation operation must be applied to reduce the size of the holes located within each of the ten rings. Application of the dilation operation alone will also increase the overall diameter of each of the ten rings, which is not desired. Binary closing, on the other hand, dilates an object first, followed by erosion. The dilation operation of closing with a circular structuring element slightly larger in diameter than the holes will fill in the centers of the holes, producing ten solid circular regions. Erosion using the same circular structuring element will yield ten filled circular regions that are of the same diameter as the rings in the original image. Figure 7.34(b) shows the use of closing on Figure 7.34(a) with a 29 pixel diameter circular structuring element. The diameters of these ten circular regions are the same size as the original ten rings.

The morphological operations of erosion and dilation can also be used to detect edges in a binary image. If a circular structuring element is used to dilate an object, it produces a new object that is larger in size by the diameter of the structuring element. The set difference between the dilated object and the original object produces a contour that straddles the outside of the contour of the original object. The *outside contour* or edge detector is defined as

$$C_O = (A \oplus B) - A \ , \tag{7.77}$$

where "−" means set difference, not subtraction. The generation of the outside contour for a rectangular region is shown in the upper right-hand plot given in Figure 7.35. So that the rolling wheel concept can be used to easily visualize the dilated image, a circular structuring function is used to dilate object A. The shaded area between the original object A and the dilated object $A \oplus B$ represents

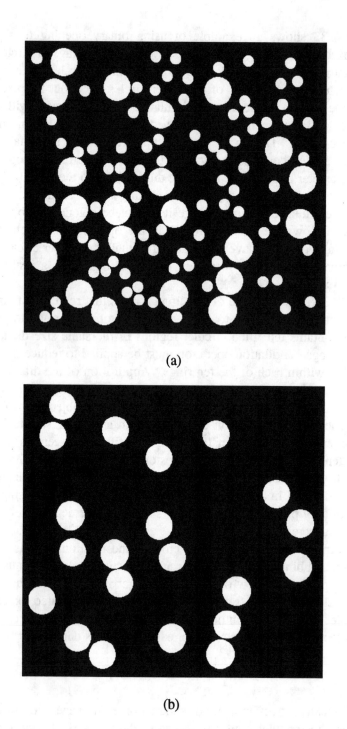

(a)

(b)

Figure 7.33: An example of using binary opening to separate two
              different sized objects: (a) the original image and (b) the
              opened image (adapted from Dougherty, 1992).

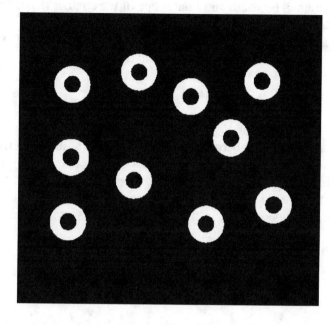

(a)

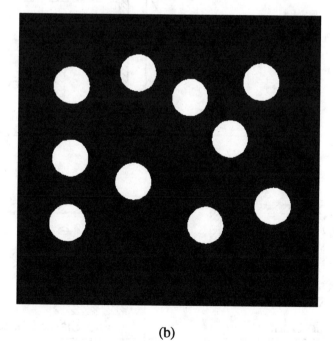

(b)

Figure 7.34: An example of using binary closing to fill in a hole
within an object: (a) the original image and (b)
the closed image.

the outside contour. The width and final shape of the contour depend on the size and shape of the structuring element used in the dilation operation.

A similar approach can be used to produce a contour that straddles the inside of an object's contour using binary erosion and the original object :

$$C_i = A - (A \ominus B).$$

(7.78)

The lower left-hand plot in Figure 7.35 shows the generation of the inside contour. The shaded region between the original object A and the eroded object $A \ominus B$ forms the inside contour. The shape of this contour is slightly different from the outside contour due to the differences between the erosion and dilation operations. The outside edge detector has produced rounded edges at places where there are outside corners in the original object A, while the inside edge detector produced sharp corners at these locations. As with the outside edge detector, the final shape and size of the inside contour depend on the size and shape of the structuring element.

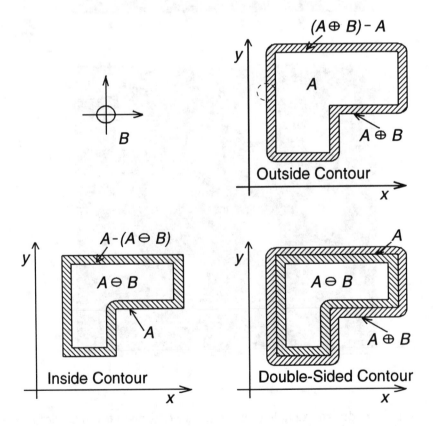

Figure 7.35: The inside, outside, and double sided morphological edge detectors.

(a)

(b)

Figure 7.36: An example of edge detection: (a) the original binary image and (b) the double sided edge image.

Combining both the inside and outside contours produces a new contour that is centered on the contour of the original image. The double-sided edge detector is defined as the set difference between the dilated and eroded objects:

$$C_d = C_o - C_i = (A \oplus B) - (A \ominus B) . \qquad (7.79)$$

Shown in the lower right-hand plot of Figure 7.35 is the double sided contour generated from the dilated and eroded objects of object A. The width of the contour produced from a double sided edge detector will be twice that of the inside and outside edge detectors. Figure 7.36 illustrates using the double sided edge detector to locate the edges of the letters in a binary image. Figure 7.36(a) is the original binary image of the letters "A", "O", "B", and "T". Figure 7.36(b) is the double sided edge detected image.

The final binary morphological operation presented in this section is that of skeletonization. In many image processing applications, it is desired to identify objects located within an image, for example, the identification of washers versus bolts in an image taken from a manufacturing assembly line. Using every pixel in each object to identify an object increases the complexity of the identification process. The edge detection example given in Figure 7.36 clearly illustrates that all the pixels within the four letters are not needed to identify each letter. The letters in the edge detected image given in Figure 7.36(b) are still completely identifiable. One method of reducing the number of pixels required to represent an object is to compute the skeleton of the object. The skeleton of an object is a unique geometrical description that requires far fewer pixels than the original image itself. Some classic examples of skeletonization are the stick figure representation of letters and geometrical shapes shown in Figure 7.37.

Figure 7.37: Several examples of skeletonization using stick figures.

The mathematical derivation that describes the skeletonization of a geometrical shape is due to Blum and is commonly referred to as the *medial axis transform* (*MAT*) as named by Blum. The approach he took was to consider the area comprising an object or geometrical shape as a field of grass, where the borders of this field are given by the contour of the geometrical shape. He next proposed starting a fire along the entire border of this grass field. During the igniting of the fire along the border, Blum assumed that the wind velocity was zero for each point along the border and the fire was ignited at exactly the same time. As the fire continued to grow toward the center of the geometrical shape, the points at which the fire fronts cross and the fire diminishes compose the medial axis transform or skeleton of that geometrical shape.

The skeleton of an object can also be defined by placing a circular disc within an object and tracing its center to form the skeleton. For every point within an object, there exists a set of maximum sized discs that just fit inside the object. The centers of these discs form the skeleton of the object. Figure 7.38 shows several maximal discs located within a rectangle and their corresponding centers. The union of all of the maximal discs' centers forms the skeleton of the rectangle, as shown by the lines inside the rectangle on the right.

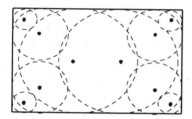 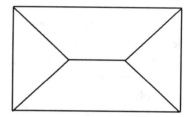

Figure 7.38: Skeletonization of a rectangle using maximal sized discs.

The skeleton of an object can be derived using the binary morphological operations of opening and erosion. Let

$$ER_n(A) = A \ominus Bn \tag{7.80}$$

be the *nth* erosion of object A with structuring element B, where Bn is defined as the *nth* dilation of structuring element B and is given by Equation (7.60). For $n = 0$, Equation (7.80) reduces to the original uneroded object A. The *nth*-order set difference between the *nth* eroded object and the opening of the *nth* eroded object becomes

$$D_n(A) = ER_n(A) - [ER_n(A)]_B \ . \tag{7.81}$$

The union of all of the set differences as defined in Equation (7.81) forms the skeleton of the object:

$$Skel(A, B) = \cup \{D_n(A): n=0, 1, 2 \cdots N_t\} \ , \tag{7.82}$$

where N_t is the total number of erosions required to completely eliminate the object. In other words,

$$ER_{n_t}(A) = A \ominus BN_t = \varnothing . \tag{7.83}$$

The process of skeletonization of a digital image using Equations (7.80) through (7.83) can be explained with the use of the 7×7 sub-image of a 5×5 square and the 3×3 structuring element given in Figure 7.39. The skeletonization operation begins by applying the opening operator to the original 7×7 sub-image using the 3×3 structuring element. The result of this operation is shown in Figure 7.40 as $f(x, y)_B$. The square in this open image is the same except for the removal of the four corner pixels, as shown by the four shaded pixels in Figure 7.40. The set difference between the opened image and the original image forms $D_0[f(x, y)]$, which contains the four corner pixels.

The next step is to repeat this process but this time using $ER_1[f(x, y)]$ formed by eroding $f(x, y)$ with the same structuring element B. This eroded image is then opened to produce a new 3×3 square, also with rounded corners, as shown by the four shaded pixels of $ER_1[f(x, y)]_B$. The set difference between $ER_1[f(x, y)]$ and this opened image yields $D_1[f(x, y)]$. This erosion followed by the opening sequence is implemented once more to yield $ER_2[f(x, y)]$ and $D_2[f(x, y)]$. This last erosion, $ER_2[f(x, y)]$, produced a 7×7 sub-image containing a single nonzero pixel. The set difference between this eroded image and the opening of this image yields $D_2[f(x, y)]$, which is also a 7×7 sub-image containing one nonzero pixel. The next erosion $ER_3[f(x, y)]$ in the erosion and open sequence produces the null image. The erosion and open sequence is halted and the union of the set difference is formed to yield the skeleton of the square, as shown by the shaded pixels in the final sub-image of Figure 7.40.

$$ER_0[f(x, y)] = f(x, y)$$

0	0	0	0	0	0	0
0	1	1	1	1	1	0
0	1	1	1	1	1	0
0	1	1	1	1	1	0
0	1	1	1	1	1	0
0	1	1	1	1	1	0
0	0	0	0	0	0	0

B

0	1	0
1	1	1
0	1	0

Figure 7.39: The structuring element and original 5×5 square object used to explain skeletonization (adapted from Myler, 1993).

Figure 7.41 shows a typical application of skeletonization in the identification of airplanes from their silhouettes. Figure 7.41(a) is a silhouette of a jet fighter showing the placement of rockets at the end of the wings. Figure 7.41(b) is the corresponding skeleton generated using the 3×3 structuring element given in Figure 7.39. It took 69 iterations of Equations (7.80) through (7.83) to compute the skeleton of the jet airplane. The angles between the

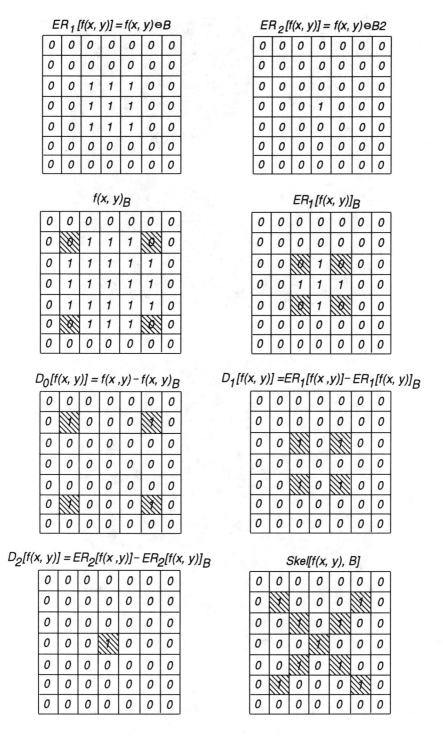

Figure 7.40: An example of skeletonization on a 5 × 5 square binary object (adapted from Myler, 1993).

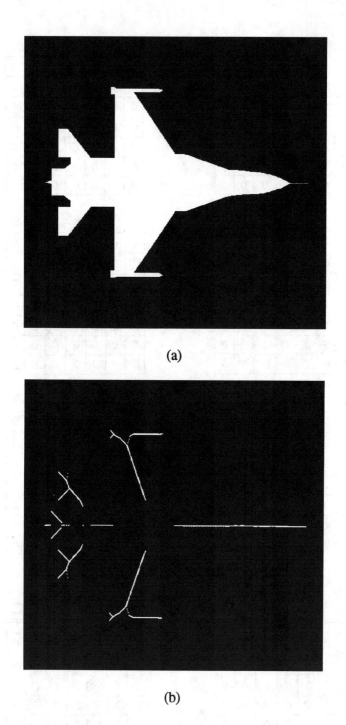

(a)

(b)

Figure 7.41: An example of using skeletonization on an airplane silhouette:
(a) the airplane silhouette image and (b) the skeletonization
image (images, © New Vision Technologies).

skeleton lines of the wings and that of the fuselage is a key feature in identifying different airplanes using their skeletons.

7.5 Binary Morphology - Hit-Miss, Thinning, Thickening, and Pruning

Binary erosion can be used as a means to locate desired geometrical objects within an image. Consider the locating and marking of three geometrical shapes that are of equal area, as shown in Figure 7.42. Neither the square, triangle, nor circle fits completely inside the others' area. Performing erosion using a structuring element equal to the circle removes the square and triangle and leaves a single point marking the location of the circle. If the square structuring element were used instead of the circle, the square would be marked and the other two shapes eliminated. In this simple example, erosion has been successfully used as a means of performing feature extraction. Now consider an image containing two different diameter circles. Performing erosion using the larger diameter circle removes the smaller circle and marks the larger diameter circle. The difficulty in using erosion for object recognition comes in trying to locate and mark the smaller diameter circle. Erosion using the smaller diameter circle as the structuring element marks this object but does not remove the larger. The only thing erosion has done to the larger diameter circle is reduce its size.

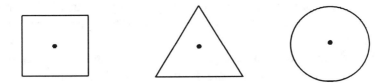

Figure 7.42: The three different geometrical shapes.

To properly perform object recognition using erosion, two different tests are performed using the binary erosion operator. Using the object to be recognized as the structuring element, erosion is performed to find all locations in which this structuring element fits into or "*hits*" the various geometrical shapes. This first erosion removes all objects from an image that are smaller than the object to be recognized. The second erosion removes all objects within the image that are larger than the object to be recognized. This erosion is performed on the complement image while using the complement of the object to be recognized as the structuring element. The result of this second erosion is to mark objects that are exactly equal to the structuring element or larger in size. In other words, this erosion has marked all objects that have been eliminated or "*missed*" by the first erosion. The intersection of the first eroded image and the second eroded image produces a single marker at the locations within the image where objects that perfectly match the structuring element are located.

Using the "*hit*" and "*miss*" approach on the two circle images mentioned above would produce two eroded images. The first eroded image, using the smaller diameter circle as the structuring element, would contain a single point marking the location of the center of the smaller diameter circle and a smaller diameter version of the larger circle. The second erosion, using the complement of the original image and the complement of the structuring element used in the first erosion, would yield a single point marker of the smaller diameter circle. The intersection of these two eroded images produces a single marker at the location of the smaller circle.

The approach of using two erosions to eliminate objects that hit or miss the structuring element is known as the *hit-or-miss binary morphological transform* and is defined in general as

$$A \circledast B = (A \ominus C) \cap (A \ominus D) \, , \qquad (7.84)$$

where the structuring element B is defined as two structuring elements C and D with

$$C \cap D = \varnothing \, . \qquad (7.85)$$

Equation (7.85) is required if the two eroded images of the hit-or-miss transform are to generate two disjointed images except at the location of objects that perfectly match the structuring element used. In many instances, the structuring elements C and D are complements of each other to guarantee that Equation (7.85) is satisfied.

The hit-or-miss transform can also be used to extract key features from objects within an image. For example, Figure 7.43(a) shows a pair of structuring elements that find the top edges of objects within an image. The use of this pair of structuring elements is easily understood using the 5×5 square in the 7×7 sub-image given in Figure 7.39. Using the translation approach for erosion requires that the two structuring functions in Figure 7.43(a) be rotated by $180°$ about their origins. The erosion of the 5×5 square by structuring element C produces a new 5×4 rectangular object, as shown in Figure 7.43(b). This object contains the same pixels as the 5×5 square but with the bottom five set of pixels set to zero.

The erosion of the complement of the square by the $180°$ rotated version of structuring element D produces a new 5×5 square of zero graylevel pixels, which is located at the bottom center of the 7×7 image given in Figure 7.43(c). In other words, this 5×5 square of zero graylevel pixels has been translated down by one vertical row with respect to the 5×5 square of ones shown in Figure 7.39. Figure 7.43(d) shows that the intersection of the first eroded image with the second eroded image yields five nonzero pixels located along the top row of the 5 × 5 square, indicating the location of the top edge of the square. By rotating the pair of structuring elements given in Figure 7.43(a) by 90° increments in the clockwise direction, the right, bottom, and left edges can also be extracted. Figure 7.44(a) shows a binary image of a complex part. Figure 7.44(b) is the hit-or-miss transformed image generated using the structuring element pair given in Figure 7.43(a). Note the extraction of the top edges of the part. The most difficult part of using the hit-miss transform is the selection of the proper structuring elements to extract the features desired.

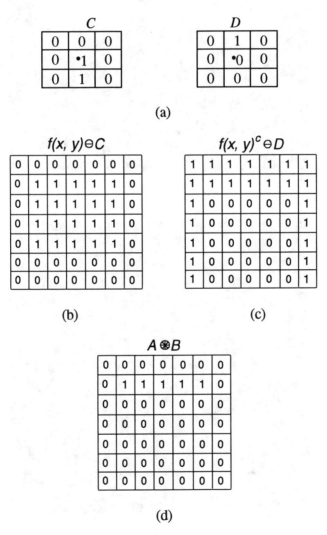

Figure 7.43: An example of using the hit-or-miss transform to find the top edges of a 5 × 5 square: (a) the structuring element pair, (b) the eroded image $f(x, y) \ominus C$, (c) the eroded image $f(x, y)^C \ominus D$, and (d) the hit-or-miss transformed image, locating the top edge of the square.

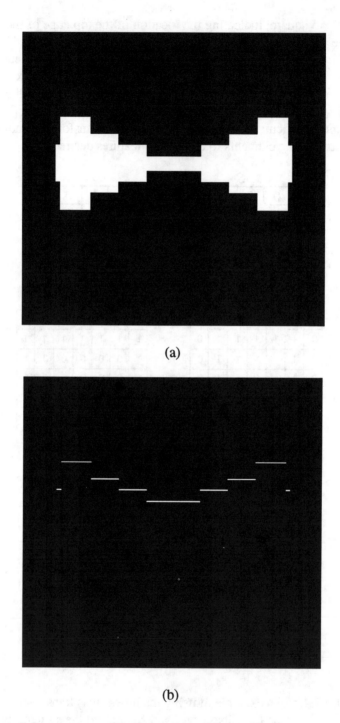

(a)

(b)

Figure 7.44: An example of extracting the top edge from a complex
                  part: (a) the original binary image and (b) the hit-or-miss
                  image using the structuring element pair in Figure 7.43(a).

The three binary morphological operations of thinning, thickening and pruning are each derived from the hit-or-miss transform. In many situations, it is desired to reduce the thickness of an object, similar to skeletonization. The *thinning* operation can be used to reduce the thickness of an object and is defined as

$$A \otimes B = A - (A \circledast B),$$

(7.86)

where "$-$" is the set difference. Equation (7.86) can also be written in terms of set intersection as

$$A \otimes B = A \cap (A \circledast B)^C.$$

(7.87)

The concept of thinning an object can be explained using the 5×5 square image given in Figure 7.39 and the structuring element pair given in Figure 7.43(a). The hit-or-miss transform of this image using these structuring elements is given in Figure 7.43(d). The set difference of this hit-or-miss transformed image and the original 5×5 square image removes the top row of five pixels from the 5×5 square, producing the 5×4 rectangle shown in Figure 7.45. Repeating the thinning operation using a $90°$ rotated version of the structuring elements given in Figure 7.43(a) removes the left-hand vertical column of pixels, yielding a 4×4 square. This process can be repeated until the desired thickness of the square is obtained. In general, there are two approaches to thinning a binary image. The first is to successively iterate over a set of thinning structuring elements until one pixel wide lines remain that describe the skeleton of a object. The other method is to select the number of thinning iterations that will yield the desired object thickness.

$$f(x, y) \otimes B$$

0	0	0	0	0	0	0
0	0	0	0	0	0	0
0	1	1	1	1	1	0
0	1	1	1	1	1	0
0	1	1	1	1	1	0
0	1	1	1	1	1	0
0	0	0	0	0	0	0

Figure 7.45: An example of morphological thinning of a 5×5 square.

Figure 7.46 gives the eight structuring element pairs that are commonly used in binary thinning, with each of these pairs performing thinning in a given orientation. The X's in each of these structuring elements are "do not cares",

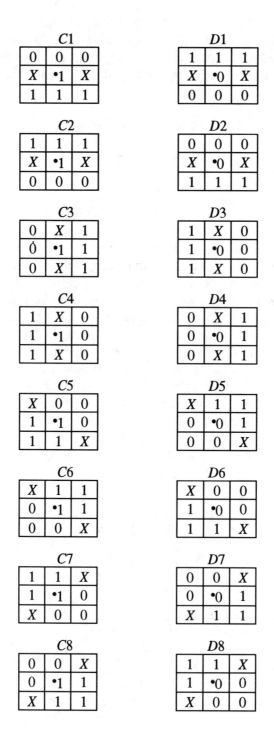

Figure 7.46: The eight masks used for morphological thinning, with the *X's* equal to "do not cares" (from Dougherty, 1992).

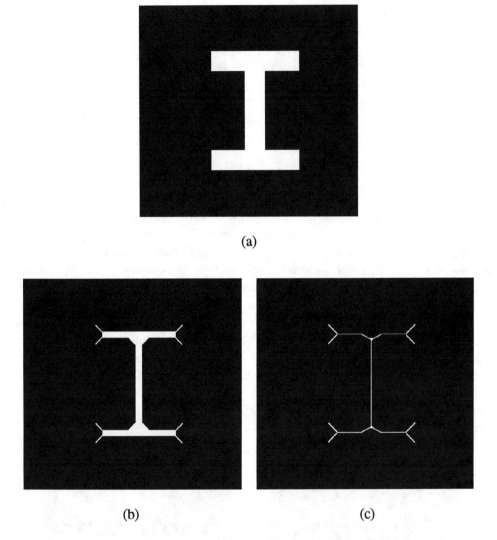

Figure 7.47: Examples of binary thinning using the eight structuring elements given in Figure 7.46: (a) the original binary image, (b) the thinned image after 8 iterations, and (c) the thinned image after 16 iterations.

which represent pixels not included in the hit-or-miss transform. There are four structuring element pairs that are used to thin the top, bottom, left-hand, and right-hand edges, with an additional four structuring element pairs used to thin edges in the direction of the four diagonals. For example, the first pair of structuring elements thins the top edge of an object, while the next pair thins the bottom edge. Figure 7.47(a) shows an example of the binary image of the letter "I". Figure 7.47(b) shows the thinning of this letter using eight iterations of thinning, with each iteration comprised of successive thinning of the object with the eight structuring elements given in Figure 7.46. The letter "I" is now half its width in the original binary image. Figure 7.47(c) is the thinned image after 16 iterations, further reducing the width of the letter. Sixteen iterations were required to reduce most of the lines to one pixel widths. The thinning process continues to cycle through each of the eight structuring elements until the desired width of the object is obtained.

The inverse of binary thinning is that of *thickening* defined as the union of the original binary object with the hit-or-miss transform of this object:

$$A \odot B = A \cup (A \circledast B). \tag{7.88}$$

The operation of thickening is the same as that of thinning. A pair of structuring elements is used to perform the hit-or-miss transform on each pixel in a binary image. The union of the hit-or-miss transformed image with the original binary image forms the thickened image. Like thinning, the operation of thickening requires several pairs of structuring element to thicken an object in several different directions. The eight structuring element pairs used in thinning can also be used for thickening. The only difference is that the structuring element pairs C and D given in Figure 7.46 are swapped. The operation of thickening is an iterative process that continues to cycle through the structuring element pairs until the desired thickness is obtained. Figure 7.48(a) is a binary image of a five pixel wide cross. Figure 7.48(b) is the thickened version of this image using the structuring element pairs given in Figure 7.46 with structuring elements C and D swapped. Eight iterations of these structuring elements were required to obtain the thickness of the cross shown.

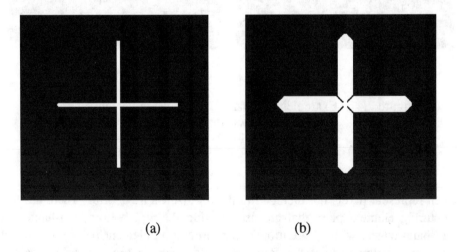

(a)   (b)

Figure 7.48: An example of binary thickening using the structuring element
            pairs given in Figure 7.46 but with (c) and (d) swapped:
            (a) the original binary image and (b) the thickened image.

The operations of thinning and skeletonization can reduce the thickness of an object down to a one pixel wide skeleton representation, which can be used for object representation. The difficulty of these two operations is that they tend to leave extra undesired "tail" pixels that must be removed prior to object recognition. The process of removing these tail pixels is known as *pruning*. Figure 7.49 shows an enlarged view of the letter "*a*" that has been thinned using eight iterations of binary thinning as defined by Equation (7.86). The thinned image of Figure 7.49 shows the presence of these undesired tail pixels. One approach for removing these tail pixels is to use binary opening. The first operation in the opening operation, erosion, would reduce the size of these tail pixels, but it would also remove the one pixel wide lines of the thinned object. The opening operation can be used only if the thinned object is more than one pixel in width. Another approach is to use a pruning algorithm to remove the end pixels of one pixel wide lines.

 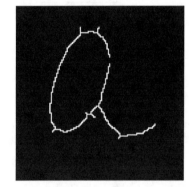

Figure 7.49: An example of thinning showing the undesired "tail" pixels:
(a) the original binary image and (b) the thinned image.

Morphological pruning using the thinning operation is defined as

$$A_{pr_i} = A_{th} \otimes B_i \ , \qquad (7.89)$$

where B_i are the eight structuring pairs given in Figure 7.50 and A_{th} is the thinned object to be pruned. One iteration of pruning requires cycling through the eight structuring pairs, similar to the approach taken for thinning. Care must be taken in the use of the pruning operation as defined in Equation (7.89). For each iteration of pruning, one pixel is removed from each end of every one pixel wide line that is present within the image, which effectively reduces the length of these tail pixels by one. Further iterations of pruning are usually needed to eliminate the presence of any tail pixels. Unfortunately, the end pixels of desired lines are also removed during the pruning process, widening any breaks that exist in the thinned contour and reducing the length of any arcs that may be present. Hence, the number of pruning iterations should be made as small as possible.

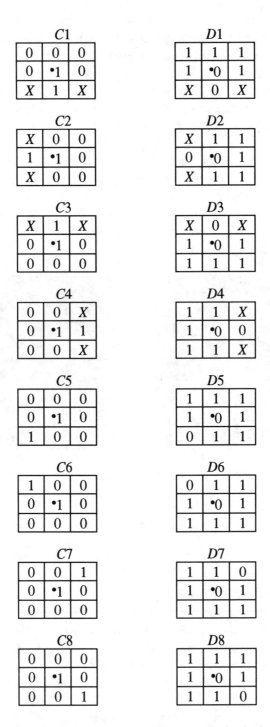

Figure 7.50: The eight masks used for morphological pruning,
with the *X*'s equal to "do not cares" (from Dougherty,
1992) .

Typically, the operation of pruning is followed by dilation to remove any breaks in the thinned contour due to the pruning process. The direct application of dilation will remove these breaks in the contour, but it will also thicken the contour beyond the desired one pixel wide description. Usually several cycles of dilation are required to fill in the breaks, increasing the width even further. A solution to this problem is to locate the end points of the lines of pruned objects and to dilate these end points using *conditional dilation* based upon the original thinned object prior to pruning. This prevents the dilation of the endpoints from exceeding the size of the original thinned object. The final step is to combine the conditional dilated end point object with the pruned object to produce a pruned image whose end points match those of the original unpruned object.

Figure 7.51 illustrates the steps involved in pruning a thinned object and using conditional dilation to fill in the ends of the contour that have been eliminated by a pruning operation. Figure 7.51(a) is the pruned image of the letter "*a*" given in Figure 7.49(b) using two iterations of pruning with the structuring element pairs given in Figure 7.50 to reduce the lengths of the tail pixels. A total of 4 iterations were required to completely remove these pixels, as shown in Figure 7.51(b). Notice the reduction in the lower contour of the letter and the enlargement of the break in the contour as a result of pruning. If the end points associated with the break and the end contour of the letter can be found, then these points can be dilated to fill in the break and the end contour. Since the structuring element pairs given in Figure 7.50 eliminate the end pixels from a contour, these structuring element pairs can also be used to automatically locate these end pixels.

An end point detector can be implemented by forming the union of the eight hit-or-miss transform objects generated using the 8 structuring elements given in Figure 7.50:

$$E = A \underset{1 \le i \le 8}{\cup} (A \circledast B_i), \qquad (7.90)$$

where A is the object in which the end points are to be located. Figure 7.51(c) shows the results of applying Equation (7.90) to locate the end points of the pruned contour given in Figure 7.51(b). The two end points associated with the break in the contour have been located as well as the end point indicating the end contour of the letter "*a*".

The next step is to dilate these three end points and then combine this result with the pruned image to fill in the break in the contour and to extend the end contour of the letter. As mentioned before, dilatation of these three end points alone would produce three new contours that are wider than one pixel. The use of conditional dilation based upon these three end points and the original thinned object will guarantee that the dilations of the three end points are one pixel wide. Conditional dilation requires that the intersection between the dilated end point image and the thinned image be formed:

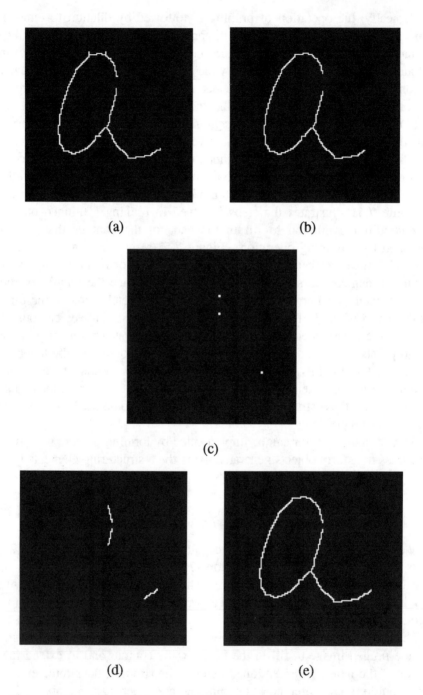

Figure 7.51: An example of pruning: (a) the pruned image using 2
              iterations, (b) the pruned image using 4 iterations, (c)
              the end point image locating the three end points of (b),
              (d) the conditional dilated end point image, and (e) the final
              pruned and restored image (adapted from Dougherty, 1992).

$$A (+) B = (E \oplus B) \cap A_{th} \ , \tag{7.91}$$

where E and A_{th} are the end point and thinned images given in Figures 7.51(c) and 7.49(b), respectively. Since the thinned image contains only one pixel wide contours, the intersection of the original image of the one pixel wide letter "*a*" with the dilated image prevents the dilation process from dilating an object beyond a width of one pixel. The conditional dilated image given in Figure 7.51(d) shows the dilation of the end points while maintaining the one pixel width that is desired. This image was obtained after ten iterations of Equation (7.91) using the 3×3 circular structuring element given in Figure 7.26. The final pruned and restored contour of the letter "*a*" is shown in Figure 7.51(e) and was produced by performing the union of the conditional dilated image [Figure 7.51(d)] with the pruned image given in Figure 7.51(b). Both the end contour and the break in the contour of the letter "*a*" have been extended to their original lengths prior to pruning.

7.6 Binary Morphology - Granulometeries and the Pattern Spectrum

The granulometric techniques developed by George Matheron can be used to measure the size and shape of objects present in a binary image. Similar to the Fourier transform, which finds the distribution of sine and cosine functions in a complex function, the concept behind granulometry is to determine the size and shape distribution of objects present within a binary image. Many of the morphological operators mentioned in this chapter can separate or eliminate different sized objects present in a binary image. For example, binary opening can remove small objects while leaving large objects. The hit-or-miss transform can remove all objects not matching the object defined by the structuring function. These morphological operations provide a means of measuring the size and shape of objects present within a binary image.

Let $\beta_{Bn}[f(x, y)]$ be one of the many morphological filters that have been mentioned in this chapter operating on a binary image $f(x, y)$ using a given structuring element B of size n. Next, let $S[\beta_{Bn}]$ be a size or shape measurement of the remaining objects that are present within the binary image after application of the morphological filter. The change in $S[\beta_{Bn}]$ as a function of the structuring element gives information about the size distribution of the objects present within the binary image. The most common measurement function that represents the size of an object is that of area. Shape measurements might include the circumference, the eccentricity, or the roundness of an object.

Consider the opening of a binary image using the fundamental structuring element Bn, where n determines the size of the structuring function. Objects into which the structuring element will not fit will be eliminated from the image, while the larger sized objects remain intact. Figure 7.52 is an example of two different sized circles $A1$ and $A2$ of diameters $3D$ and $6D$. Let $S[f(x, y)] = S_0$ be defined as the area of all of the objects prior to any morphological operations. For the example given in Figure 7.52, this reduces to $S_0 = 11.25\pi D^2$. Opening these two circles with a circular structuring element Bn of diameter nD produces exactly the same image as the original image for a structuring element with a diameter smaller than the diameter of the smaller circle $A2$. Hence, $S[f(x, y)_{Bn}] = S_0$, for $n < 3D$. Increasing the size of the structuring element to a diameter slightly larger than the diameter of the smaller circle and then performing opening by this structuring element eliminates the smaller diameter circle from the image. At this point, the structuring element no longer fits into the smaller circle. The area measure of this opened image reduces to $S[f(x, y)_{B2}] = 9\pi D^2$.

The process of increasing the size of the structuring element and then performing opening produces no change in the image until the structuring element is slightly greater than $6D$, the diameter of the larger circle $A1$. This structuring element no longer fits into the larger circle, and opening using it eliminates this circle from the image. Hence, $S[f(x, y)_{Bn}] = 0$, for $n > 6D$. No further openings are needed since all of the objects have been removed from the image. A plot of the normalized area measurement parameter $S[f(x, y)_{Bn}]$ as a function of n is given in Figure 7.53. This function is a decreasing function as a function of n. It starts at one and decreases to zero at the point where the size of the structuring function no longer fits into the objects present within the image. The two discontinuities in this plot are the locations at which the circles $A1$ and $A2$ were eliminated from the image.

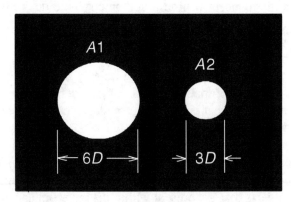

Figure 7.52: An example image containing two different diameter circles.

The normalized measure parameter can be rewritten as an increasing function using

$$D(Bn) = \frac{S[f(x, y)] - S[f(x, y)_{Bn}]}{S[f(x, y)]} = \frac{S0 - S[f(x, y)_{Bn}]}{S_0} , \qquad (7.92)$$

where $D(Bn)$ increases from zero to one as the size of the structuring element Bn increases. A close inspection of Equation (7.92) shows that $D(Bn)$ is equivalent to a probability distribution, which is also an increasing function between 0 and 1. Equation (7.92) gives the *granulometric size distribution* of objects present in a binary image. The derivative of a probability distribution yields the probability density function. Hence, the derivative of Equation (7.92) with respect to n, which determines the size of the structuring element, yields the granulometric size density function:

$$P(Bn) = \frac{-\dfrac{d\, S[\, f(x,\, y)_{Bn}]}{dBn}}{S_0} \,. \tag{7.93}$$

Equation (7.93) gives the distribution of object sizes within a binary image and is often referred to as the *pattern spectrum*.

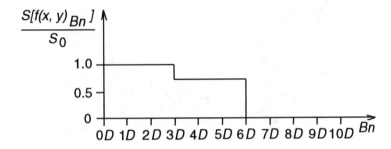

Figure 7.53: A plot of area of the opened image of Figure 7.52 as a function of structuring element size Bn.

The pattern spectrum for the two circle image of Figure 7.52 is given in Figure 7.54. The equation that describes this plot is given as

$$P(Bn) = 0.2\, \delta(Bn - 3D) + 0.8\, \delta(Bn - 6D) \,, \tag{7.94}$$

where $\delta(x)$ is the unit impulse function as defined in Chapter 2. The height of these two impulse functions gives the percentage of the area that each circle occupies relative to the total area of both circles. Figure 7.55(a) shows a complex rectangular object. Using a square structuring element of size Bn, the pattern spectrum for this image is found using Equation (7.93). There is no change in the opened image until the size of the structuring element reaches the size of 1, which is the size of the squares in the lower part of the rectangle. The opening of the complex rectangle with a structuring element slightly larger than 1 removes the two lower squares and reduces the area of the complex rectangle by 9.09%. Opening of the rectangle with the square structuring element continues again with no further change until the structuring element is slightly

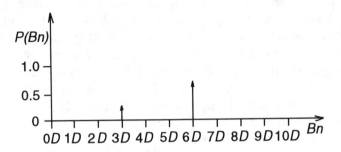

Figure 7.54: The pattern spectrum for the two circle image given in Figure 7.52.

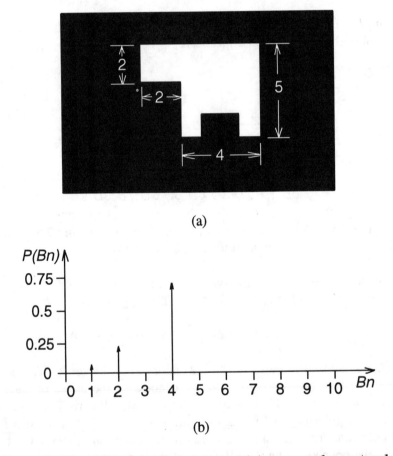

(a)

(b)

Figure 7.55: (a) An example image containing a complex rectangle
and (b) its corresponding pattern spectrum.

greater than 2. At this point, the larger square on the left of the rectangle is
removed and the area of the rectangle is further   reduced by 18.18%. Opening
of the rectangle by the square structuring element again remains the same until

the structuring element is slightly greater than 4, which completely removes the rest of the rectangle. The pattern spectrum for this complex rectangle is shown in Figure 7.55(b). The pattern spectrum shows that the rectangle is composed of three different sized rectangles.

The computation of the pattern spectrum is essentially the same for digital images, except the structuring element is no longer a continuous function of n but takes on discrete sizes. Let B be the fundamental or smallest structuring element that will be used to compute the pattern spectrum. The size of this structuring element determines the resolution of the pattern spectrum. The smaller the structuring element, the finer the resolution of the pattern spectrum. The smallest structuring element that can be used with a digital image is of size 2 $\times$ 2, with the size 3 $\times$ 3 typically chosen so that a symmetric structuring element can be used. The digital image is then opened using this structuring element using the same approach as for the continuous case, but each increase in the structuring element is computed using

$$ Bn = (\cdots(B \oplus B) \oplus B) \oplus \cdots \oplus B) \ , \qquad (7.95) $$

where there are a total of n dilations.

Figure 7.56(a) is an image of nonoverlapping circles of random diameters. A visual inspection of this image shows that there are predominantly two different sized circles. Using Equation (7.93) and a circular structuring element of size 3 $\times$ 3 for B, the pattern spectrum for this image was computed and is shown in Figure 7.56(b). This pattern spectrum shows that approximately 30% of the circles have a diameter of 10 pixels and that the minimum diameter is 8 pixels, while the maximum diameter of the circles is 26 pixels. The bimodal nature of the pattern spectrum confirms that there are two predominant sized circles present in the image of Figure 7.56(a). This example clearly shows the advantage of using the pattern spectrum to obtain the size distribution of objects within an image.

The concept of using the pattern spectrum to find the size distribution of objects within an image using a fundamental structuring element can also be expanded to decompose a complex geometrical object into a set of simple geometrical shapes. The decomposition of a complex object by a given geometrical shaped structuring element B is similar to the decomposition of a complex function into a set of cosine and sine functions using the Fourier transform. In the computation of the pattern spectrum, the image is opened by an ever increasing sized structuring element Bn. At the point at which the opening operation produces a change in the image by eliminating a component of an object, information about the size of the component is obtained. The structuring element that is slightly smaller than the structuring element that removed the component of the object corresponds to the maximum sized structuring element that will fit into that component. Using this information, a complex geometrical shaped object can be decomposed into a set of geometrical shaped objects that depend on the shape of the structuring element B.

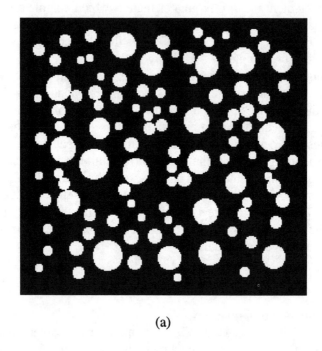

(a)

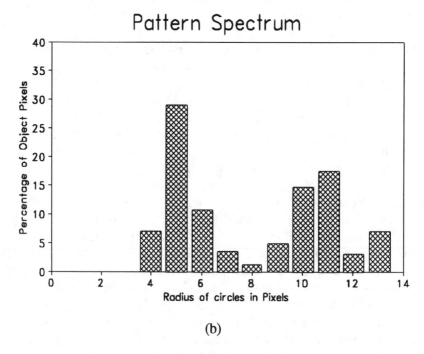

(b)

Figure 7.56: An example of using the pattern spectrum: (a) original
              binary image of randomly sized circles and (b) its
              corresponding pattern spectrum.

The concept of the geometrical decomposition of a complex shaped object into a set of simpler objects is to decompose an object based upon the set of maximal sized structuring elements. The process of shape decomposition of binary image $f_0(x, y)$ starts with a fundamental structuring element B. The image is then opened using an ever increasing structuring element Bn, until the maximum sized structuring element Bm that will fit into the object is found. Next, a new binary image $S_1(x, y)$ is formed containing this exact structuring element located within the image at the place where it was just able to fit into the object. The set difference "−" is then used to remove the area defined by this structuring element from the original binary image. Effectively, this produces a new binary image with the part of the object missing that corresponds to the area defined by the maximum sized structuring element. The process of finding the maximum structuring element is then repeated once more on this new binary image, finding the next largest structuring element that will fit into the object. Another new binary image is formed containing this structuring element.

A set of decomposition components are created by repeating the above procedure until there are no components left from the original image to decompose. An iterative process of opening and then removing the maximum sized structuring element from the object is used to create the composite components. At the *ith* iteration, the decomposition process is defined as

$$f_{i+1}(x, y) = f_i(x, y) - f_i(x, y)_{Bm_i} \quad , \tag{7.96}$$

where Bm_i is the ith maximum structuring element that will fit into object $f_i(x, y)$. Equation (7.96) is repeated until $f_i(x, y)$ equals the null set. At this point, the object has been completely decomposed into a set of objects based upon the structuring element B. The union of each maximum sized structuring element Bm_i translated within the image to the location where each was just able to fit inside the object produces a new image based upon the decomposition of its components by structuring element B.

Figure 7.57(a) is a binary image of a key. The goal of this example is to decompose this key using a set of square structuring elements, with the initial structuring element B defined as a 3×3 square. The image of Figure 7.57(a) was then opened with increasing square structuring elements 5×5, 7×7, 9×9, etc., until the maximum sized structuring element was found. Next, the image was eroded using this structuring element to locate the position where this structuring element was located within the object. If more than one nonzero pixel was present, the position of the first pixel found was used to mark the location of this structuring element. A new binary image was created containing the maximum sized square structuring element centered at the given coordinate found during the previous erosion operation. The set difference between this new image and the original image removed the area defined by the maximum sized structuring element from the object, producing the decomposition image

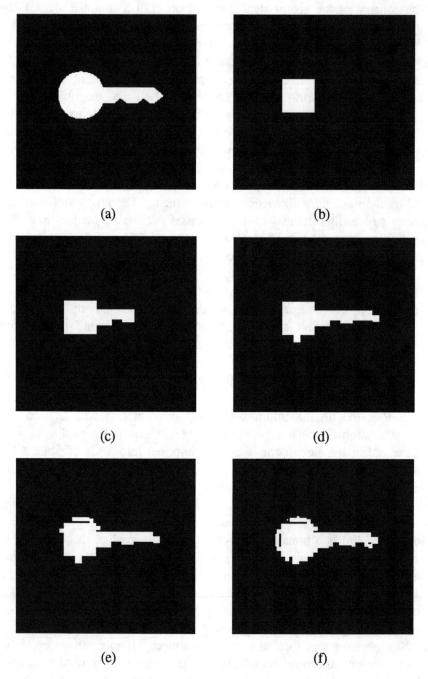

Figure 7.57: An example of geometrical decomposition of a binary
              object using a square structuring element: (a) the original
              binary image of a key, (b) 1 component, (c) 4 components,
              (d) 8 components, (e) 15 components, and (f) 30 components
              (images, © New Vision Technologies).

for the next iteration. The opening, erosion, and set difference operations were then repeated until the desired number of iterations were completed or until the object was completely decomposed. The final step was to combine all of the component images together to form a new image based upon its decomposed components.

Figure 7.57(b) is the decomposition of the key given in Figure 7.57(a) showing one iteration. The square represents the decomposition of the large circular region of the key. Figure 7.57(c) is the representation of the key using the first four square components. Each successive iteration of the decomposition process produces smaller decomposition components, as seen by the smaller squares in this image. Figure 7.57(d) is the key recomposed using the first eight components, while Figure 7.57(e) was generated from the first 15 components. Figure 7.57(f) shows an image of the key created from its first 30 square components. Note that the addition of more components produces a better composition of the original object.

The process of shape decomposition provides a means of representing complex geometrical shaped objects using a set of simpler components. Applications include object recognition and image compression. The concepts presented here can easily be expanded to include more than one shaped structuring element, such as a circle, a square, and a triangle. For example, if a circle and a square structuring element were used in the decomposition of the key as shown in Figure 7.57, better results would have been obtained. In particular, the decomposition of the circular part of the key is best performed using a circle as the structuring element rather than a square. Conversely, the shaft of the key is better decomposed using a square structuring element than a circular one.

7.7 Graylevel Morphology

The concepts and the various operations of binary morphology are easily expanded to graylevel images. Binary morphology treats objects within a binary image as the set of pixels defining the object. To describe an object, only the two-dimensional coordinates of the pixels defining the object need to be specified. The modification of the area and geometrical shape of an object is the fundamental concept behind binary morphology. In graylevel morphology, objects are no longer defined as two-dimensional objects. A third dimension is added that is proportional to the graylevels within a grayscale image. Hence, objects are now defined in three dimensions, as volumes.

The mathematical connection between binary objects as defined by their area and graylevel objects as defined by their volume is that of the *umbra*. A graylevel image can be represented by a three-dimensional plot of x, y position and graylevel value, as shown in Figure 7.58. The graylevel distribution of the pixels within the image forms a complex surface, as shown by the shaded regions. Figure 7.58 shows an image of a circular object containing a variation in graylevels. The volume occupied by this object is defined by the pixels comprising the object plus the graylevels of each pixel. The sides of this volume are defined by the pixels in the contour of the object, and the top of the volume is defined by the distribution of graylevels. The volume occupying graylevels 0 to $f(x, y)$ is shown in solid lines, while the volume defined by $-\infty$ to 0 is shown in dotted lines. The umbra of this object is then simply the union of these two different volumes.

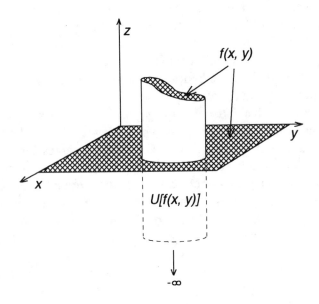

Figure 7.58: A three-dimensional plot of a graylevel image, showing the umbra.

The umbra of a graylevel image defines the total volume below the surface $f(x, y)$, from $-\infty$ to $f(x, y)$. Mathematically the umbra is defined as

$$U[f(x, y)] = \{x, y \in O^2, z < f(x, y), 0 \leq f(x, y) \leq G_{max}\} , \qquad (7.97)$$

where O^2 is the set of coordinates defining the pixels within the object and G_{max} is the maximum allowable graylevel of the image. The graylevels of a grayscale image can be extracted from the umbra using

$$f(x, y) = max[z] \qquad (7.98)$$

where $(x, y, z) \in U[f(x, y)]$. Since z gives the height of the volume occupied by the umbra, the maximum of z is simply the top surface of the volume, which is the graylevels of the image. Related to Equation (7.98) is the *top surface* operator $TS[U[f(x, y)]]$, which extracts the top surface of the umbra. The top surface operator yields the graylevels of the original image or object and is defined as

$$TS[U[f(x, y)]] = \{(x, y) \in I^2, \max(z), (x, y, z) \in U[f(x, y)]\} . \qquad (7.99)$$

The umbra function can be interpreted as the mapping of a two-dimensional surface to a three-dimensional volume and the top surface function as the inverse mapping of the umbra to the graylevels of the image or object.

Figure 7.59(a) contains two overlapping objects defined with graylevels $r(x, y)$, with $x, y \in R$, and $s(x, y)$, with $x, y \in S$, where R and S define the pixels contained within each corresponding object. The umbra of each object is also shown in Figure 7.59(a). As with the set union and intersection of binary objects, the union and set intersection of the umbras of several objects can also be found. Two functions of interest between two objects $r(x, y)$ and $s(x, y)$ are the minimum and maximum functions:

$$\min[r(x, y), s(x, y)] \qquad (7.100)$$

and

$$\max[r(x, y), s(x, y)] . \qquad (7.101)$$

The umbra of Equation (7.100) is equal to the set intersection of the umbra of each function,

$$U[\min[r(x, y), s(x, y)]] = U[r(x, y)] \cap U[s(x, y)] , \qquad (7.102)$$

while the umbra of Equation (7.101) is equal to the union of the umbra of each object,

$$U[\max[r(x, y), s(x, y)]] = U[r(x, y)] \cup U[s(x, y)] . \qquad (7.103)$$

Figure 7.59(b) shows the intersection of the umbra of $r(x, y)$ with the umbra of $s(x, y)$. Note, how the top surface of this umbra is different from the top surface of $U[s(x, y)]$. Figure 7.59(c) shows the use of Equation (7.103), with the union of the umbra of $r(x, y)$ with the umbra of $s(x, y)$ given in the figure.

Equations (7.100) and (7.101) provide a means of computing the union and intersection of two umbras using the minimum and maximum operators. The modification of the umbra of an object is the concept behind graylevel erosion and dilation. Graylevel erosion of a graylevel object $r(x, y)$ with another graylevel object $s(x, y)$ is the process of reducing the umbra of $r(x, y)$, while graylevel dilation of a graylevel object $r(x, y)$ with another graylevel object $s(x, y)$ enlarges the umbra of $f(x, y)$.

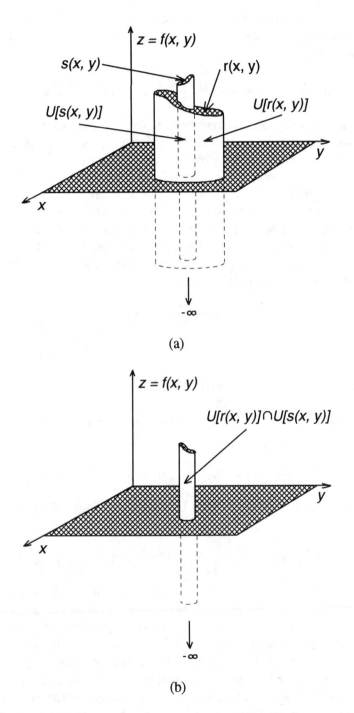

Figure 7.59: Three-dimensional plots of a graylevel image showing:
(a) the umbra of two objects and (b) the intersection of
$U[r(x, y)]$ with $U[s(x, y)]$.

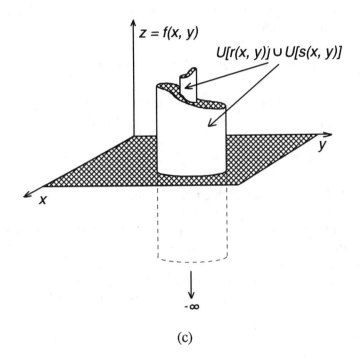

(c)

Figure 7.59: Three-dimensional plots of a graylevel image showing:
(c) the union of $U[r(x, y)]$ with $U[s(x, y)]$.

Similar to binary erosion, graylevel erosion of a graylevel object is defined as the top surface of the intersections of all possible umbras of $r(x, y)$ created by translating the umbra of $r(x, y)$ in both the x and y spatial directions and translating the graylevels z of $r(x, y)$ by the umbra of $s(x, y)$. The main difference between binary erosion and graylevel erosion is that the translations are now in three dimensions, as compared to two dimensions for binary erosion. But the translation of the umbra of $r(x, y)$ by the umbra of $s(x, y)$ is the erosion of the two umbras. Hence, graylevel erosion can be defined as

$$(r \ominus s)(x, y) = TS[U[r(x, y)] \ominus U[s(x, y)]] . \tag{7.104}$$

Graylevel erosion can also be written in terms of the graylevels of the two objects using the minimum operator:

$$(r \ominus s)(x, y) = \min[r(u, v) - s(u - x, v - y) : u, v \in O^2; u - x, v - y \in S], \tag{7.105}$$

where S defines the pixels included in the object $s(x, y)$. Similarly, graylevel dilation is defined as the top surface of the union of all umbras of $r(x, y)$ translated by the umbras of $s(x, y)$, as opposed to the intersection for graylevel erosion. But, the union of the translated umbras of $r(x, y)$ is the dilation of the umbra of $r(x, y)$ by the umbra $s(x, y)$:

$$(r \oplus s)(x, y) = TS[U[\ r(x, y)] \oplus U[s(x, y)]]\ . \qquad (7.106)$$

Like erosion, graylevel dilation can also be written in terms of the graylevels of the two objects $r(x, y)$ and $s(x, y)$ using the maximum operator:

$$(r \oplus s)(x, y) = \max[r(u, v) + s(u - x, v - y): u, v \in O^2; u - x, v - y \in S]\ . \quad (7.107)$$

Similar to binary erosion and dilation, the image or object $s(x, y)$ is typically referred to as the *graylevel structuring element*. The main difference between the graylevel structuring element and its binary counterpart is that the graylevel structuring element not only includes the domain of the x, y coordinates, but it also includes its graylevels $s(x, y)$ for each location. A special case of Equations (7.105) and (7.107) is when the top surface of the structuring element is flat and set to zero. Under this condition, Equations (7.105) and (7.107) reduce to

$$(r \ominus s)(x, y) = \min[r(u, v): u, v \in O^2; u - x, v - y \in S] \qquad (7.108)$$

and

$$(r \oplus s)(x, y) = \max[\ r(u, v): u, v \in O^2; u - x, v - y \in S\ ]. \qquad (7.109)$$

Equations (7.108) and (7.109) are simply the nonlinear minimum and maximum filters using the filter mask given by S.

Like binary erosion and dilation, graylevel erosion and dilation can be interpreted using the concept of the rolling wheel. Instead of using a two-dimensional wheel and rolling it about the contour to form the eroded or dilated binary object, a three-dimensional sphere is used. The graylevels of an image form a complex surface structure in which this sphere is rolled about to generate a new surface by tracing the center of the sphere as it is rolled about the image's surface. Graylevel dilation is then interpreted as rolling the sphere about on the top side of an image's surface, increasing the average graylevel of the image. On the other hand, graylevel erosion is defined by rolling the sphere below the image's surface and tracing the center of the sphere. This reduces the average graylevels of the eroded image. Graylevel erosion can now be interpreted as reducing positive going graylevel peaks, while increasing negative going graylevel valleys, with the opposite true for dilation.

Figure 7.60(a) shows a 7×7 sub-image with a 3×3 square graylevel object located in the center. For every pixel within this 7×7 sub-image, Equations (7.105) and (7.107) are used to perform graylevel erosion and dilation. As done previously in this chapter, erosion and dilation of the boundary pixels of the

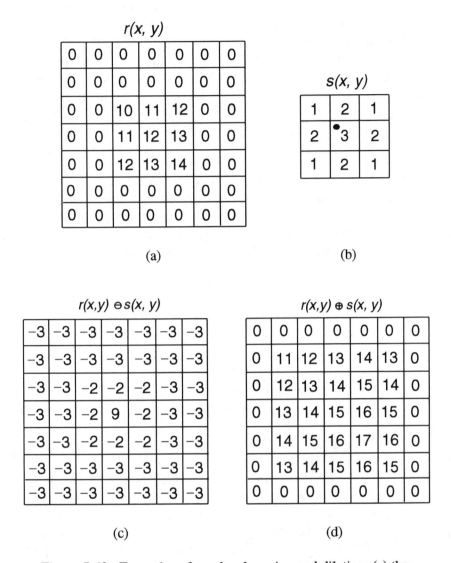

Figure 7.60: Examples of graylevel erosion and dilation: (a) the original graylevel image, (b) the graylevel structuring element, (c) the graylevel eroded image, and (d) the graylevel dilated image.

7×7 sub-image are computed by replicating these boundary pixels to form a new 7×7 sub-image. Figure 7.60(c) shows the result of graylevel erosion using the structuring element given in Figure 7.60(b). For the most part, the 3×3 square graylevel object has been reduced to one pixel with a graylevel of 9. Figure 7.60(c) shows that the zero graylevel has been reduced to a value of -3 by the erosion operation. Since negative graylevels cannot exist in a true digital image, there are several options to remove negative graylevels introduced by the erosion operation. The first is to truncate all negative graylevels to 0. The other method is to rescale the minimum and the maximum graylevels within the eroded image to the range of allowable graylevels. Figure 7.60(d) shows the result of graylevel dilation of Figure 7.60(a) using the same structuring element given in Figure 7.60(b). Note how the size of the square graylevel object has been increased from 3×3 to 5×5. Also notice the increase in the graylevels of the 5×5 square as compared to the original 3×3 square. As with grayscale erosion, care must be taken to scale the dilated graylevels to the range of allowable graylevels using either graylevel scaling or truncation as described above.

Even though every member of the 3×3 structuring element given in this example was used in the graylevel erosion and dilation operations, not every member must be included in these operations. Typically, excluded members are shown within a structuring element by using an asterisk, as shown in Figure 7.61. As previously defined in this text, the dot represents the origin of the structuring element.

-1	3	•1
-2	3	*
2	2	*

Figure 7.61: An example of a 3×3 structuring element with excluded members.

Examples of graylevel erosion and dilation are shown in Figure 7.62. Figure 7.62(a) is an image of a Hewlett Packard 35 calculator showing both white and black letters. The black letters against a light background can be interpreted as a valley in the complex surface as defined by this image's graylevels, while the white lettering against a dark background can be viewed as peaks in this surface. Figure 7.62(b) is the corresponding graylevel eroded image using the graylevel structuring element given in Figure 7.60(b). Since the white letters appear as peaks and the black letters appear as valleys in the complex graylevel surface of the image, graylevel erosion has removed these positive peaks while enhancing these negative valleys. Figure 7.62(c) is the graylevel dilated image of Figure 7.63(a) using the same structuring element given in Figure 7.60(b). Graylevel dilation has eliminated the black letters while enhancing the white letters. Contrary to erosion, the operation dilation has enhanced the positive peaks and reduced the valleys within the graylevel surface of the image.

Both graylevel erosion and dilation operations have similar properties to their binary counterparts. The first property is that graylevel dilation is *commutative:*

$$(r \oplus s)(x, y) = (s \oplus r)(x, y) .$$ (7.110)

(a)

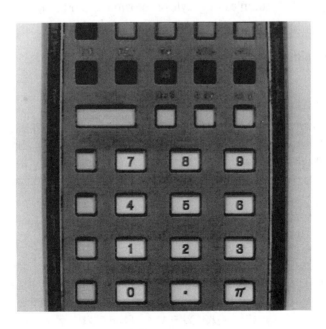

(b)

Figure 7.62: Examples of graylevel erosion and dilation: (a) the original grayscale image and (b) the corresponding graylevel eroded image.

(c)

Figure 7.62: Examples of graylevel erosion and dilation: (c) the
corresponding graylevel dilated image.

Dilation also follows the *associative* property of

$$[(r \oplus s) \oplus q](x, y) = [r \oplus (s \oplus q)](x, y) \quad . \tag{7.111}$$

Both graylevel erosion and dilation are *translation invariant*. It does not
make a difference whether the object is translated first and then eroded or dilated
or eroded or dilated first and then translated. Let r_t define the translation of an
object $r(x, y)$ by t. The translation invariant property for erosion and dilation
states that

$$(r_t \ominus s)(x, y) = (r \ominus s)_t(x, y)$$
$$(r_t \oplus s)(x, y) = (r \oplus s)_t(x, y) \quad . \tag{7.112}$$

Translation of the structuring element $s(x, y)$ yields

$$(r \ominus s_t)(x, y) = (r \ominus s)_{-t}(x, y)$$
$$(r \oplus s_t)(x, y) = (r \oplus s)_t(x, y) \quad . \tag{7.113}$$

The several properties of graylevel erosion and dilation that are important are

$$[r \oplus \max[s(x, y), q(x, y)]](x, y) = \max[(r \oplus s)(x, y), (r \oplus q)(x, y)] \; , \tag{7.114}$$

$$[\min[r(x, y), s(x, y)] \ominus q](x, y) = \min[(r \ominus q)(x, y), (s \ominus q)(x, y)] , \quad (7.115)$$

and

$$[r \ominus \max[s(x, y), q(x, y)]](x, y) = \min[(r \ominus s)(x, y), (r \ominus q)(x, y)] . \quad (7.116)$$

Both graylevel erosion and dilation follow the *increasing* property. If $r(x, y) < q(x, y)$ then

$$\begin{aligned}
(r \ominus s)(x, y) &< (q \ominus s)](x, y) \\
(r \oplus s)(x, y) &< (q \oplus s)](x, y) \quad .
\end{aligned} \quad (7.117)$$

For structuring elements, if $s(x, y) < q(x, y)$, the increasing property yields

$$\begin{aligned}
(r \ominus s)(x, y) &> (r \ominus q)](x, y) \\
(r \oplus s)(x, y) &< (r \oplus q)](x, y) \quad .
\end{aligned} \quad (7.118)$$

Another important property is the *distributive* property between erosion and dilation:

$$[(r \ominus s) \ominus q](x, y) = [r \ominus (s \oplus q)](x, y) \quad . \quad (7.119)$$

The erosion of object $r(x, y)$ by two sequential erosions with two structuring elements $s(x, y)$ and $q(x, y)$ is equivalent to dilating $s(x, y)$ with $q(x, y)$ and then performing the erosion on $r(x, y)$ with this dilated structuring element. Equation (7.119) is an important property in that it enables structuring elements to be decomposed into smaller structuring elements.

Let sT be the overall structuring element composed of smaller structuring elements:

$$sT(x, y) = (\cdots(s1 \oplus s2) \oplus s3) \oplus \cdots \oplus sN)(x, y) . \quad (7.120)$$

Then for erosion

$$(r \ominus sT)(x, y) = (\cdots(r \ominus s1) \ominus s2) \ominus s3) \ominus \cdots \ominus sN)(x, y) \quad (7.121)$$

and for dilation

$$r \oplus sT(x, y) = (\cdots(r \oplus s1) \oplus s2) \oplus s3) \oplus \cdots \oplus sN)(x, y) . \quad (7.122)$$

A special case of Equations (7.121) and (7.122) is the erosion or dilation of an object by the same structuring function n times:

$$(r \ominus sn)(x, y) = (r \ominus s) \ominus s) \ominus s) \ominus \cdots \ominus s)(x, y) \quad (7.123)$$

and

$$(r \oplus sn)(x, y) = (r \oplus s) \oplus s) \oplus s) \oplus \cdots \oplus s)(x, y) , \qquad (7.124)$$

where

$$sn(x, y) = (\cdots(s \oplus s) \oplus s) \oplus \cdots \oplus s) , \qquad (7.125)$$

yielding a total of *n* dilations.

The graylevel operations of opening and closing are defined the same as their binary versions:

$$r_S(x, y) = [(r \ominus s) \oplus s)](x, y) \qquad (7.126)$$

and

$$r^S(x, y) = [(r \oplus s) \ominus s)](x, y) \quad . \qquad (7.127)$$

The relationship between the graylevel operations of erosion, dilation, opening, and closing is given by

$$(r \ominus s)(x, y) \leq r_S(x, y) \leq r(x, y) \leq r^S(x, y) \leq (r \oplus s)(x, y) \quad . \qquad (7.128)$$

As with graylevel erosion and dilation, the graylevel operations of opening and closing are also translation invariant. It does not make a difference if the translation is performed before or after the operations of closing or opening.

The graylevel operations of opening and closing also obey the increasing property. If $r(x, y) \leq q(x, y)$ then

$$r_S(x, y) \leq q_S(x, y) \qquad (7.129)$$

and

$$r^S(x, y) \leq q^S(x, y) . \qquad (7.130)$$

Finally, both operations follow the *idempotent* property in that once either opening or closing has been applied, the opened or closed image is invariant to further openings or closings:

$$[r_S(x, y)]_S = r_S(x, y) \qquad (7.131)$$

and

$$[r^S(x, y)]^S = r^S(x, y) . \qquad (7.132)$$

Both graylevel operations of opening and closing are fundamental morphological filters used on graylevel images. Since opening is erosion followed by dilation, the erosion operation first eliminates positive peaks, enhances valleys, and decreases the graylevels of the graylevel surface structure of an image, and then the dilation operation increases the graylevels of the image back to those of the original image. Hence, the average brightness of an opened image is approximately the same as the original image. Graylevel opening is typically used to highlight dark regions of an image, while reducing or eliminating bright regions. The opposite holds true for graylevel closing. Graylevel closing is best at enhancing bright regions while reducing dark regions. In terms of impulse noise removal, graylevel opening is good at removing positive impulsive or salt noise and graylevel closing is good at removing negative impulsive or pepper noise.

Graylevel opening and closing can be combined to form two new morphological filters of *open-close* and *close-open*

$$\text{OPENCLOSE}[r(x, y), s(x, y)] = [r(x, y)_S]^S \qquad (7.133)$$

and

$$\text{CLOSEOPEN}[r(x, y), s(x, y)] = [r(x, y)^S]_S \; . \qquad (7.134)$$

Since opening is good at removing salt noise and closing is good at removing pepper noise, both the close-open and open-close filters are good at removing both types of noise.

The open-close and close-open filters obey some interesting properties in relationship to the median filter. Consider successive applications of a median filter to an image. Eventually, after n iterations, application of the median filter no longer produces any changes in the output median filtered image. At this point, the image has become invariant to the median filter and no longer changes. An image that is invariant to the median filter is known as the *root image* of the median filter. Let $\text{MED}_R(r(x, y))$ be the output root image of a median filter obtained after n iterations of the median filter. Then

$$[r(x, y)_S]^S \le \text{MED}_R[r(x, y)] \le [r(x, y)^S]_S \; . \qquad (7.135)$$

Equation (7.135) states that the root image of a median filter is bounded by the graylevel open-close and close-open filter operations. Also, both open-close and close-open filtered images are root images of a median filter:

$$[r(x, y)_S]^S = \text{MED}_R[[r(x, y)_S]^S] \qquad (7.136)$$

and

$$[r(x, y)^S]_S = \text{MED}_R[[r(x, y)^S]_S] \; . \qquad (7.137)$$

Another graylevel morphological filter that has similar properties to the open-close and close-open filters is the *hybrid* filter:

$$\text{HYBRID}[r(x, y), s(x, y)] = \frac{(r \ominus s)(x, y) + (r \oplus s)(x, y)}{2} \ . \qquad (7.138)$$

The concept behind this filter is that erosion lowers the graylevels of an image, while dilation increases the graylevels. The average of the eroded image with the dilated image should produce a new image with its graylevels near those of the original image. For the special case in which the structuring element is flat, the hybrid filter reduces to the average of the minimum and maximum filters. This is simply the nonlinear midpoint filter discussed in Chapter 5.

Figure 7.63 gives examples of using grayscale morphology to remove salt-and-pepper noise from an image. Figure 7.63(a) is the original grayscale image and Figure 7.63(b) is the corresponding 10% salt-and-pepper noise degraded image. The structuring element used for each morphological filtering operation presented in Figure 7.63 is the 3 × 3 structuring element given in Figure 7.60(b). Figure 7.63(c) is the opened image, showing that most of the salt noise has been removed. Since the first operation in the opening filter is erosion, this operation removes the salt noise, but it also enhances the pepper noise. The best the dilation of the eroded filtered image can do is reduce the enhanced pepper noise. Hence, the opening filter is unsuccessful at removing the pepper noise from the image. Figure 7.63(d) is the closed filtered image, showing the removal of the pepper noise. As expected, the closing filter was unable to remove the salt noise. Figure 7.63(e) and (f) are the open-close and close-open filtered images. The closing operation of the open-close filter was able to remove most of the pepper noise that remained in the opened image. Likewise, the opening operation in the close-open filter was able to remove most of the salt noise that remained in the closed image.

Another important grayscale morphological filter is the *gradient* filter, defined as the difference between the dilated image and the eroded image:

$$\text{GRADIENT}[r(x, y), s(x, y)] = (r \oplus s)(x, y) - (r \ominus s)(x, y) \ . \qquad (7.138)$$

The output of the gradient filter is an edge detected image. Figure 7.64(a) is the gradient filtered image of Figure 7.63(a) using the 3 × 3 structuring element given Figure 7.60(b). Notice how well the gradient filter was able to detect the edges present within the original grayscale image. The output of the gradient filter is very similar to the nonlinear range filter given in Chapter 5. In fact, for a flat structuring element, the gradient filter reduces to the difference between the maximum and minimum filters, which is the nonlinear range filter.

A grayscale morphological filter that can detect positive peaks in the graylevel distribution of an image is the *tophat* filter. The tophat filter is defined

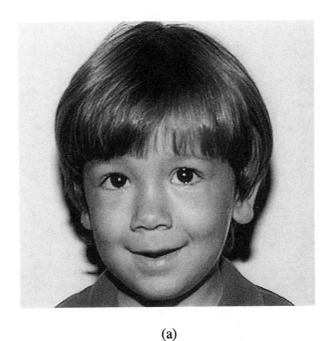

(a)

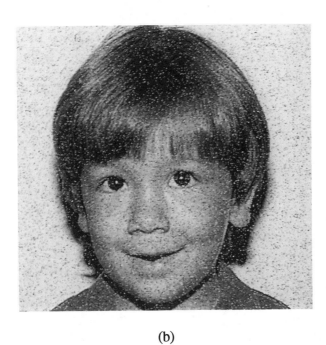

(b)

Figure 7.63: Examples of graylevel morphological filtering: (a) the original grayscale image and (b) the 10% salt-and-pepper noise degraded image.

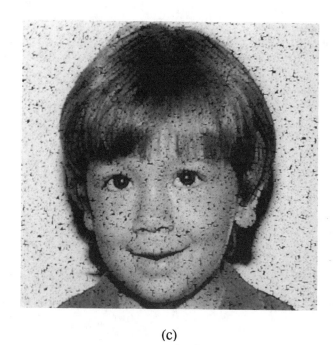

(c)

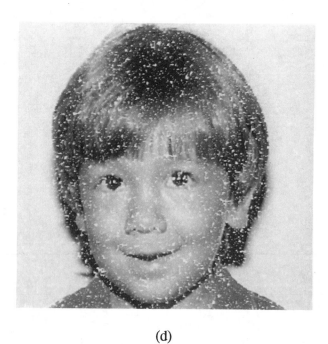

(d)

Figure 7.63: Examples of graylevel morphological filtering: (c) the
opened image of (b) and (d) the closed image of (b).

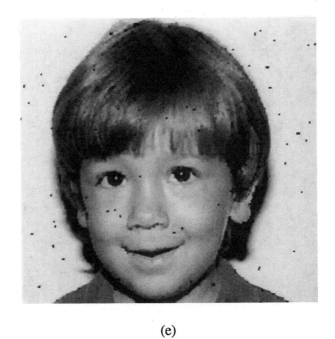

(e)

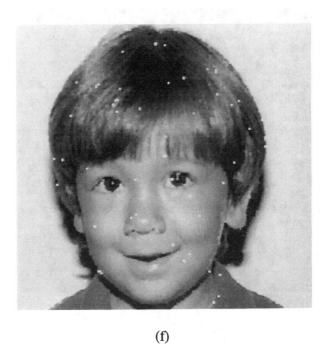

(f)

Figure 7.63: Examples of graylevel morphological filtering: (e) the open-close image of (b) and (f) the close-open image of (b).

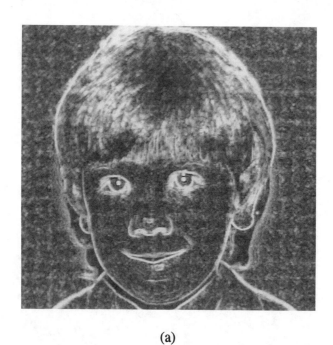

(a)

(b)

Figure 7.64: Examples of graylevel morphological edge detection:
(a) the gradient filtered image of Figure 7.63(a) and (b)
the tophat filtered image of Figure 7.62(a).

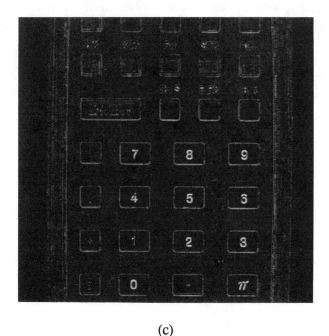

(c)

Figure 7.64: Examples of graylevel morphological edge detection:
(c) the valley filtered image of Figure 7.62(a).

as the difference between the original grayscale image and the opened image:

$$\text{TOPHAT}[r(x, y), s(x, y)] = r(x, y) - r(x, y)_s \cdot \quad (7.139)$$

Figure 7.64(b) is the tophat filtered image of Figure 7.62(a) using the structuring element of Figure 7.60(b). The tophat filter was able to detect the white letters, while removing all the black letters and the background. The opposite of the tophat filter is the *valley* filter. This filter detects negative going peaks in the image and is defined as the difference between the closed image and the original image:

$$\text{VALLEY}[r(x, y), s(x, y)] = r(x, y)^s - r(x, y) \cdot \quad (7.140)$$

Figure 7.64(c) is the valley filtered image of Figure 7.62(a), showing the detection of the black letters and the removal of the background and the white letters. This valley filtered image was also obtained using the 3×3 structuring element given in Figure 7.60(b).

The goal of this chapter was to introduce the concepts of morphological image processing and to give some examples of how the various types of morphological filters might be used. There are many more uses and applications of morphological processing that can be applied to digital images that have not been mentioned in this chapter. The interested reader is referred to the two texts *An Introduction to Morphological Image Processing* by Edward Dougherty and *Nonlinear Digital Filters: Principles and Applications* by I. Pitas and A. N. Venetsanopoulos. These two texts give excellent treatments of morphological image processing.

CHAPTER 8

Image Segmentation and Representation

An important area of electronic image processing is the segmentation of an image into various regions to separate the objects from the background. Image segmentation can be categorized by three methods. The first method is based upon a technique called image thresholding, which uses a predetermined graylevel as a decision criteria to separate an image into different regions based upon the graylevels of the pixels. The second method uses the discontinuities between graylevel regions to detect edges/contours within an image. Edges play a very important role in the extraction of features for object recognition and identification. The final method of image segmentation is to separate an image into several different regions based upon a desired criteria. For example, pixels that are connected and that have the same graylevel are grouped together to form one region.

After an image has been segmented into different objects, it is often desired to describe these objects using a small set of descriptors, thus reducing the complexity of the image recognition process. Since edges play an important role in the recognition of objects within an image, contour description methods have been developed that completely describe an object based upon its contour. The three most common methods are one based upon a coding scheme called chain codes, the use of higher order polynomials to fit a smooth curve to an object's contour, and the use of the Fourier transform and its coefficients to describe the coordinates of an object's contour. An object within an image can also be described using several region descriptors such as its area, perimeter, curvature, height, and width. An object can also be described based upon its surface texture. For example, images of a smooth circular object and a rough circular object can be separated and described solely on the texture difference between them. A set of parameters has been developed that quantifies the description of an object's surface texture.

8.1 Image Thresholding

Thresholding is the process of separating an image into different regions based upon its graylevel distribution. Consider an image containing a homogeneous object and background, as shown in Figure 8.1, with graylevels 200 and 80, respectively. Separation of the object pixels from the background pixels is accomplished by selecting a graylevel value K such that all pixels within the image with $f(x, y) > K$ will be classified as pixels belonging to the object. For the example image given in Figure 8.1, any threshold value K between 80 and 199 will separate the object pixels from the background pixels. The goal of thresholding is to select a threshold value that separates an image into two distinct graylevels:

$$g(x, y) = \begin{cases} G_a & f(x, y) \leq K \\ G_b & f(x, y) > K \end{cases}, \qquad (8.1)$$

where G_a and G_b are the desired two graylevels in the thresholded image. The process of thresholding as described by Equation (8.1) reduces a multilevel image to a two graylevel image containing graylevels of G_a and G_b. Typically, this two graylevel image is referred to as a binary or binarized image.

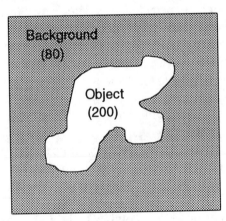

Figure 8.1: An example of a homogeneous object and background.

Key to the selection of a threshold value is an image's histogram, which defines the graylevel distribution of its pixels. The observation of an image's histogram can immediately locate the best threshold value. Figure 8.2(a) is an image of a light square object against a dark background that has been corrupted by Gaussian noise. Figure 8.2(b) is its corresponding histogram, showing the distribution of graylevels for this image. The bimodal nature of this histogram is typical of images containing two predominant regions of two different graylevels. The first mode of this histogram gives the distribution of pixels associated with the background, while the second mode gives the distribution of pixels associated with the square object. Inspection of this histogram shows that the graylevel distribution of pixels for both the background region and the square object appear to be Gaussian distributed with the same variance but with different average graylevels of 64 and 192. In fact, the original undegraded image was a square with a graylevel of 192 against a dark background with a graylevel of 64. The histogram for this image contains two peaks located at graylevels 64 and 192, with the height of these peaks equal to the percentage of background and object pixels present within the image.

The degradation of this image by additive Gaussian noise has redistributed the graylevels of the pixels associated with the square object and its background. The valley between the two peaks in the histogram indicates a separation point between the graylevels associated with the square object and those associated with the background. Thresholding Figure 8.2(a) using Equation (8.1) with G_a = 0, G_b = 255, and K = 128 yields the binarized image given in Figure 8.2(c). For the most part, the pixels associated with the square object have been separated from the pixels associated with the background.

(a)

Figure 8.2: An example of thresholding: (a) the original graylevel image.

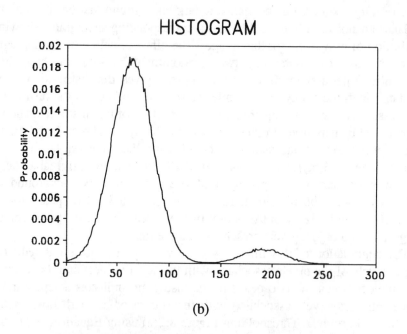

(b)

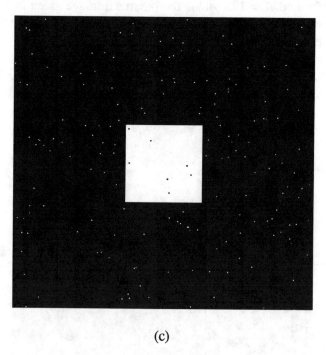

(c)

Figure 8.2: An example of thresholding: (b) the graylevel histogram
            of (a) and (c) the thresholded image with $G_a = 0$,
            $G_b = 255$, and $K = 128$.

The few black noise pixels that are present within the square object of this binarized image are due to pixels that have graylevels below 128. The degradation of this square by additive Gaussian noise has redistributed the graylevel of these pixels from 192 to graylevels below 128. Hence, they have been misclassified as background pixels when they really belong to the square object. Similarly, the white noise pixels present in the background are due to the redistribution of the graylevels of these pixels from a graylevel of 64 to a graylevel greater than 128. Selecting a threshold value less than 128 would eliminate some of the black pixels present in the white square but would also increase the misclassification of background pixels, resulting in an increase in the amount of white pixels present in the background. Likewise, increasing the threshold value would increase the number of black pixels present in the square, while decreasing the number of white pixels present in the background. The goal is to select a threshold value that minimizes the misclassification of background and object pixels. This is the approach that is taken in the implementation of optimum thresholding that will be discussed later in this section.

The thresholding of the example image given in Figure 8.2 was easily accomplished, since the graylevels of the pixels associated with the square were well separated from those associated with the background. Figure 8.3(a) shows an image of two objects with different graylevels against a dark background that has been corrupted with additive Gaussian noise. Figure 8.3(b) is the corresponding histogram of Figure 8.3(a), showing the presence of the three modes. The first mode, centered around graylevel 64, corresponds to the pixels associated with the background, while the other two modes, centered at 128 and 192, correspond to the pixels associated with the rectangular and circular objects. The selection of one threshold value using Equation (8.1) to segment this image would eliminate one of the three regions present, as shown in Figure 8.3(c). Figure 8.3(c) was obtained using Equation (8.1) with $G_a = 0$, $G_b = 255$, and a threshold value of $K = 160$, which is halfway between the two peaks of the last two modes present in the histogram. This thresholding operation for the most part has eliminated the rectangular object, except for a few pixels that have obtained graylevels above 160 due to the degradation of the image by the additive Gaussian noise.

A better method of thresholding the graylevel image given in Figure 8.3(a) is to use *multilevel thresholding*. The approach taken by multilevel thresholding is to expand Equation (8.1) to include more than one threshold value:

$$g(x, y) = \begin{cases} G_a & 0 \le f(x, y) < K_1 \\ G_b & K_1 \le f(x, y) < K_2 \\ G_c & K_2 \le f(x, y) \le G_{\max} \end{cases}, \qquad (8.2)$$

where $G_{\max}$ is the maximum allowable graylevel of the image $f(x, y)$. Equation (8.2) segments the image into three graylevel regions G_a, G_b, and G_c.

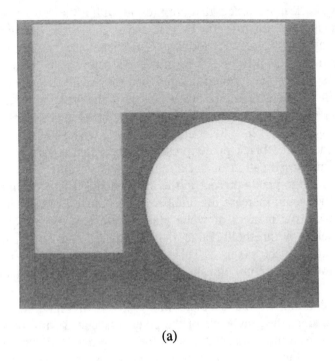

(a)

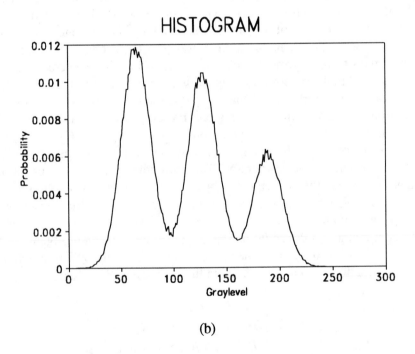

(b)

Figure 8.3: Examples of thresholding: (a) the original graylevel
image containing two objects at different graylevels
and (b) its corresponding histogram.

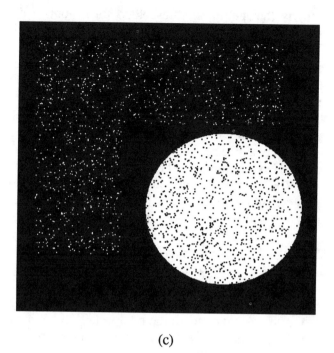

(c)

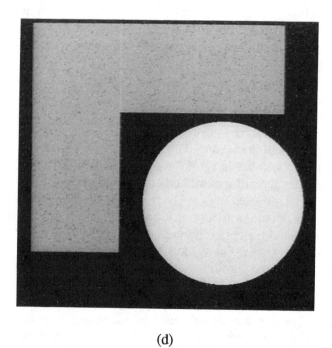

(d)

Figure 8.3: Examples of thresholding: (c) the single level thresholded image with $K = 160$ and (d) the multilevel thresholded image with $K_1 = 96$ and $K_2 = 160$.

Figure 8.3(d) is the multilevel thresholded image obtained with the two threshold values $K_1 = 96$ and $K_2 = 160$ and the three graylevels $G_a = 0$, $G_b = 128$, and $G_b = 255$. The two threshold values $K_1 = 96$ and $K_2 = 160$ were chosen because they corresponded to the two valleys separating the three objects. The multilevel thresholding approach was able to segment the two different graylevel objects from the background. The noise pixels that are present in the rectangular object are the result of misclassifying these pixels as belonging to either the background or to the circular object. The expected graylevel of 128 (the graylevel of the noise free rectangular object) for these pixels was degraded by the addition of Gaussian noise to below the graylevel of 96 or above the graylevel of 160. The same holds true for the noise pixels present in the circular object. The degradation of the image by Gaussian noise has lowered the graylevel values of these pixels below 160, resulting in these pixels being classified as belonging to either the background or the rectangular object.

The thresholding of a graylevel image is easy if the graylevels of the pixels defining an object are well separated from the graylevels defining the background, as in the thresholding example given in Figure 8.1. Under this condition, the histogram of the image will be bimodal, and the best threshold value is selected in the valley between the two peaks. In a complex image, the graylevels of the objects may not be well separated, making the selection of the threshold value difficult. Typically, images acquired using a poor illumination source will result in images that are no longer bimodal, which makes it difficult to find the best threshold value. As given by Equation (1.4), the model for an image $f(x, y)$ is the product of its illumination component $i(x, y)$ and its reflectivity component $r(x, y)$:

$$f(x, y) = i(x, y) \cdot r(x, y) . \qquad (8.3)$$

Figure 8.4(a) is the same image of Figure 8.2(a) of the rectangular object but acquired under poor illumination. The brightness of this image ranges from a low level at the top of the image to a high level at the bottom of the image. Figure 8.4(b) shows the histogram of this image. Comparing it to the histogram of the uniformly illuminated image, given in Figure 8.2(b), one can see that the introduction of a nonuniform illumination has blended the two modes of the uniformly illuminated histogram into one mode. The well separated graylevels for the rectangular object have been blended with the graylevels of the background, making the thresholding of this image difficult. The minimum valley between the two peaks that describes the separation between the rectangular object and the background is no longer present. Figure 8.4 shows the importance of obtaining a uniformly illuminated image prior to thresholding of the image. If the illumination component alone can be determined, the poorly illuminated image can be multiplied by the inverse of the illumination component to produce a uniformly illuminated image. Alternatively, if the illumination component cannot be determined, then the homomorphic filter described in Chapter 5 should be used to remove the shading from the image.

(a)

HISTOGRAM

(b)

Figure 8.4: The effect the illumination component has on image thresholding: (a) the original shaded image and (b) its corresponding histogram.

Typically, the histogram of a complex image is anything but bimodal, and the selection of the best threshold is at best a compromise between the segmentation of the different features that are present within the image. Figure 8.5(a) is an image of a young girl and Figure 8.5(b) is its corresponding histogram. Note the lack of a clear separation between the graylevels of the girl and the graylevels of the background behind the girl. Figures 8.5(c) and (d) are the thresholding of this image with the threshold values of $K = 67$ and $K = 180$. Depending on the threshold value selected, the facial features of the girl present in the thresholded image varies. For a threshold value of $K = 67$, the features of the girl's hair and eyes are still present in the binary image. On the other hand, a threshold value of $K = 180$ has eliminated these features and has highlighted the features associated with the girl's nose and mouth. In thresholding this example image, there is no best threshold value. The value chosen depends on the desired features to be segmented from the other features present in the graylevel image. Usually, a trial and error method is used to select the threshold value that involves a human interpretation of the thresholded image to determine if the threshold value selected has extracted the desired object features.

There are several adaptive thresholding methods that attempt to select a threshold value that either maximizes the information present in the thresholded image or attempts to minimize the error associated with the selection of a threshold value. The first of these adaptive thresholding algorithms is that of *optimum thresholding*. Optimum thresholding is based upon a two object model in which an image is separated into a set of object pixels and a set of background pixels. Essentially, optimum thresholding assumes an image histogram that is bimodal, with a valley separating the two modal peaks. Optimum thresholding is based upon the Minimum Probability of Error MPE classifier/detection model used in the detection of a binary signal in the presence of noise. The MPE classifier model selects a threshold value that minimizes the error of detecting a 0 or 1 from a binary signal. Let $P1$ be the probability of an object in an image and $P2$ be the probability of the background; then

$$P1 = \frac{n_o}{n_t} \tag{8.4}$$

and

$$P2 = \frac{n_b}{n_t}, \tag{8.5}$$

where n_o is the number of object pixels, n_b is the number of background pixels, and n_t is the total number of pixels within the image. From Equations (8.4) and (8.5)

$$P1 + P2 = 1. \tag{8.6}$$

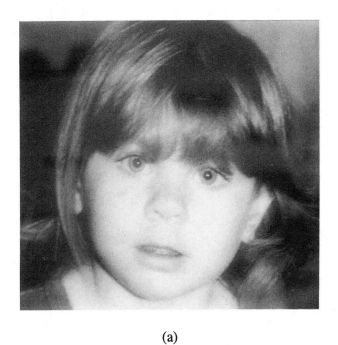

(a)

HISTOGRAM

(b)

Figure 8.5: The thresholding of a complex graylevel image: (a) the
original graylevel image and (b) its corresponding
histogram.

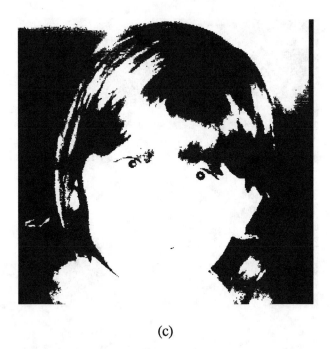

(c)

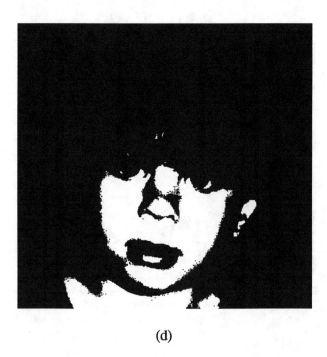

(d)

Figure 8.5: The thresholding of a complex graylevel image: (c) the
            thresholded image with $K = 67$ and (d) the thresholded
            image with $K = 180$.

The sum of the two probabilities must add up to one since the assumption is that there are only background and object pixels present within the image.

Consider an object pixel within an image with graylevel G_o and background pixels with graylevel G_b. The histogram for this image contains two nonzero values, at G_o and G_b. Next, consider a new image consisting of this noise-free image containing the object and its background with the addition of zero mean additive noise. As shown in Figures 8.2(a) and (b), the degradation of an image by additive noise redistributes the pixels associated with the graylevels within the image. In fact, the histogram of the object pixels has several similarities to the histogram of Gaussian noise. Figure 8.6(a) shows the histogram of a noise free image containing background and object pixels with graylevels G_b and G_o. Figure 8.6(b) is the histogram of a zero mean noise process that will be used to degrade this noise free image. Figure 8.6(c) shows the noise degraded image's histogram. The degradation of an image by additive noise produces a new image in which its histogram contains two versions of the histogram of the noise, one centrally located about G_o and the other about G_b. In addition, the distribution of pixels about the background and object obtain the same variance as that of the noise degradation process.

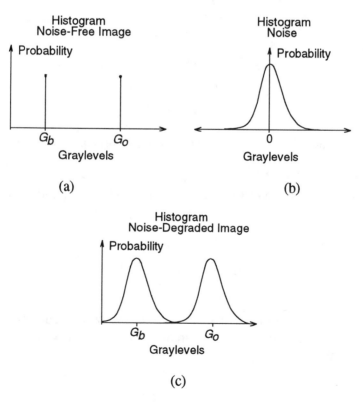

Figure 8.6 Example of degrading an image by additive noise: (a) the histogram for a noise free image, (b) the histogram for a noise process, and (c) the degraded image's histogram.

Figure 8.6 shows how a noise degraded image can obtain a distribution of graylevels that is bimodally distributed.   In the derivation of optimum thresholding it is not required that the distribution of graylevels for the background be equal to the distribution of graylevels for the object. Let $h_2(G_i)$ be the histogram for the background pixels and $h_1(G_i)$ be the histogram for the object pixels. Figure 8.7 shows in detail the two histograms of the distribution of graylevels for the background and the object. Thresholding this image with a value of K produces two errors. The first error is the misclassification of object pixels with graylevels below K as background pixels. This is shown in Figure 8.7 as the shaded area given by $E_1(K)$. The second error is the misclassification of background pixels with graylevels greater than K as object pixels. This error corresponds to the second shaded region $E_2(K)$ given in Figure 8.7.

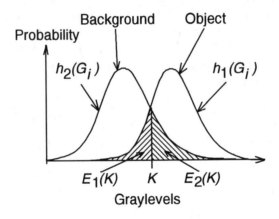

Figure 8.7: A detailed plot showing a bimodal histogram.

The total error $E_1(K)$ of assigning object pixels as background pixels is defined as the sum of the histogram $h_1(G_i)$ of the object from 0 to the graylevel $K-1$:

$$E_1(K) = P1 \cdot \sum_{i=0}^{K-1} h_1(G_i) \; . \qquad (8.7)$$

Likewise, the error $E_2(K)$ associated with selecting background pixels as object pixels is equal to the sum of the histogram $h_2(G_i)$ of the background pixels from graylevels K to G_{max}

$$E_2(K) = P2 \cdot \sum_{i=K}^{G_{max}} h_2(G_i) \, , \qquad (8.8)$$

where G_{max} is defined as the maximum graylevel contained within the image. The total error of thresholding this bimodal image is then equal to the sum of the two errors:

$$E_T(K) = E_1(K) + E_2(K) \ , \tag{8.9}$$

or

$$E_T(K) = P1 \cdot \sum_{i=0}^{K-1} h_1(G_i) + P2 \cdot \sum_{i=K}^{G_{max}} h_2(G_i) \ . \tag{8.10}$$

The goal of optimum thresholding is to select a threshold value K that minimizes the error function given in Equation (8.10). The relationship between the total error as defined by Equation (8.10) and the threshold value K can be understood from Figure 8.7. As the threshold value is decreased, the error $E_1(K)$ associated with misclassifying object pixels decreases but the error $E_2(K)$ of misclassifying background pixels as object pixels *increases*. If the threshold value increases, the error $E_1(K)$ associated with misclassifying object pixels as background pixels increases, while the error $E_2(K)$ of misclassifying background pixels as object pixels decreases. There is an optimum threshold value that minimizes these two errors. Taking the derivative of Equation (8.10) and setting it to zero yields the minimum error when

$$P1 \cdot h_1(K) = P2 \cdot h_2(K) \ . \tag{8.11}$$

Equation (8.11) states that the minimum classification error occurs when the product of the probability of occurrence of the object and its histogram evaluated at graylevel K is equal to the product of the probability of occurrence of the background and its histogram evaluated at graylevel K.

The solution of Equation (8.11) to compute the optimum threshold value assumes that the histograms of the background and object pixels are known. In many cases, if an image has been degraded by a known type of noise, $h_1(G_i)$ and $h_2(G_i)$ can be estimated directly from the noise histogram. The probabilities $P1$ and $P2$ are then found from the percentage of area that the object and background occupy within the image. For the case in which the object pixels have a mean graylevel of G_o, the background pixels have a mean graylevel of G_b, and the noise process can be approximated as Gaussian distributed with the same variance σ^2, the histograms for the object and background become

$$h_1(G_i) = \frac{e^{-(G_i - G_o)^2/\sigma^2}}{\sigma\sqrt{2\pi}} \tag{8.12}$$

and

$$h_2(G_i) = \frac{e^{-(G_i - G_b)^2/\sigma^2}}{\sigma\sqrt{2\pi}} \qquad . \tag{8.13}$$

Substituting Equations (8.13) and (8.14) into Equation (8.11) and solving for the optimum threshold value K yields

$$K = \frac{G_o + G_b}{2} + \frac{\sigma^2}{G_o - G_b} \ln\left(\frac{P2}{P1}\right) . \tag{8.14}$$

Equation (8.14) represents the solution of the two class discrimination problem under the presence of Gaussian noise, which is known as the MPE classifier. For an equal probability of occurrence for the object and the background (they both occupy the same area within the image), the optimum threshold value is equal to the average of the graylevel of the object with that of the background. If the probability of the object exceeds that of the background, the optimum threshold value is reduced. This reduces the error associated with the misclassification of the object pixels as background pixels. The bias toward a lower threshold value should be expected since the probability of an object being higher than the background states that there will be fewer background pixels available to be misclassified as object pixels. Hence, the threshold value can be lowered to reduce the error of misclassifying object pixels as background pixels. The inverse is true when the probability of the background is greater than the probability of the object. Under this condition, the threshold value is increased to reduce the error associated with misclassifying background pixels as object pixels.

Optimum thresholding has been successfully used in thresholding cardioangiograms to highlight the flow of blood in an artery. The main limitation with optimum thresholding is that it requires modeling the histogram of an image as a bimodal histogram, which can then be separated into two known histograms that represent the distribution of graylevels for the object and the background. If these two histograms cannot be determined, and if the histogram of the image to be thresholded is approximately bimodal, a search algorithm can be used to locate the minimum of the valley between the two modal peaks. This graylevel then becomes the threshold value to segment the object pixels from the background pixels. Typically, an image's histogram is noisy due to the finite number of samples used to compute the histogram. Figure 8.5(b) shows an example of an image histogram that is noisy. Prior to searching for the minimum of the valley, the histogram is typically lowpass filtered to smooth its curve and reduce the noise present in the histogram, making it easier to locate the minimum between the two modal peaks. A small windowed mean filter such as a 3×1 or a 5×1 window can be applied to the image's histogram to perform the smoothing operation.

There are several thresholding methods that are not based upon a prior knowledge of the histogram of the image as required by optimum thresholding. A method proposed by Otsu treats the segmenting of a grayscale image into a binary image as a classification problem in which two classes are generated from the set of pixels within the grayscale image. The first class defines the set of pixels within the image with graylevels below or equal to the threshold value K, while the second class defines the set of pixels with graylevels above the threshold value K. The concept of grouping data into classes based upon similarity is one of the fundamental concepts of pattern recognition. A short discussion here will explain the concepts required to understand Otsu's adaptive thresholding method. The interested reader is referred to the book *Pattern Recognition Principles* by J. T. Tou and R. C. Gonzalez listed in the bibliography section for a detailed discussion of classes and their use in pattern recognition.

The concept of grouping data in classes based upon the similarly of the data is easily explained with the use of Figure 8.8, which shows a two-dimensional plot of scattered data described by their x, y coordinates. Inspection of Figure 8.8 shows that these scattered data have clustered into two different regions. A decision function (line) can then be drawn between the two clusters to separate these two regions into separate groupings or classes. All the scattered data below this decision function are grouped together to form Class 1, while the scattered data above this decision function are grouped together to form Class 2. For example, the unknown point A will be classified as belonging to Class 1. Figure 8.8 shows that associated with each class is its centroid, or central location, and the area of the class. The tighter the scattered data are clustered together, the smaller the area of the class will be. The process of segmenting these clustered data into two distinct classes is easily accomplished if the area occupied by each class is small and the separation between the centroids of the classes is large. Maximizing the class separation while minimizing the areas of the classes is the approach taken by Otsu's method of adaptive thresholding to select the most optimum threshold value.

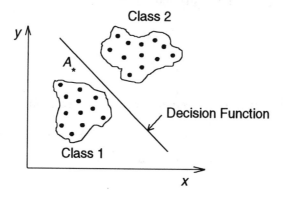

Figure 8.8: An example plot showing the concept of placing data in classes.

The two-dimensional classification example given in Figure 8.8 can be applied using any number of dimensions desired. For the case of adaptive thresholding using Otsu's method, only one dimension is needed to group pixels together into two classes based upon their graylevels. The goal of Otsu's thresholding method is to select a threshold value that minimizes the areas of the two classes while maximizing the distance (variance) between these two classes. For example, the unknown point A in Figure 8.8 is selected as belonging to the class in which its area increases the least. Obviously, if this unknown point A were selected as belonging to Class 2, that would increase the area of Class 2 by a much larger extent than selecting point A as belonging to Class 1, which increases the area of Class 1 only slightly.

The first step in implementing Otsu's method is to compute the graylevel histogram $h(G_i)$ of the image to be thresholded. Associated with the set of pixels belonging to the class (Class 1) with graylevels below or equal to the threshold value K is its average graylevel:

$$m_1(K) = \sum_{i=0}^{K} i \cdot h(G_i).$$ (8.15)

Equation (8.15) is just the one-dimensional equivalent of the centroid given in Figure 8.8 and describes the central location of the graylevels for this class. Likewise, the average graylevel for the second class (Class 2), describing pixels with graylevels above the threshold value K, is

$$m_2(K) = \sum_{i=K+1}^{G_{max}} i \cdot h(G_i).$$ (8.16)

It should be noted that Equation (8.16) is the average over graylevels $K + 1$ through G_{max}, while Equation (8.15) is the average over graylevels 0 through K. Also, the average graylevel for the entire image is

$$m_T = \sum_{i=0}^{G_{max}} i \cdot h(G_i).$$ (8.17)

The probability of Class 1 occurring within the grayscale image is

$$P1(K) = \sum_{i=0}^{K} h(G_i),$$ (8.18)

while the probability of Class 2 occurring is

$$P2(K) = \sum_{i=K+1}^{G_{\text{max}}} h(G_i) . \tag{8.19}$$

The total of these two probabilities must equal one:

$$P1(K) + P2(K) = 1 . \tag{8.20}$$

The distance between the two classes is given by the between-class variance,

$$\sigma^2 = \frac{[m_1(K) \cdot P2(K) - m_2(K) \cdot P1(K)]^2}{P1(K) \, P2(K)} . \tag{8.21}$$

The maximization of Equation (8.21) with respect to the threshold value K gives the maximum separation between and minimizes the sizes of the two classes and therefore yields the optimum threshold value.

The maximization of Equation (8.21) to determine the optimum threshold value defines the adaptive thresholding method as proposed by Otsu. The adaptive thresholding of a grayscale image starts with an initial threshold value of 1. Equations (8.15), (8.16), (8.18), and (8.19) are used to compute $m_1(K)$, $m_2(K)$, $P1(K)$, and $P2(K)$, and these values are then used to compute the between-class variance given in Equation (8.21). The process continues with a new threshold value of 2, which is used to compute the between-class variance as given by Equation (8.21). The between-class variance is computed for each threshold value between 1 and $G_{\text{max}} - 1$. The threshold value that yields the maximum between-class variance is chosen as the optimum threshold value. The final step in Otsu's adaptive thresholding algorithm is to threshold the graylevel image at this threshold value.

Figure 8.9(a) is the thresholded image of Figure 8.5(a) obtained using Otsu's adaptive thresholding algorithm, which calculated a threshold value of 107. A plot of the between-class variance as a function of threshold value is given in Figure 8.9(b). The maximum between-class variance occurred for a threshold value of 107. This thresholded image is sort of a compromise between the two thresholded images of the young girl given in Figures 8.5(c) and (d). Thresholding of Figure 8.5(a) with a value of 67 eliminated all of the features of the face including the mouth and the nose but preserved the features of the hair. On the other hand, thresholding of this image using a threshold value of 180 preserved the nose and mouth of the young girl but removed all of the features associated with the girl's hair and most of the features of her eyes. The adaptive threshold value of 107 chosen by Otsu's method preserved some of the features of the girl's eyes as well as some of the features associated with her mouth. The selection of the optimum threshold value for this image is difficult since there

(a)

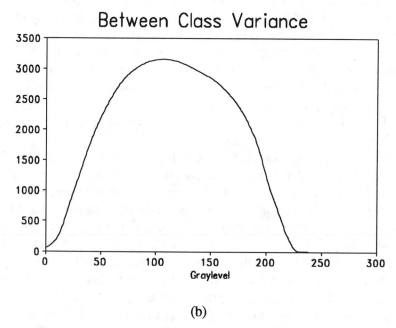

(b)

Figure 8.9: The adaptive thresholding of Figure 8.5(a) using Otsu's
method: (a) the thresholded image and (b) the between-
class variance.

is no clear-cut separation between the graylevels in the features of the girl. This is typical of most complex images in that the best thresholded image that can be obtained is a compromise between preserving various features. Figure 8.10(a) is the example image of the Hewlett Packard model 35 calculator, and Figure 8.10(b) is its corresponding adaptive thresholded image using Otsu's method with $K = 123$. Notice how well the thresholding of this image preserved both the white and black letters present in the grayscale image; unfortunately the white letters on the gray colored keys [on the left side of Figure 8.10(a)] were removed from the thresholded image.

Another adaptive thresholding method selects a threshold value that maximizes the contours present within a binarized image, as presented by Weeks et al. The goal of this adaptive thresholding algorithm is that edges play a key role in the identification of objects within an image. By maximizing the number of contours that are present in the binary image, a large number of key recognizable features of the original graylevel image will be observable, making the recognition of the objects present within the thresholded image easier. An example of computer recognition in which key features are directly related to the object's contours is that of the automatic character recognition of handwritten characters using binary images. Here, it is desirable to produce a binary image containing the maximum number of contours possible. The importance of edges or contours in the recognition of objects within an image was also illustrated with the range filter given in Chapter 5 and the morphological gradient filter given in Chapter 7. The goal then is to select a optimum threshold value that preserves within the thresholded image the most contours (features) that are present within the original grayscale image. The recognition of features is directly related to the contours that remain after thresholding.

Perceptual psychologists have shown that the visual information within an image is concentrated at the contours and that these contours and their nearby neighboring pixels are essential for image perception. A good example of image recognition based upon contours is that of caricatures, where an artist accentuates facial features of a subject in simple line drawings that are readily recognizable. Other studies by perceptual psychologists have also shown that the perception of complex shapes is determined largely from regions of concavity and high contour change. This finding provided a means of decomposing an image into simpler components for recognition. Other psychologists support the finding that a complex object can be recognized using a simpler set of components and proposed a set of visual primitives that are essential for visual recognition. Among the visual primitives proposed are color, brightness, line ends or terminators, blobness or closure in the sense of convexity, tilt, and curvature.

The research findings of Resnikoff showed that the change in the angular direction of a contour is a reasonable measure of the visual information present in the contour. This measurement determines the information gained as a measure of an angular change along a contour of an image relative to other

(a)

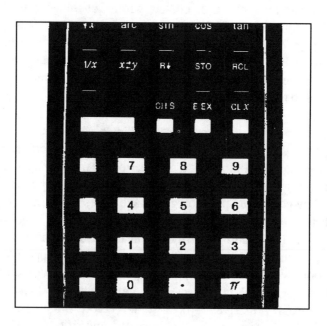

(b)

Figure 8.10: An example of adaptive thresholding using Otsu's
method: (a) the original image and (b) the
thresholded image with $K = 123$.

angular measurements along the contour. Consider the sequence of points $\{(x_0, y_0), (x_1, y_1), (x_2, y_2), (x_3, y_3) ...\}$ that comprise a contour, with the ordering of the sequence the same as the occurrence of the points in the contour as shown in Figure 8.11. Let (x_0, y_0) and (x_1, y_1) form one straight line segment and (x_1, y_1) and (x_2, y_2) form another straight line segment, with the point (x_1, y_1) as the vertex between the two line segments. The angle between the two straight line segments is defined as the single angular measurement α at the point (x_1, y_1) in a range of $-\pi$ to π. The absolute angular information from this single angular measurement is then defined as

$$I(\alpha) = \log_2\left(\frac{\pi}{|\alpha|}\right) \text{ bits} \quad , \qquad (8.22)$$

where $|\alpha|$ is the absolute value of α.

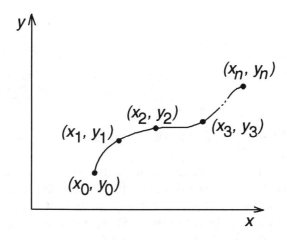

Figure 8.11: A diagram of a contour containing n points.

Next, consider a second angular measurement β ($-\pi$ to π) on the contour at the vertex point (x_2, y_2) formed between the two straight line segments (x_1, y_1) to (x_2, y_2) and (x_2, y_2) to (x_3, y_3). The information gained by traversing the contour by the second angular measurement β is defined as

$$I(\beta) - I(\alpha) = \log_2\left(\frac{\pi}{|\beta|}\right) - \log_2\left(\frac{\pi}{|\alpha|}\right) \text{ bits}, \qquad (8.23)$$

or

$$I(\beta) - I(\alpha) = \log_2\left(\frac{|\alpha|}{|\beta|}\right) \text{ bits}. \qquad (8.24)$$

The information gained (entropy) by traversing a contour is relative to a previous information gain (entropy) measurement. Figure 8.12 shows a diagram of an angle formed by the interconnection of two straight lines. Acute angles less than 45° tend to yield more information than acute angles larger than 45° and obtuse angles less than 180°. Furthermore, a contour with rapid changes in angular direction per unit distance indicates the presence of high contour information. The measure of the total contour information (entropy) is then the accumulation of information gain along a complete contour containing n angular comparisons:

$$I_t = \sum_{i=1}^{n} \left| \log_2\left(\frac{|\alpha_i|}{|\beta_i|}\right) \right| \text{ bits.} \qquad (8.25)$$

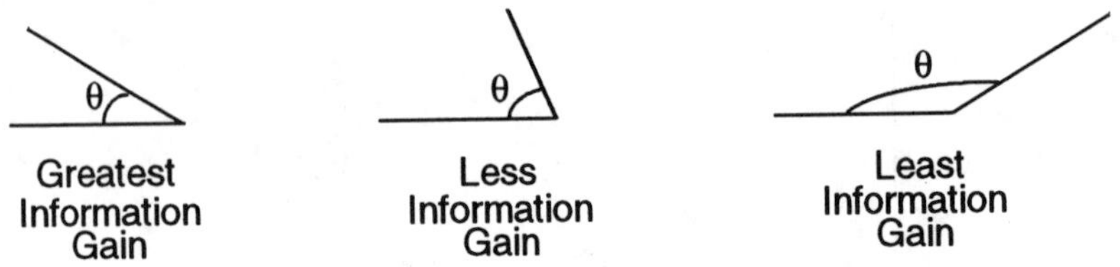

Grediest   Less   Least
Information   Information   Information
Gain   Gain   Gain

Figure 8.12: The amount of information contained within various contours.

The threshold value that produces the maximum total contour entropy in a grayscale image is obtained dynamically by varying the threshold value between the minimum graylevel 0 and the maximum graylevel G_{max}. The threshold value starts at zero and is incremented until the maximum threshold value is reached. For each threshold value, a series of steps is performed. First, the entire image $f(x, y)$ is thresholded at a given graylevel K to produce a binary image $g(x, y)$ as given by Equation (8.1). Next, one of the many edge detection algorithms that exist (presented in Section 8.2) is used to obtain a contour image containing the contours present in the thresholded image.

For each contour present in the contour image, its corresponding contour information is computed using Equation (8.25). This contour is then removed from the contour image, and the contour information for the next contour is computed. The process of computing the contour information for each contour continues until all of the contours have been traced. At this point, the sum of the contour information of each individual contour yields the total contour information present within the entire thresholded image. The threshold value K is then incremented to the next threshold value and the process of determining the total contour information for this thresholded image is repeated. The threshold value that yields the maximum total contour information is selected as the optimum threshold value to threshold the grayscale image with.

Figure 8.13(a), entitled "Sultry", is a picture of the girl's face containing low spatial detail within the facial region and high spatial detail within the region of the hair. The corresponding histogram for this image is given in Figure 8.13(b). Note the difficulty in using an optimum thresholding algorithm based upon the graylevel histogram for this image. The histogram contains many modes due to several bins within the histogram containing 0 pixels at that graylevel as a result of a malfunctioning digital video camera. Figure 8.13(c) shows a plot of the total contour entropy as a function of every fifth threshold value. The maximum contour entropy for this image occurs at a threshold value of $K = 77$. Figure 8.13(d) shows the binary image for this threshold value. Notice how the high spatial detail of the hair remains in the binary image. Unfortunately, the facial details of the nose and part of the mouth were lost. Choosing a threshold that optimizes these features would eliminate most of the identifiable features such as the eyes and the facial outline, reducing the spatial information in the binary image. Figure 8.13(e) gives the thresholded image using Otsu's method with a threshold value of $K = 140$. In comparing the adaptive thresholded image using the contour entropy method given in Figure 8.13(d) with Otsu's method, the contour entropy algorithm yielded more spatial features. This is especially true in the eyes and the hair region.

(a)

Figure 8.13: Examples of adaptive thresholding using the contour
entropy method: (a) the original grayscale image
(from Weeks et al., 1993).

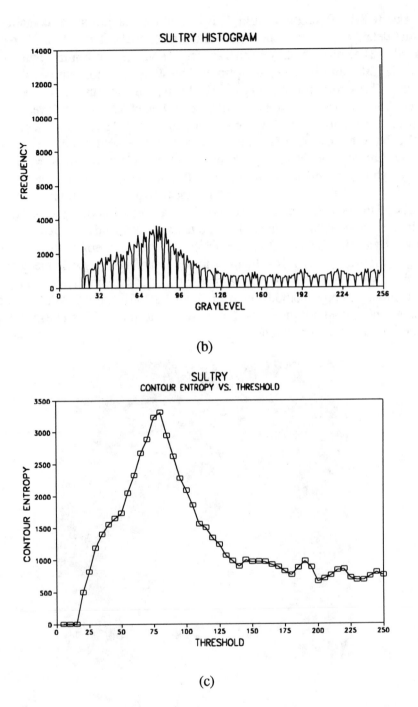

Figure 8.13: Examples of adaptive thresholding using the contour
           entropy method: (b) the histogram of (a) and (c) the
           contour entropy as a function of threshold value K
           (from Weeks et al., 1993).

(d)

(e)

Figure 8.13: Examples of adaptive thresholding using the contour
entropy method: (d) the contour entropy thresholded
image with $K = 77$ and (e) the thresholded image using
Otsu's method with $K = 149$ (from Weeks et al., 1993).

The advantage of Otsu's method over the contour entropy method is its speed. The adaptive thresholding of an image using the contour entropy algorithm is extremely computationally intensive because of the requirement of tracing and computing the contour information for each contour present in each of the thresholded images. In fact, the total time required to threshold an image using the contour entropy algorithm depends on the total number of contours that are present within the grayscale image. The larger the number of contours, the longer it takes to threshold the image. It took approximately 7000 times longer to compute the adaptive thresholded image using the contour entropy method than using Otsu's method. An application of adaptive thresholding using the contour entropy method is the thresholding of images containing handwritten characters for character recognition. Since the contour entropy method selects a threshold based upon maximizing the contours present within the thresholded image, the contour entropy method produces a thresholded image of the handwritten characters with the most contours guaranteeing that the largest number of character features possible is available for recognition.

8.2 Edge, Line, and Point Detection

The detection of graylevel discontinuities play an important part in the recognition of objects present within an image. In this section, several filters will be presented that detect edges, lines, and points that are present within a grayscale image. The importance of graylevel discontinuities in the recognition of objects is easily illustrated with Figure 8.14. Figure 8.14(a) starts with one circular contour and additional contours are added to each successive image of Figure 8.14 until the caricature is recognizable. Figure 8.14(b) shows the addition of two ellipses oriented and placed within the first circular contour to resemble two eyes. Figure 8.14(b) could now be interpreted as a face. The addition of a vertically oriented ellipse in Figure 8.14(c), which could be interpreted as a nose, gives a higher confidence that the caricature is a face. But there is a finite probability that Figure 8.14(c) could be viewed as a cartoon drawing of a bowling ball. In Figure 8.14(d), with the addition of a circular arc, which resembles a mouth, the inclination is to dismiss the bowling ball assumption and confirm the presence of a face. The addition of six straight lines placed adjacent to the elliptical contour representing a nose, as shown in Figure 8.14(e), gives the impression that this face contains whiskers and that this image might be a caricature of an animal. Finally, with the addition of two sets of contours that resemble ears, Figure 8.14(f) yields a recognizable caricature of a cat's face. Figures 8.14(a) through 8.14(f) show the importance of using contours (graylevel discontinuities) in the recognition of images and how features are easily recognized from just the contours.

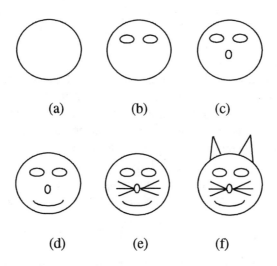

Figure 8.14: The importance of contours in the recognition of
objects within an image (from Weeks et al., 1993).

Typically, points, lines, and edges are detected using neighborhood spatial filtering as presented in Chapter 4. The type of filter mask chosen determines the type of detector implemented. From Chapter 4, the spatial filtering of an image using a 3 × 3 filter mask *A*, as shown in Figure 8.15, is simply the multiplication of every pixel within the image that is under the filter mask with the corresponding filter mask coefficients.

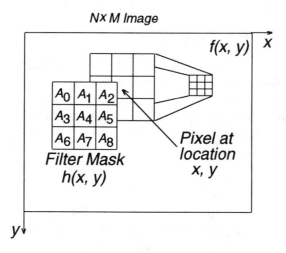

Figure 8.15: An illustration of spatial filtering using a 3 × 3 filter mask.

For the 3 × 3 filter mask given in Figure 8.15, the spatial filter operation as defined in Chapter 4 is

$$g(x, y) = \sum_{i=0}^{2} \sum_{j=0}^{2} f(x-1+j, y-1+i) \, A_{(j+3i)} \; . \tag{8.26}$$

The output of Equation (8.26) is simply the sum of the multiplication of each pixel under the mask with the corresponding filter weight. Spatial filtering of an image is performed by scanning the image pixel by pixel and then using Equation (8.26) to perform the spatial filtering operation. Single point discontinuities are detected within a grayscale image using the 3×3 filter mask given in Figure 8.16. As stated in Chapter 7, the central pixel of this mask is given by the black dot. Point detection using this filter mask is given by the absolute value of the spatial filtering operation given in Equation (8.26):

$$\text{POINT}[x, y] = |g(x, y)| \; . \tag{8.27}$$

In homogeneous graylevel regions of an image, consisting of a constant graylevel, the output of the point detector reduces to zero. When a single pixel with a graylevel different from its neighboring pixels is spatially filtered with the mask of Figure 8.16, the output of Equation (8.27) will be nonzero, detecting the presence of a single point graylevel discontinuity.

−1	−1	−1
−1	•8	−1
−1	−1	−1

Figure 8.16: The filter mask used to detect single points within an image.

The output of this point detector is a graylevel value that is directly proportional to the graylevel difference between the center pixel and its surrounding eight neighbors. In many applications, it is desired to make a decision if a single point graylevel discontinuity has been detected within an image. Under these applications, the output of the point detector is thresholded at a given threshold value K. If

$$\text{POINT}[x, y] > K \; , \tag{8.28}$$

then a single point discontinuity is present. Otherwise, there is no point discontinuity present. Equation (8.28) sets the minimum graylevel difference between a single point discontinuity and its neighboring pixel before a single point is detected. Figure 8.17(a) shows an example image containing the words "IMAGE Processing". Figure 8.17(b) is its corresponding point detected image obtained using Equation (8.27). Notice how well the contours of the two words have been detected.

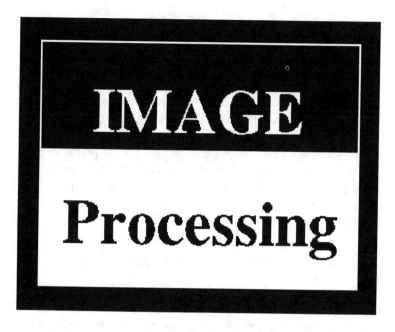

(a)

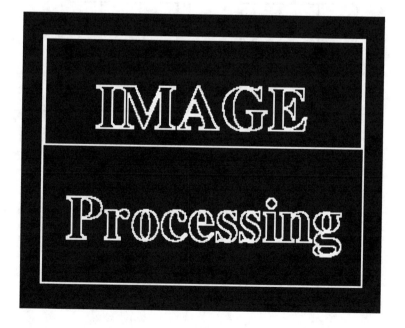

(b)

Figure 8.17: An example of using the point detector to detect graylevel discontinuities within an image.

Similar to the detection of point discontinuities is that of detecting line discontinuities. Figure 8.18 gives the four 3×3 spatial filter masks that detect horizontal and vertical lines and diagonal lines at 45° and 135°. The direction of the three weight factors of 2 gives the direction of the line detector. For example, Figure 8.18(a) detects horizontal lines, while Figure 8.18(c) detects diagonal lines at 45°. The detection of lines using these four masks is accomplished using Equation (8.26) and substituting this result into Equation (8.27). Consider the application of Equation (8.27) using the four line detection masks given in Figure 8.18 to a graylevel image $f(x, y)$ producing four output images L_1, L_2, L_3, and L_4, corresponding to vertical and horizontal lines and diagonal lines at 45° and 135°, respectively. The relative magnitude between these four line images describes the type of line that is present at a given pixel. If at a given coordinate position x, y

$$L_n(x, y) > L_m(x, y) \text{ for } n \neq m \text{ and } n, m = 1, 2, 3, 4 , \qquad (8.29)$$

then the pixel at this coordinate is best described as being associated with the line given by $L_n(x, y)$. For example, if $L_3(x, y)$ is the maximum of the four output images at x, y, then this pixel is best described as belonging to a diagonal line at 45°.

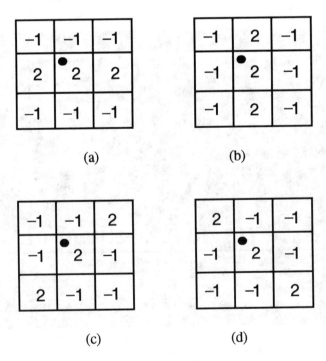

(a)   (b)

(c)   (d)

Figure 8.18: The four 3×3 filter masks to detect lines: (a) horizontal, (b) vertical, (c) 45° diagonal lines, and (d) 135° diagonal lines.

Figure 8.19(a) gives an example image containing horizontal and vertical lines as well as diagonal lines in both directions. Figure 8.19(b) is the corresponding horizontal line detected image. Note the presence of the horizontal lines and the lack of vertical lines. Diagonal lines are also detected by this horizontal line detection mask [from Figure 8.18(a)] but with a much lower graylevel value than a horizontal line as the result of diagonal lines containing a horizontal component. Figure 8.19(c) is the vertical line detected image, showing the detection of the vertical lines and the removal of the horizontal lines. As with the horizontal line detected image, the diagonal lines are also present as a result of diagonal lines containing a vertical component.

Figure 8.19(d) is the 45° diagonal line detected image, while Figure 8.19(e) is the 135° line detected image. Notice the lack of vertical and horizontal lines in both images. However, the 45° line detector also detected diagonal lines at 135°, detected as two lines but with one-third the amplitude of a 45° line. This is easily seen by considering that when a 135° line overlies the top-rightmost and the bottom-leftmost mask elements of Figure 8.18(c), it produces a nonzero pixel at these locations, thus producing two lines at 135° in the 45° line detected image. Figure 8.19(d) shows the detection of the 135° lines as sets of two thin lines also at 135°. Figure 8.19(e) is the 135° line detected image; the same reasoning describes why the 45° diagonal lines on the left side of this image are detected as two double lines.

(a)

Figure 8.19: Examples of line detection: (a) the original image.

(b)

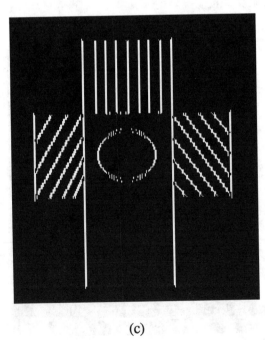

(c)

Figure 8.19: Examples of line detection: (b) the horizontal
line detected image and (c) the vertical line
detected image.

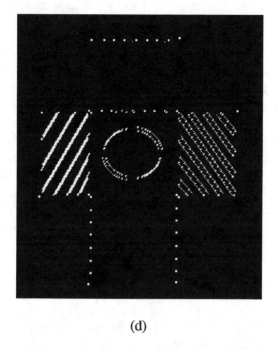

(d)

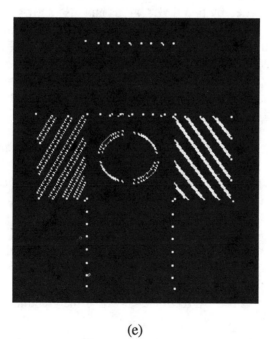

(e)

Figure 8.19: Examples of line detection: (d) the 45° diagonal
line detected image and (e) the 135° diagonal line
detected image.

The concept of using a set of filter masks as templates to extract lines in a given orientation can also be used to detect edges in a given direction. The eight filters masks given in Figure 8.20, which are known as the *Kirsch edge detector* masks, provide a set of templates to detect the presence of edges in the two horizontal, two vertical, and four diagonal directions. The direction of the 5's in each of these masks determines the orientation of the edge that is detected with them. The first four masks, in Figures 8.20(a)-(d), detect the presence of the top, bottom, right, and left edges, while the last four masks, given in Figures 8.20(e)-(g), detect the presence of diagonal edges at 45°, 135°, 225°, and 315°. These eight filter masks are used with Equation (8.26) eight times to produce eight template images. Next, for each pixel within the eight template images, the maximum value is determined and used as the output for the Kirsch edge detector. Like the single point and line detectors, the sum of each element from each of these masks also sums to zero. Hence, in constant graylevel regions, the output of Equation (8.26) for all of the filter masks is zero.

The presence of an edge within a grayscale image indicates that there is a change in the graylevels from one region to another. The derivative of the graylevel change within an image as a function of the x, y position provides a means of detecting the presence of an edge. Figure 8.21(a) shows a one-dimensional plot profiling the graylevel distribution within a 128×128 image at the vertical coordinate $y = 64$. This plot represents a white region in the center of the image surrounded by two dark regions and also shows the presence of two edges. Figure 8.21(b) shows the derivative in the x direction for this graylevel distribution with the height of the derivative proportional to the sharpness of the edge. The magnitude of this derivative can be used to detect the presence of edges. Figure 8.21(c) shows the relationship between the two edges in Figure 8.21(a) and the second derivative. For dark to light graylevel transitions, the sign of the second derivative goes from positive to negative, while for light to dark graylevel transitions the sign of the second derivative goes from negative to positive. The magnitude of the second derivative can also be used to detect the presence of an edge. Unlike the magnitude of the first derivative, which produces only one pulse per edge, the second derivative produces two pulses per edge. Of these two derivatives, the second is much more sensitive to the detection of edges. Unfortunately, it is also very sensitive to the presence of noise.

In two dimensions, the derivative edge detection example given in Figure 8.21 requires the computation of the derivative of the graylevel distribution in both the x and y directions. This forms a two-dimensional vector of derivative values in each of the x and y directions, known as the gradient:

$$\nabla f(x, y) = \begin{bmatrix} f_x(x, y) \\ f_y(x, y) \end{bmatrix} = \begin{bmatrix} \dfrac{\partial f(x, y)}{\partial x} \\ \dfrac{\partial f(x, y)}{\partial y} \end{bmatrix} . \tag{8.30}$$

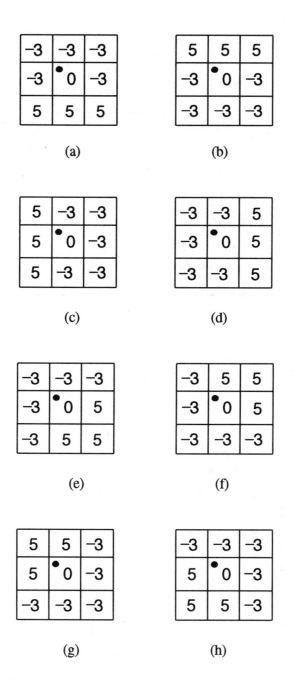

Figure 8.20: The eight Kirsch edge filter masks: (a) top, (b) bottom, (c) right vertical, (d) left vertical, (e) top-left diagonal, (f) bottom-left diagonal, (g) bottom-right diagonal, and (h) top-right diagonal edges.

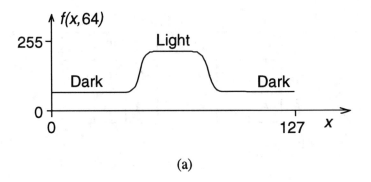

(a)

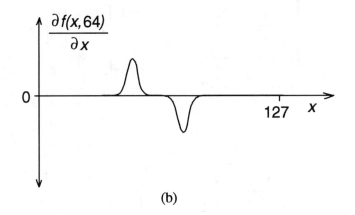

(b)

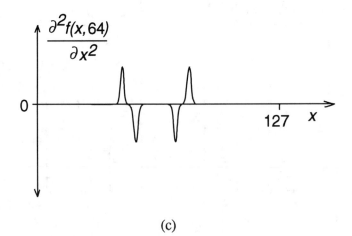

(c)

Figure 8.21: Edge detection using derivatives: (a) a plot of the
             graylevel distribution of a 128×128 image in the
             x direction for $y = 64$, (b) its first derivative,
             and (c) its second derivative.

Associated with the gradient is its magnitude,

$$\text{mag}[\nabla f(x, y)] = \sqrt{f_x(x, y)^2 + f_y(x, y)^2} \ , \tag{8.31}$$

and its direction,

$$\theta(x, y) = \tan^{-1}\left[\frac{f_y(x, y)}{f_x(x, y)}\right] , \tag{8.32}$$

given as an angle in radians or degrees relative to the x axis. In other words, an angle of $0°$ corresponds to the gradient vector in the positive x direction. To eliminate the requirement that the square root be used in the computation of the magnitude of the gradient, Equation (8.31) is typically approximated as

$$\text{mag}[\nabla f(x, y)] \approx |f_x(x, y)| + |f_y(x, y)| \ . \tag{8.33}$$

There are several different ways to approximate the gradient and its magnitude, given by Equations (8.30) and (8.31). Figure 8.22 shows the eight neighboring pixels surrounding the pixel at the coordinate x, y. The difference between graylevels of adjacent pixels can be used to approximate the derivative of the graylevel distribution in the x direction:

$$f_x(x, y) = f(x, y) - f(x - 1, y) . \tag{8.34}$$

Likewise, the approximation for the derivative of the graylevel distribution in the y direction can be written as

$$f_y(x, y) = f(x, y) - f(x, y - 1) . \tag{8.35}$$

Figure 8.23 shows the two 2×2 spatial filter masks that implement Equations (8.34) and (8.35). Once Equations (8.34) and (8.35) are computed for each pixel within the grayscale image, Equation (8.33) is used to compute the final gradient magnitude image, which corresponds to the gradient edge image.

$f(x-1, y-1)$	$f(x, y-1)$	$f(x+1, y-1)$
$f(x-1, y)$	$f(x, y)$	$f(x+1, y)$
$f(x-1, y+1)$	$f(x, y+1)$	$f(x+1, y+1)$

Figure 8.22: The eight neighboring pixels of $f(x, y)$.

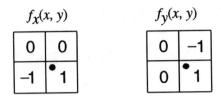

$$f_x(x, y) \qquad\qquad f_y(x, y)$$

Figure 8.23: The 2×2 spatial filter masks that implement
Equations (8.34) and (8.35).

Another common form for the x and y gradient components are the two cross
differences of

$$f_1(x, y) = f(x + 1, y + 1) - f(x, y) \tag{8.36}$$

and

$$f_2(x, y) = f(x, y + 1) - f(x + 1, y) . \tag{8.37}$$

Figure 8.24 gives the 2×2 spatial filtering masks used to compute the cross
differences as given in Equations (8.36) and (8.37). The final gradient edge
image is then given by

$$\text{mag}[\nabla f(x, y)] \approx |f_1(x, y)| + |f_2(x, y)| . \tag{8.38}$$

The output gradient edge image using Equations (8.36) and (8.37) is commonly
referred to as the Robert's cross operator or Robert's edge detected image.

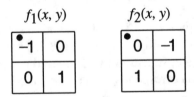

$$f_1(x, y) \qquad\qquad f_2(x, y)$$

Figure 8.24: The 2×2 Robert's spatial masks used to detect edges
within an image.

Figure 8.25 gives several other examples of spatial 3×3 filtering masks that
implement several types of gradient edge detectors, of which the most popular
are the Prewitt and Sobel gradient edge detectors. The Sobel edge filter provides
good edge detection and is somewhat insensitive to noise present within the
image. This is due to the averaging that is performed by this edge detector
during the computation of the gradient. For example, the Sobel x gradient mask
computes the graylevel differences in the x direction while averaging these

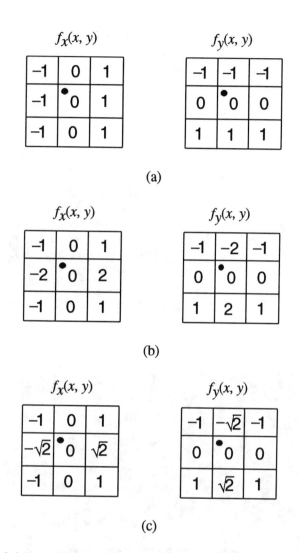

Figure 8.25: Example of several 3×3 gradient edge detector masks:
(a) the Prewitt gradient masks, (b) the Sobel gradient
masks, and (c) the Frei and Chen gradient masks.

differences in the y direction. Likewise, the Sobel y gradient mask computes the
graylevel differences in the y direction while performing averaging in the x
direction. The Frei and Chen gradient masks given in Figure 8.25 are a subset of
the nine template masks that were proposed by Frei and Chen. These two masks
give the gradient components in the x and y directions, respectively.

Figure 8.26(a) is an image of the Forth suspension bridge located in
Scotland. Figure 8.26(b) is the magnitude of the Sobel vertical gradient
$|f_y(x, y)|$. Notice the detection of the horizontal edges associated with the bridge
span and the horizontal edge of the shoreline located just below the bridge. Also

(a)

(b)

Figure 8.26: An example of using the Sobel gradient edge detector:
(a) the original grayscale image and (b) the magnitude
of the vertical Sobel gradient component $| f_y(x, y) |$.

(c)

(d)

Figure 8.26: An example of using the Sobel gradient edge detector:
(c) the magnitude of the horizontal Sobel gradient
component $|f_x(x, y)|$ and (d) the total Sobel gradient
edge image.

present in this image are the horizontal edges due to the clouds in the sky. Missing from this image are the vertical edges, which is noticeable in the support pillar of the bridge on the right-hand side of the figure. Figure 8.26(c) is the Sobel horizontal gradient $|f_x(x, y)|$ image. This image shows the vertical edges present in the original grayscale image. The vertical edges of the support pillars for the bridge that were missing in Figure 8.26(b) are now present, as are the vertical supports that run the length of the bridge. Missing from this image are the horizontal edges. In particular, the horizontal edges associated with the shoreline and the clouds are missing. Figure 8.26(d) is the total Sobel gradient edge image given by Equation (8.33). The complete edge information of the bridge is observable in this image, as well as the cloud structure of the sky.

Figure 8.21(c) illustrates the use of the second derivative as a means of detecting graylevel discontinuities within an image. The one-dimensional example given in Figure 8.21(c) can be easily expanded to two dimensions using the Laplacian operator

$$\nabla^2 f(x, y) = \frac{\partial^2 f(x, y)}{\partial x^2} + \frac{\partial^2 f(x, y)}{\partial y^2} . \qquad (8.39)$$

The discrete version of Equation (8.39) can be derived with the use of Figure 8.22, which describes the eight neighbors of the pixel at the coordinate x, y. The first partial derivative in the x direction can be approximated as

$$\frac{\partial f(x, y)}{\partial x} \approx f(x + 1, y) - f(x, y) \qquad (8.40)$$

and

$$\frac{\partial f(x, y)}{\partial x} \approx f(x, y) - f(x - 1, y) . \qquad (8.41)$$

Taking the difference between Equations (8.41) and (8.40) gives the discrete version of the second partial derivative of $f(x, y)$ in the x direction:

$$\frac{\partial^2 f(x, y)}{\partial x^2} \approx 2f(x, y) - f(x - 1, y) - f(x + 1, y) . \qquad (8.42)$$

The same approach can be used to derive the first and second partial derivatives of $f(x, y)$ in the y direction as

$$\frac{\partial f(x, y)}{\partial y} \approx f(x, y + 1) - f(x, y) , \qquad (8.43)$$

$$\frac{\partial f(x, y)}{\partial y} \approx f(x, y) - f(x, y - 1) \, , \tag{8.44}$$

and

$$\frac{\partial^2 f(x, y)}{\partial y^2} \approx 2f(x, y) - f(x, y - 1) - f(x, y + 1) \, . \tag{8.45}$$

Combining Equations (8.42) and (8.45) yields the Laplacian operator as

$$\frac{\partial f(x, y)}{\partial x} \approx 4f(x, y) - f(x - 1, y) - f(x + 1, y) - f(x, y - 1) - f(x, y + 1) \, . \tag{8.46}$$

Figure 8.27(a) gives the 3×3 spatial filter mask that implements Equation (8.46). The edge detection of a grayscale image using the Laplacian operator is accomplished by computing the Laplacian at every pixel within the image and then taking the magnitude of this result:

$$\text{LAPLACIAN}[f(x, y)] = |\nabla^2 f(x, y)| \, . \tag{8.47}$$

Another common Laplacian edge mask is derived from the Gaussian spatial filter mask given in Equation (4.21):

$$h(x, y) = e^{-\pi(x^2 + y^2)/a^2} \, . \tag{8.48}$$

The second derivatives of this filter in the x and y directions are

$$\frac{\partial h(x, y)}{\partial x} = \left(\frac{4\pi^2 x^2}{a^4} - \frac{2\pi}{a^2} \right) e^{-\pi(x^2 + y^2)/a^2} \, , \tag{8.49}$$

and

$$\frac{\partial h(x, y)}{\partial y} = \left(\frac{4\pi^2 y^2}{a^4} - \frac{2\pi}{a^2} \right) e^{-\pi(x^2 + y^2)/a^2} \, . \tag{8.50}$$

Substituting Equations (8.49) and (8.50) into Equation (8.39) gives the Laplacian filter as

$$\nabla^2 h(x, y) = \left(\frac{4\pi^2 (x^2 + y^2)}{a^4} - \frac{4\pi}{a^2} \right) e^{-\pi(x^2 + y^2)/a^2} \, . \tag{8.51}$$

Figure 8.27(b) gives the 5 × 5 Laplacian spatial filter mask computed from Equation (8.51) for $a = 2$. In using Equation (8.51) to generate a spatial filter mask, the size of the mask should be at least the next larger integer in size above $2.25a$ and the sum of all of its elements should add up to zero. In the generation of the 5 × 5 Laplacian filter mask given in Figure 8.27(b), the rounding of each element required that the center element be modified from $-78/25$ to $-80/25$ so that the sum of the filter elements added up to zero.

$\frac{1}{25}$	$\frac{5}{25}$	$\frac{7}{25}$	$\frac{5}{25}$	$\frac{1}{25}$
$\frac{5}{25}$	$\frac{9}{25}$	$\frac{-7}{25}$	$\frac{9}{25}$	$\frac{5}{25}$
$\frac{7}{25}$	$\frac{-7}{25}$	$\frac{-80}{25}$	$\frac{-7}{25}$	$\frac{7}{25}$
$\frac{5}{25}$	$\frac{9}{25}$	$\frac{-7}{25}$	$\frac{9}{25}$	$\frac{5}{25}$
$\frac{1}{25}$	$\frac{5}{25}$	$\frac{7}{25}$	$\frac{5}{25}$	$\frac{1}{25}$

0	−1	0
−1	4	−1
0	−1	0

(a)   (b)

Figure 8.27: The Laplacian filter masks: (a) the 3 × 3 mask as defined by Equation (8.46) and (b) the 5 × 5 mask as defined by Equation (8.52).

Figure 8.28(a) is an image of an old building against a mountain peak in the Swiss Alps taken during midafternoon. Figure 8.28(b) is the Laplacian edge detected image obtained using the 5 × 5 Laplacian spatial filter mask given in Figure 8.27(b). Both vertical and horizontal edges are present in this image. In particular, notice how well the edges of the bricks defining the wall of the building on the far left have been preserved during the Laplacian edge detection. The Laplacian edge detector did not perform as well within the mountain regions of the grayscale image. The surface texture of the mountain regions appear as noise in the edge detected image. Even though the Laplacian edge detector did a reasonable job at edge detection of the building in Figure 8.28, the Laplacian filter is seldom used as an edge detector due to its poor response in the presence of noise. Since the Laplacian edge detector is derived from the second derivative of the graylevel distribution of an image, any graylevel variations that are due to noise are easily detected by this edge detector. The Laplacian edge filter is usually used in conjunction with another edge filter such as the Sobel to determine the direction of the graylevel transition, such as dark to light or light to dark.

(a)

(b)

Figure 8.28: An example of using the Laplacian operator for
edge detection: (a) the original grayscale image and
(b) the Laplacian edge detected image using the 5×5
filter mask given in Figure 8.27(b).

Edge detection can also be accomplished in the Fourier frequency domain using a highpass spatial filter. The Fourier transform components $F(n, m)$ of the grayscale image $f(x, y)$ are first obtained and then filtered using a highpass filter $H(n, m)$ to yield

$$G(n, m) = F(n, m) \cdot H(n, m) \ . \tag{8.52}$$

Finally, the inverse Fourier transform is applied to the filtered frequency components to produce the edge detected image. Described in Chapter 4 was the circularly symmetric Butterworth highpass filter:

$$H(\rho) = \frac{1}{\sqrt{1 + \left(\dfrac{\omega_c}{\rho}\right)^{2N}}} \ , \tag{8.53}$$

where $\rho^2 = n^2 + m^2$. Since this filter is a highpass filter, it can be used as an edge detector. Figure 8.29 is the edge detected image of Figure 8.26(a) using this highpass filter with $N = 4$ and $\omega_c = 60$. Even though the edges are not as pronounced as with the Sobel edge detection of this image, given in Figure 8.26(d), the Butterworth highpass filter did a reasonable job detecting both the vertical and horizontal edges of the bridge. It also should be noted that increasing the cutoff frequency of this filter beyond 60 decreased the amplitude of the edges associated with the vertical supports of the bridge, making them barely observable.

Figure 8.29: The edge image of Figure 8.26(a) obtained using the
Butterworth highpass filter given in Equation (8.53).

As mentioned previously, the process of edge detection is usually followed by the thresholding of the edge detected image. The edge detected image contains a set of graylevels that are proportional to the slope of the graylevel discontinuities present. The process of thresholding an edge image provides a means of separating weak edges from strong edges. Let EDGE(x, y) define an edge image. The thresholding of this image yields a new image that contains only the strong edges:

$$g_e(x, y) = \begin{cases} 0 & \text{for EDGE}(x, y) \leq K \\ 1 & \text{for EDGE}(x, y) > K \end{cases}, \tag{8.54}$$

where the strength of the edges that are present in the threshold edge image is determined by the threshold value K. The larger the threshold value K, the larger the strength of the edge must be before an output of 1 occurs.

Another commonly used edge detector is based upon image thresholding. This edge detector is a two pass edge detection algorithm that locates edges separately in the x and y directions in the first pass and then combines these detected edges together in the second pass to form a composite edge detected image. Consider two pixels $f(x, y)$ and $f(x - 1, y)$ that are adjacent horizontal neighbors. If these pixels have very dissimilar graylevels, then there is a good probability that they are separated by a graylevel discontinuity. On the other hand, if they have very similar graylevels, then it is highly likely that they belong in the same region, and hence the probability is high that an edge does not separate these two pixels. Equation (8.55) gives the decision process to detect the presence of a vertical edge using $f(x, y)$ and its adjacent horizontal neighbor $f(x - 1, y)$

$$g_v(x, y) = \begin{cases} 1 & \text{for } f(x, y) \leq K \text{ and } f(x - 1, y) > K \\ & \text{for } f(x, y) > K \text{ and } f(x - 1, y) \leq K \\ 0 & \text{otherwise} \end{cases}, \tag{8.55}$$

where K is the threshold value that separates the two different graylevels. The same decision process given in Equation (8.55) can be used to detect horizontal edges:

$$g_h(x, y) = \begin{cases} 1 & \text{for } f(x, y) \leq K \text{ and } f(x, y - 1) > K \\ & \text{for } f(x, y) > K \text{ and } f(x, y - 1) \leq K \\ 0 & \text{otherwise} \end{cases}. \tag{8.56}$$

Equations (8.55) and (8.56) implement the first pass of this edge detector, locating edges in both the vertical and horizontal directions. The second pass combines Equations (8.55) and (8.56) to form one composite edge detected image:

$$g_e(x, y) = \begin{cases} 1 & \text{if } g_v(x, y) = 1 \text{ or } g_h(x, y) = 1 \\ 0 & \text{otherwise} \end{cases} \quad . \tag{8.57}$$

After edges have been detected in an image, using any one of the edge detectors that have been mentioned in this section, it is often desired to trace the edges within the image either for object description or edge description. For example, to compute the perimeter of an object, its contour must be traced so that its length can be determined. A simple algorithm based upon the magnitude of the edges that have been detected can be used. Consider an image $f_e(x, y)$ that has been edge detected using any one of the gradient edge detectors mentioned in this section. The edge image is a grayscale image in which the graylevels represent the strength of the edge. If the graylevels are above a given threshold value K,

$$f_e(x, y) > K \ , \tag{8.58}$$

then these edges are strong enough to be considered for tracing.

The edge tracing algorithm begins by locating the starting pixel for the first edge to be traced. This is accomplished by scanning the edge image and locating the pixel that satisfies Equation (8.58) and that also has the highest graylevel value. If more than one edge pixel is present with this magnitude, then the starting point is picked arbitrarily from these pixels. Clearly, choosing the starting point in this way guarantees the edge tracing algorithm is starting with the strongest edge. Once the starting pixel has been located, its eight neighbors are examined to determine the next pixel of the edge to be traced. Only eight neighbor pixels that have not been previously traced are considered. If more than one edge pixel is present within the eight neighboring pixels, then the selection of this next edge pixel can be based upon the similarity between the gradient magnitude of the present pixel on the edge and its eight neighbors:

$$\min[|\, f_e(x + i, y + j) - f_e(x, y)\,|] < K_m \ , \tag{8.59}$$

where $f_e(x + i, y + j)$ is one of the eight neighbors of $f_e(x, y)$. The threshold value in Equation (8.59) also guarantees that the next edge pixel to be traced is similar enough in magnitude to be considered as the same edge. Once the next edge pixel has been determined, the previously traced edge pixel is removed from the edge image. This prevents edge pixels located on a closed contour from being traced twice. This edge tracing process continues until all the pixels associated with the edge have been traced or the starting point for the edge has been reached, defining a closed contour. The process of tracing edges continues until all edges have been traced within the image.

Figure 8.30(a) gives an example of a grayscale image of four arrows. Figure 8.30(b) shows the first edge traced by the edge tracing algorithm with $K = 90$, and $K_m = 156$. The starting point for this edge was located in the upper left-

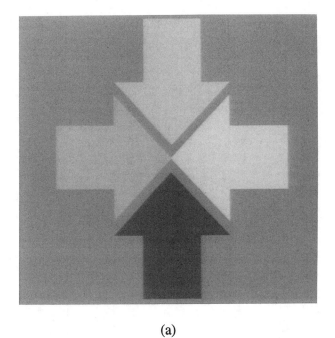

(a)

(b)

Figure 8.30: An example of edge tracing: (a) the original grayscale
image and (b) the edge traced image showing the
first edge traced.

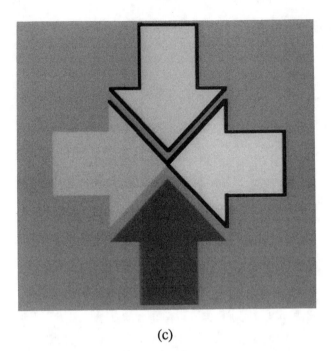

(c)

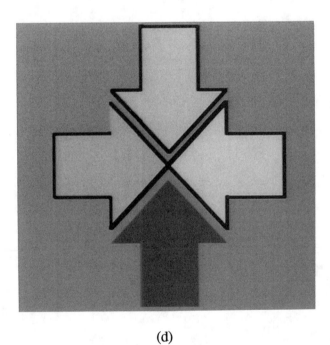

(d)

Figure 8.30: An example of edge tracing: the edge traced images
showing (c) the second edge traced and (d) the
third edge traced.

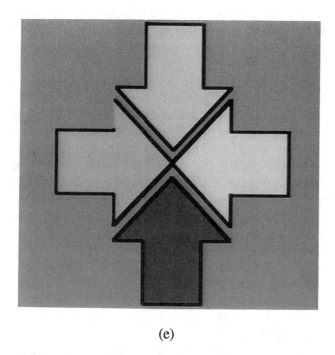

(e)

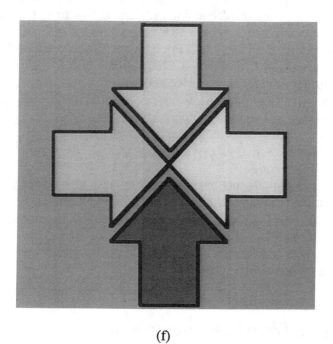

(f)

Figure 8.30: An example of edge tracing: the edge traced image
showing (e) the fourth edge traced and (f) the fifth
and final edge traced.

hand corner of the top arrow. When the complete closed contour of the top arrow was located, the tracing of this edge discontinued. The edge was then tagged as being traced and the edge image was then scanned to find the next nontagged edge that contained the highest magnitude to be traced. The next starting point was located at the top corner of the left arrow. Figure 8.30(c) shows the tracing of this edge. Notice how the edge tracing algorithm did not trace the contour of the left arrow but continued straight at the point where the left arrow intersected the right arrow. At this point, there were several edge points to choose from to be traced next. The edge algorithm chose to follow the straight contour. The tracing continued until the closed contour of the right arrow was traced. Figure 8.30(d) shows the tracing of the next edge. Unfortunately, the start location of this traced edge was at a different location than the start point of the previously traced edge. Tracing of this third edge continued until the intersection of the left and right arrows were reached. At this point, there were no more contour pixels to trace. This left a small vertical line segment of the left arrow to be defined as a separate edge and to be traced later by the algorithm. Figure 8.30(e) shows the tracing of the lower arrow, leaving only the small vertical edge associated with the left arrow to be traced. Figure 8.30(f) shows the tracing of all of the edges. A total of five edges were located and traced in this example.

Care must be taken in using the simple edge tracing algorithm presented here. How the edges are divided and traced depends on the initial starting points as well as the thickness of the edges. This algorithm has difficulty tracing edges that are more than 1 pixel in width. For example, if the width of the edge to be traced is wide enough to produce a small homogeneous region (3×3), this algorithm will trace the region as a closed edge and terminate tracing of the edge at this location. This problem is especially true at corners, where small homogeneous regions can exist in edges with widths wider than one. For more advanced edge tracing algorithms based upon graph theoretic methods, the interested reader is referred to *Digital Image Processing Algorithms* by I. Pitas given in the bibliography.

Not mentioned in this section were the nonlinear edge detectors based on the order statistic filters (range filter) and those that are derived from the several nonlinear mean filters given in Chapter 5, as well as the morphological edge detectors given in the last chapter. These edge detectors are just as useful as the ones that have been mentioned here and should not be overlooked in determining what edge detector should be used.

8.3 Region Based Segmentation

Another method of image segmentation is to consider the similarity of neighboring pixels so that pixels can be grouped together to form regions in which there is a common factor between each pixel within the region. The basic concept behind region segmentation is that an image can be divided into smaller nonoverlapping homogeneous regions. Consider the spatial decomposition of an image $f(x, y)$ defined into N smaller nonoverlapping regions $R_1, R_2 \cdots R_i \cdots R_N$, as shown in Figure 8.31 for 4 separate regions. The union of each of these regions must be equal to the spatial domain R_T of the original image $f(x, y)$:

$$R_T = \bigcup_{i=1}^{n} R_i \quad . \tag{8.60}$$

Since the regions are to be non overlapping, the intersection of each region must be the null set

$$\varnothing = R_i \cap R_j \quad \text{for } i \neq j . \tag{8.61}$$

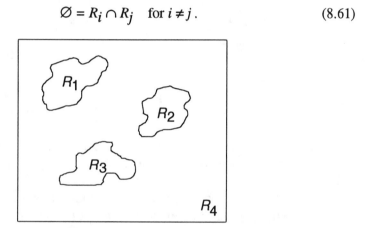

Figure 8.31: The spatial decomposition of an image into smaller regions.

The process of dividing an image into smaller regions is based upon some predetermined rule on how to group the pixels into one logical region. Typically, this rule is based upon a logical predicate $P(R)$. For example, all pixels within the region must be exactly the same graylevel. Another example might be that all pixels greater than a given threshold are combined together to form one region. Hence, the predicate is true for all pixels within the region R_i:

$$P(R_i) = \text{true} \quad \text{for } i = 1, 2, \cdots N \quad . \tag{8.62}$$

In other words, all pixels within a region follow the same logical predicate rule. To prevent the decomposition of an image into smaller regions than necessary, the predicate of the union of two different regions should be false:

$$P(R_i \cup R_j) = \text{false} \quad \text{for } i \neq j \quad . \tag{8.63}$$

This guarantees that each region within the image corresponds to a different homogeneous area with a different predicate that is true.

Only pixels that are connected can be combined to form one region R_i. Two pixels P_1 and P_2 can be considered connected if a straight line can be drawn between them and if for every pixel on this line the same logical predicate $P(R_i)$ is true. In addition, the pixels defining this straight line form a sequence of adjacent neighbors from P_1 to P_2. Figure 8.32(a) shows an example of two nonconnected pixels in separate regions, while Figure 8.32(b) shows two connected pixels that form one region.

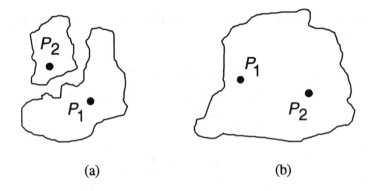

(a)   (b)

Figure 8.32: An example of (a) two nonconnected pixels and (b) two connected pixels.

The first of the region segmentation methods, known as region growing by pixel aggregation, is based upon the selection of a seed pixel, followed by the determination of whether adjacent neighboring pixels belong to the same region as the seed pixel. The process starts with one pixel and continues to grow until no more pixels can be added to the region. The algorithm starts with the selection of the seed pixel and of the desired predicate $P(R_1)$ to compare neighboring pixels with. This initial seed pixel is then assigned to region R_1. Next, the eight neighboring pixels of the seed pixel are assigned to region R_1 if these pixels meet the predicate requirement. The process of comparing the eight neighboring pixels continues until no further pixels can be added to the region. At this point, region R_1 has been completely defined. Pixels belonging to this region are removed from the original image to prevent further classification of them. A new seed pixel is located within another area of the image and the region growing procedure is repeated until all pixels defining region R_2 are found. This process of placing a seed pixel and growing a region continues until all regions have been defined.

There are several methods to automatically determine the seed pixels within an image. The first is to scan the image until the first pixel that meets the desired predicate is found. Another method is to scan the image pixel by pixel and to use the first pixel found that has either the minimum or maximum graylevel. After the seed pixel is located, the region growing process must search for all connected pixels. There have been recursive algorithms proposed that perform this task, but a simpler method is to scan all pixels from the seed in increasing order of either 4-distance r_4 or 8-distance r_8 as defined in Chapter 1. As shown in Figure 1.23, using a constant 4-distance compares adjacent pixels following a diamond shape, while a constant 8-distance compares adjacent neighboring pixels using a square shape. This approach guarantees that each pixel checked during each iteration of increasing distances will always be an eight neighbor to a previously defined pixel. The process of increasing the distance from the seed pixel to determine whether unknown pixels belong to the same region as the seed pixel continues until no further pixels can be assigned to the region.

Figure 8.33(a) is a graylevel map of the United States with the regions of the states defined by a graylevel of 199 and the state boundaries with a graylevel of 0. The goal is to perform region segmentation using the region growing by pixel aggregation algorithm. A seed pixel was located at the pixel coordinate of 251, 265, which corresponds to the location of Oklahoma City, the capital of the state of Oklahoma. The growing process is based upon the logical predicate of grouping all pixels that have the same graylevel as the initial seed pixel into one common region. The 4-distance measure was used to locate and assign pixels to the same region as the seed pixel. The graylevels of the immediate four-neighbors of this seed pixel ($r_4 = 1$) were compared to the seed pixel's graylevel, and those pixels having the same graylevel as the seed pixel were assigned to the same region as the seed pixel. All pixels belonging to this region were given a graylevel of 70 to separate them from the other pixels present in the image.

The 4-distance was then increased to two and the graylevels of those pixels with this distance were compared to the seed pixel's graylevel value. If at least one of its eight neighbors was previously assigned to the same region as the seed pixel and its graylevel is the same as the seed pixel's graylevel, this pixel is then assigned to the same region as the seed pixel. The requirement that at least one of its eight neighbors belongs to the region guarantees that all pixels are connected within the region. Figure 8.33(b) shows the region growing process after 7 iterations for 4-distance scans from 1 to 7. Notice the diamond shape of the region due to using the 4-distance as a means of locating and assigning pixels to it. At this point in the region growing process, if the 8-distance was used to scan the pixels, the shape of this region would be square instead of the diamond shape shown in Figure 8.33(b). Figure 8.33(c) shows the region growing process after 25 iterations. The diamond shaped region of Figure 8.33(b) has continued to grow, but the top has reached the state boundary and has stopped growing. Figure 8.33(d) is the region segmentation of the state of Oklahoma after self-termination of the growing process after 60 iterations.

Another method of region segmentation is based upon the dividing of an image into smaller and smaller regions until all the pixels within each region

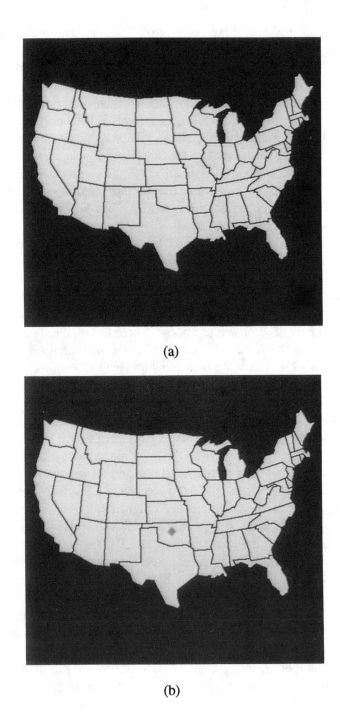

(a)

(b)

Figure 8.33: Region segmentation using region growing by pixel aggregation: (a) a map of the United States and (b) the region segmentation of the state of Oklahoma after 7 iterations (images, © New Vision Technologies and adapted from Gonzalez and Woods, 1992).

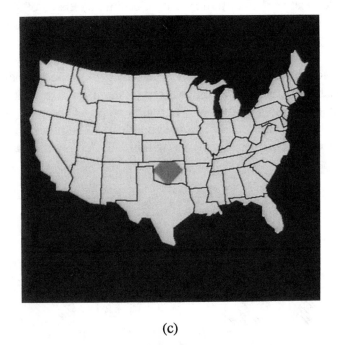

(c)

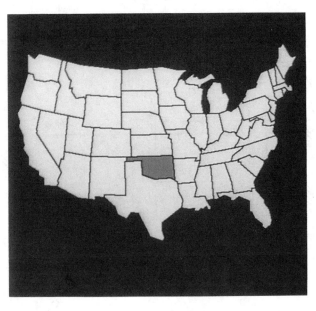

(d)

Figure 8.33: Region segmentation using region growing by pixel aggregation:
the region segmentation of the state of Oklahoma after (c) 25
and (d) 60 iterations (images, © New Vision Technologies
and adapted from Gonzalez and Woods, 1992).

satisfy the predicate $P(R_i)$ for that region. Consider an image of size $N \times M$ that is divided into four equally sized regions of $N/2 \times M/2$. For each of the four regions, every pixel is scanned to determine if it meets the logical predicate of the region. For every region in which every pixel does not satisfy the predicate, these $N/2 \times M/2$ regions are again subdivided by 2 to produce 4 new $N/4 \times M/4$ regions. Each new region is again checked to see if its pixels satisfy the given predicate. If so, the subdividing of this region stops. Otherwise, all regions in which the predicate fails are once more subdivided. The process of subdividing the image stops when all the pixels within all the regions produce a true predicate.

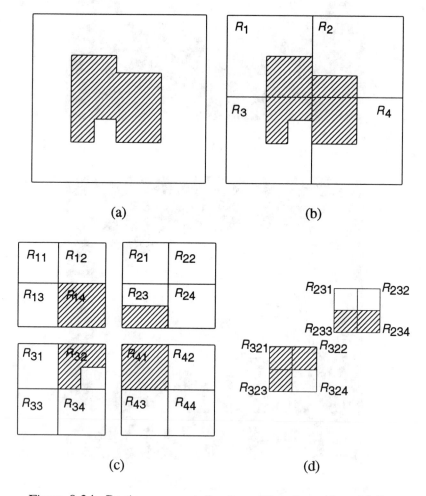

(a)   (b)

(c)   (d)

Figure 8.34: Region segmentation by split and merging: (a) the original image, (b) initial subdivision of the image into four regions, (c) the division of the image into sixteen regions, and (d) the further subdivision of regions R_{23} and R_{32}.

Figure 8.34(a) shows an example of a simple rectangular shaped object. Figure 8.34(b) shows the initial subdivision of the image into four equally spaced regions R_1, R_2, R_3, and R_4. Using a logical predicate that the graylevel of each region should be a constant homogeneous value shows that each of these four regions need to be further subdivided into four smaller regions. Figure 8.34(c) shows the subdivision of each region into four smaller regions. There are now a total of sixteen nonoverlapping regions that comprise the original image. Inspection of regions R_{23} and R_{32} shows that these two regions need to be further subdivided into four equally sized regions, as shown in Figure 8.34(d). At this point, each region contains a constant homogeneous graylevel and the splitting process ends.

Figure 8.35 shows a tree representation known as *quadtrees* that gives the region division of the image shown in Figure 8.34(a). At the top of the tree is the original image $f(x, y)$. At the next level is the initial division of the image into four nonoverlapping regions. The next level below this is the subdivision of these four regions into four smaller regions, decomposing the image into sixteen regions. The final level of the quadtree shows the subdivision of regions R_{23} and R_{32} into four smaller regions.

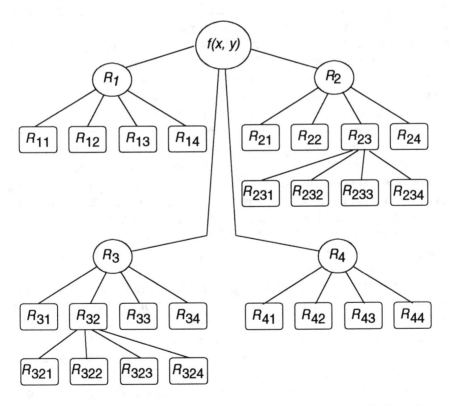

Figure 8.35: The quadtree representation of the region splitting of Figure 8.34(a).

After the image has been split into regions that meet the logical predicate requirement, regions that are connected and have the same predicate requirement are merged to form a new enlarged region. The process of merging connected regions continues until no further merging is possible. The process of splitting an image into smaller regions and then merging connected regions together is called region segmentation by *splitting and merging*.

Figure 8.36(a) is an image of twelve squares, some of which are at different graylevels, that together form a large square object. The initial splitting of these squares based upon the predicate that the graylevel within each region must be a constant separates each one of 12 squares into different regions. For the merging process, the logical predicate was selected so that connected regions with graylevel differences of less than 50 would be merged together to form one region. Figure 8.36(b) is the merged image showing the five different regions that remain after the merging process, where each region has been assigned a unique graylevel for illustration purposes. It is interesting to note that the four squares in the lower left-hand side of the image have been merged into one big square and the three squares on the right side of the object have also been merged into one long vertical rectangle. The most difficult task in the split and merge algorithm is the selection of the splitting and merging predicates. Once these predicates are found, the difficult part of the splitting and merging process is keeping track of all the different nonoverlapping regions. Usually, a recursive algorithm with a link list data structure is used to perform the splitting process. This algorithm continues to call itself until no further splitting can be accomplished. The merging algorithm is also a recursive algorithm that calls itself until no further connected regions can be combined.

A common method of region segmentation of binary images is based upon a process called *connected component labeling*. Connected component labeling involves the grouping into regions of nonzero pixels that are connected. The process begins by scanning a binary image for the first nonzero pixel and assigning this pixel to region 1. Next, a search algorithm is performed to locate all pixels connected to this initial pixel so that these pixels can be assigned to the same region. All labeled pixels are removed from the image and the process of finding another nonzero pixel and its connected neighbors is repeated so that the next region can be labeled. This process repeats until all nonzero pixels have been assigned to a labeled region.

The region growing by pixel aggregation is one of the many algorithms that can be used to perform connected component labeling of a binary image. Instead of placing the seed manually within the image, the image is scanned until the first nonzero pixel is located. This pixel becomes the seed for the growing algorithm, which is initiated to locate all of the pixels connected to the seed pixel. Once the growing algorithm is terminated, all pixels that have been assigned to the region given by the seed pixel are labeled as region 1. All pixels belonging to this region are removed from the binary image and the next seed location is found by scanning the binary image a second time until the first

(a)

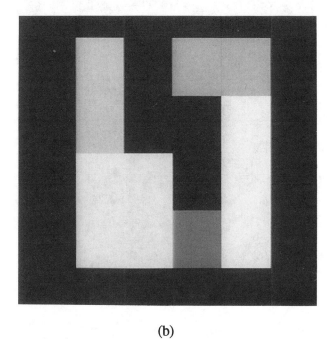

(b)

Figure 8.36: An example of region segmentation using the split-merge algorithm: (a) the original grayscale image and (b) the segmented image.

nonzero pixel is located. This second seed pixel becomes the first member of region 2. The growing process is initiated and all connected pixels are assigned to this second region. These labeled pixels are then removed from the binary image and the process of locating a new seed pixel and its connected pixels repeats. The locating of a seed pixel followed by region growing terminates when all nonzero pixels within the image have been assigned to a region.

Figure 8.37(a) shows a binary image of a clock face. There are a total of 17 unconnected regions present within this image. Using the region growing method for component labeling, the first region that was labeled corresponded to the outer ring of the clock face. With this outer ring removed from the binary image, the next two regions that were labeled were the numbers one and two associated with the number twelve on the clock face. Figure 8.37(b) gives the first ten regions that were labeled using connected component labeling. Each region is shown using a different graylevel as a means of labeling each region. The first region is assigned a graylevel of 255 and each successive region is assigned a graylevel decreased by a factor of 11. Figure 8.37(c) gives the total labeled image, showing all 17 unconnected regions. Notice how the number six has the lowest graylevel of all the regions. This was due to this region being labeled last and hence being assigned the lowest graylevel.

(a)

Figure 8.37: An example of component labeling: (a) the original binary image of a clock face (images, © New Vision Technologies).

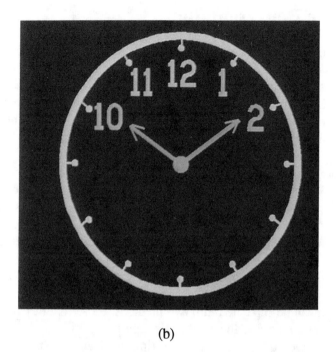

(b)

(c)

Figure 8.37: An example of component labeling: (b) the labeling of the first 10 objects and (c) the labeling of all 17 objects (images, © New Vision Technologies).

8.4 Image Representation

The process of image segmentation divides an image into separate regions so that objects present within an image can be identified. As presented in the last three sections, an image can be segmented using thresholding, edge, or region techniques. Once an image has been segmented, it is often desirable to represent an object using other than the coordinates of the pixels defining the object. For many image recognition algorithms, the description of an object's shape is all that is needed to recognize one object from another.   For example, the discrimination of a rectangular object from a circular object is easily accomplished from the shapes of the object contours. In other instances, an object is best described by the pixels composing the object's region.   For example, an object with a smooth reflective surface is easily separated from an object with a rough surface by comparing their surface textures.   In the processing of fruit using grayscale images, apples can be easily separated from oranges based upon their textures. These examples illustrate the importance of region and contour methods of describing objects within an image.   Image representation can be separated into two methods, those that describe an object's contour and those that describe an object's region.

The importance of edges in the description of an object was illustrated in Chapter 7 during the discussion of morphological filters. The edges of an object immediately give its size and shape. The use of edges to define an object also gives a compressed means of describing it, as shown in the skeletonization of an object given in Chapter 7.   Once the edges associated with an object are determined, there are several means of describing these edges. The first is based upon an orientation code that describes the tracing of the edge from its beginning point to its ending point through a set of connected pixels. Figure 8.38 shows two orientation codes known as *chain codes* that are commonly used. As an edge is traced from its beginning point to its ending point, the direction that must be taken to move from one pixel to the next is given by the number present in either the 4-*chain code* or 8-*chain code*. An edge can be completely described in terms of its starting coordinate and its sequence of chain code descriptors.

Of the two chain codes, the 4-chain code is the simplest requiring only four different code values. Consider an edge pixel at the location x, y. If the next connected pixel is located at $x, y - 1$, the 4-chain code description of this movement is given as a 1. Likewise, if the next adjacent pixel of the edge is given by $x - 1, y$, this direction is given by 2. The 4-chain code describes the tracing of contour using only vertical and horizontal descriptors. On the other hand, the 8-chain code of a contour as it is traced from x, y to $x, y - 1$ yields the code value of 6. The difference between the 4-chain code and the 8-chain code is that the 8-chain code allows an edge to be described using vertical and horizontal descriptors as well as diagonal descriptors at 45°, 135°, 225°, and 315°. The advantage of the 4-chain code description of an edge over the 8-chain code description is that it is more compact, requiring only two bits per edge tracing direction compared to three bits per edge tracing direction. The 4-chain code description of an edge is used when storage is an issue and an edge can be satisfactorily described in terms of its vertical and horizontal directions. Usually though, the 8-chain code is used because of its better edge description.

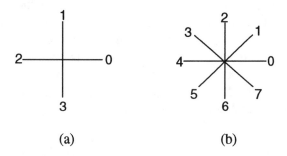

(a)   (b)

Figure 8.38: Chain code definitions: (a) 4-chain code and (b) 8-chain code.

Figure 8.39 shows an example of an edge containing 17 pixels, with the beginning pixel marked with the letter "B" and the ending pixel marked with the letter "E". Using the 8-chain code description for this edge and starting at the pixel identified with the letter "B", the direction of the next connected pixel is oriented in the direction given by the value of 7. This value becomes the first value in the chain code description. Moving to this connected pixel and determining the direction of the next connected pixel along the edge also yields a chain code value of 7. Again moving to this connected pixel and then determining its adjacent pixel produces the next chain code value of 0. This process continues until every pixel within the edge has been traced. The final chain coded for the edge given in Figure 8.39 is

$$7, 7, 0, 1, 1, 2, 1, 1, 1, 7, 0, 1, 7, 0, 7, 0 \ . \tag{8.64}$$

The implementation of either the 4-chain code or the 8-chain code requires the tracing of the edge from its beginning pixel to its ending pixel. This can be accomplished using the edge following algorithm given in the previous section. As the edge is traced, the values given by Figure (8.38) are stored in an array representing the edge. These array values and the initial starting point completely define the edge.

Figure 8.39: An example of an edge.

Because an image is composed of finite spatially sampled pixels, during the edge detection process it is not uncommon to find pixels shifted over by one position, as shown in the top horizontal line of Figure 8.39, which should be a straight line. This line has been distorted by two pixels shifted up by one pixel position. Noise pixels like these can drastically change the final chain code description of the edge, making it difficult to use the chain code description of an edge as a means of object recognition. A simple method of smoothing an edge is to resample the edge using a lower resolution. Figure 8.40(a) is the edge of Figure 8.39 placed against a grid that is half the resolution of the original edge pixels. Figure 8.40(b) is the new lower resolution edge. Starting at the beginning pixel given by "B", the 8-chain code for this edge is

$$7, 1, 2, 0, 2, 0, 0, 0, 7 \ . \tag{8.65}$$

Reducing the spatial resolution of the edge has smoothed the edge removing the two noise pixels from the top horizontal line and has reduced the number of chain-code descriptors required to define this edge.

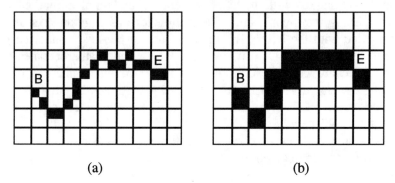

(a)   (b)

Figure 8.40: Reducing the resolution of an edge for chain coded description.

Another method of describing the contour of an object is through the use of the one-dimensional discrete Fourier transform. Each pixel on an edge can be defined by its x_j and y_j coordinate values. Given an edge with N pixels, these two coordinate values can be used to generate a complex function of the form

$$f(r) = x_r + j\, y_r \quad \text{for } r = 0, 1, 2, \cdots N - 1 \ . \tag{8.66}$$

The Fourier transform of this function yields the frequency components that describe the edge. One of the advantages of describing an edge using Equation (8.66) is that it reduces the edge description problem from two dimensions to one dimension. The discrete one-dimensional Fourier transform of $f(r)$ is

$$F(n) \ = \ \frac{1}{N} \sum_{r=0}^{N-1} f(r) \cdot e^{-j2\pi nr/N} \ , \tag{8.67}$$

and its inverse is

$$f(r) = \sum_{n=0}^{N-1} F(n) \cdot e^{j2\pi nr/N} \; . \tag{8.68}$$

The Fourier components computed from Equation (8.67) give several interesting properties about the contour. For example, $F(0)$ yields the geometrical centroid of the contour, with the x value given by the real part of $F(n)$ and the y centroid given by the imaginary part of $F(n)$. This property is easily shown by substituting Equation (8.66) into Equation (8.67) and setting $n = 0$. The real part of $F(n)$ becomes

$$x_{centroid} = \frac{1}{N} \sum_{r=0}^{N-1} x_r \; , \tag{8.69}$$

and the imaginary part of $F(n)$ becomes

$$y_{centroid} = \frac{1}{N} \sum_{r=0}^{N-1} y_r \; . \tag{8.70}$$

The translation of the edge by x_o, y_o to a new location,

$$f(r) = (x_r + x_o) + j(y_r + x_o) \quad \text{for } r = 0, 1, 2, \cdots N-1 \; , \tag{8.71}$$

changes only the first Fourier coefficient $F(0)$; all other coefficients remain unchanged. For a closed contour, the changing of the starting point of the contour simply results in a translation of $f(r)$ to $f(r - r_o)$. The two-dimensional Fourier property of Equation (2.55) can be rewritten in one dimension and results in a change of the starting point of the contour:

$$e^{-j2r_o n/N} F(n) \; . \tag{8.72}$$

Equation (8.72) describes the modulation of the Fourier coefficients as a result of changing the starting point of the contour.

Changing the size of an object by scaling the image results in the scaling of the magnitude components. Given the original edge image $f(x, y)$, which produces real and imaginary Fourier components of $F(n) = Re(n) + j \, Im(n)$, the scaling of this edge image by $f(ax, by)$ produces Fourier components $F_s(n)$ of

$$F_s(n) = a \cdot Re(n) + j \, b \cdot Im(n) \; . \tag{8.73}$$

Finally, rotating an object by θ is equivalent to writing the rotated version of $f(r)$ in polar form as

$$f_r(r) = \sqrt{x_r^2 + y_r^2}\, e^{j(\alpha_r + \theta)} \quad , \tag{8.74}$$

where α_r is defined as

$$\alpha_r = \tan^{-1}\!\left(\frac{y_r}{x_r}\right) . \tag{8.75}$$

The Fourier transform of Equation (8.74) yields Fourier coefficients that are phase shifted by the same angle θ:

$$F_r(n) = \sqrt{Re(n)^2 + Im(n)^2}\, e^{j(\beta(n) + \theta)} \quad , \tag{8.76}$$

where $\beta(n)$ is defined as

$$\beta(n) = \tan^{-1}\!\left(\frac{Im(n)}{Re(n)}\right) . \tag{8.77}$$

One common use of the Fourier components of a contour is in the number of elements required to define the object as compared to using the coordinates of every pixel on its contour. Consider the image of the capital letter "E" shown in Figure 8.41(a). This letter was scaled so that there were exactly 1024 edge pixels so that the fast Fourier transform algorithm could be used instead of Equation (8.67). If the number of edge points N is not equal to a power of 2, then $f(r)$ is padded with zeros to produce a new function in which the number of elements is equal to a power of 2. The reconstruction of the edge is then simply the first N unpadded values of $f(r)$. Figure 8.41(b) is the reconstruction of the letter "E" obtained by setting all of the Fourier components of the edge $F(n)$ as computed from the FFT to zero except the first component $F(0)$, which gives the centroid of the letter and the first positive and negative frequency components $F(1)$ and $F(-1)$. Eliminating all but three components completely destroyed the information contained in the edge pertaining to the shape of the object. In fact, truncating all of the Fourier components to zero except $F(0)$, $F(-1)$, and $F(1)$ will always produce a circle:

$$f(r) = F(1)\,[\cos(2\pi r/N) + j\,\sin(2\pi r/N)] + F(0) \ \text{ for } r = 0, 1, 2 \ldots N - 1 . \tag{8.78}$$

Equation (8.68) is simply the equation of a circle in polar coordinates with a radius equal to $F(1)$ and centered at $F(0)$.

Figure 8.41(c) is the edge image reconstructed using only the first 21 Fourier components and setting the other 1003 Fourier components to zero. Notice the

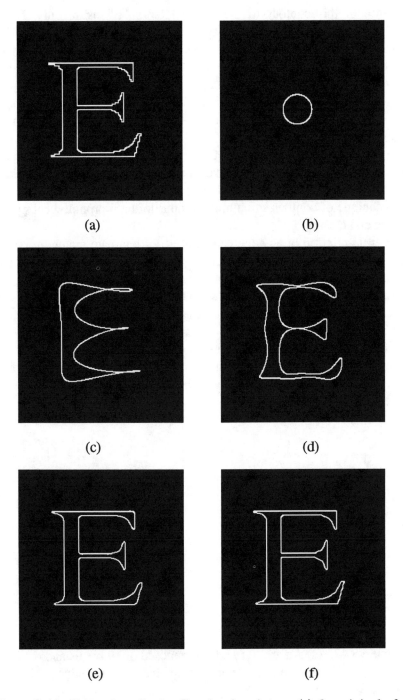

Figure 8.41: Examples of using Fourier descriptors: (a) the original edge
image, (b) 3 Fourier coefficients, (c) 21 Fourier coefficients,
(d) 61 Fourier coefficients, (e) 201 Fourier coefficients, and
(f) 401 Fourier coefficients.

resemblance of this contour to the original letter "E" using only 21 values. Figure 8.41(d) is the reconstructed image generated with only the first 61 Fourier components, showing even a better resemblance to the original letter "E". Figures 8.41(e) and (f) are the reconstructed images using 201 and 401 out of the 1024 Fourier components. With only 201 elements, there is very little difference between the reconstructed letter and the original letter. This corresponds to approximately a 5 to 1 compression ratio. A close comparison of Figures 8.41(e) and (f) shows that the addition of the high frequency components used to generate Figure 8.41(f) has added the fine variations in the contour that is present in the curved part of the bottom stroke of the letter "E". Lowpass filtering the Fourier components to remove the fine spatial structure of an edge provides a simple technique of contour smoothing, similar to the morphological operations of opening and closing.

Fourier description of an edge is also used for template matching. Since all the Fourier components except the first $F(0)$ do not depend on the location of the edge within an image, this provides a convenient method of classifying objects using template matching of an object's contour. A set of Fourier components is computed for the contours of each of the known objects. Ignoring the first Fourier component of each known object, the other Fourier components are compared against the Fourier components of the unknown object. The known object, whose Fourier components are the most similar to the unknown object's Fourier components, is the object the unknown object is classified as.

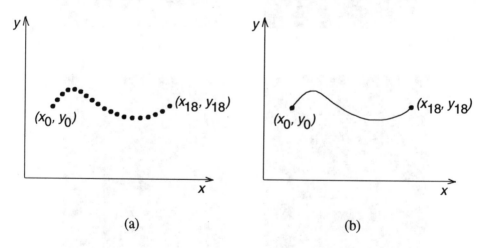

(a)   (b)

Figure 8.42: (a) An arbitrary edge composed of 19 pixels and (b) a smooth polynomial approximation to the edge.

The edge of an object can also be described using polynomial curves to define short smooth contour segments that intersect the pixels defining the edge. Figure 8.42(a) shows an arbitrary edge composed of 19 pixels and Figure 8.42(b) shows the corresponding polynomial curve fit of this edge. This polynomial curve provides a smooth continuous curve from points x_0, y_0 to x_{18}, y_{18} that matches the original curve defined by the 19 pixels. There exist two major techniques of approximating contours using short straight line segments. The first method, known as edge merging, starts with a straight line defined by the beginning pixel of the edge and the next adjacent pixel on the edge. Figure 8.43(a) shows the initial definition of this straight line in relationship to an arbitrary edge composed of seven pixels. The slope of this line is then changed so that the line intersects the first and third edge pixels. The deviation between this line and the second edge pixel is computed. If this deviation is less than the maximum allowable deviation, the linear edge is then redefined using the first and fourth edge pixels. The maximum deviation determines how much the edge can deviate from a straight line before a new line segment must be defined.

The deviation between this new line segment and the second and the third edge pixel is computed. Figure 8.43(b) shows the difference between this line and the second and third pixels as the lengths of the perpendicular lines connecting the pixels to the line. If one of these values exceeds the maximum allowable deviation, this line is terminated and the first line segment is defined using the beginning edge pixel and the third edge pixel. Otherwise, the straight line is redefined using the beginning edge pixel and the fifth edge pixel. If the maximum deviation of the straight line defined between the first and fifth edge pixels exceeds the given threshold value, the line segment is then defined from the beginning pixel to the fourth edge pixel. A new beginning point at the fifth pixel is defined and the process of moving the line from one pixel to the next and checking the maximum allowable deviation is repeated until the deviation from the actual edge pixels to this line exceeds the predetermined threshold. At this point, the second straight line of the edge has been defined. The process of line fitting continues until the end of the edge is reached. Figure 8.43(c) shows the linear curve fit of Figure 8.43(a) after the generation of two line segments and Figure 8.43(d) is the complete definition of the seven pixel edge using a total of three line segments.

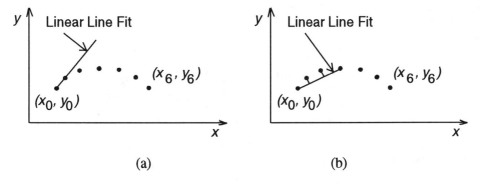

(a)   (b)

Figure 8.43: Line fitting using edge merging: (a) the initial line description using the first and second edge points and (b) the straight line fit to the first four edge points.

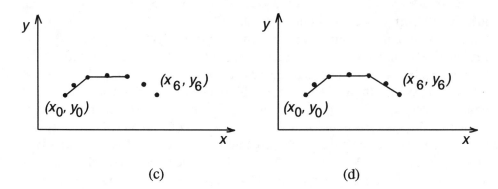

(c)   (d)

Figure 8.43: Line fitting using edge merging: (c) the straight line
            descriptions of the first two line segments and (d)
            the edge description using a total of three straight lines.

The inverse technique of edge merging is edge splitting. This technique
starts with a long straight line and divides it into smaller line segments until all
the edge pixels that are being approximated by this line segment are within the
maximum allowable deviation. Figure 8.44(a) shows the first straight line
segment of the seven edge contour given in Figure 8.43(a). This line segment
was defined using the first and last edge pixels. The fourth edge pixel is the
farthest away from this straight line and the distance from this point to the
straight line exceeds the maximum allowable deviation. Hence, this line is split
into two lines at the fourth edge pixel, as shown in Figure 8.44(b). The
deviation between these two straight line segments and the actual edge is
computed and if the deviation from the edge pixels to the line segments exceeds
the maximum allowable deviation, either line segment is split again. The
process of line splitting continues until the distance from edge pixels to the
corresponding line segment is within the maximum allowable deviation.

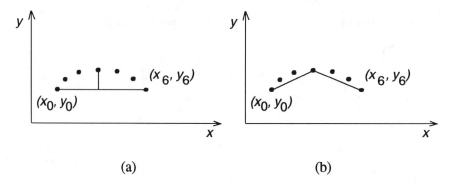

(a)   (b)

Figure 8.44: Line description of an edge using edge splitting: (a) the
            initial line description using the first and last edge points
            and (b) the splitting of the line given in (a) into two lines.

Figure 8.45 shows an example of straight line fitting of the edge of the letter "E" given in Figure 8.41(a) using the merging technique. The initial starting point of the edge was located in the upper left-hand corner of the letter, which was used as the starting location for the line following algorithm presented in the previous section. As the line was traced from this beginning pixel, the previous location at which the actual edge of the letter deviated from the straight line beyond the selected threshold became the ending point of this line segment and the start of a new line segment was initiated. This process continued until the upper left-hand corner of the letter was reached by the edge following algorithm. Figure 8.45(a) is the linear edge fit using a maximum allowable deviation value of 60 pixels. This deviation value produced a line fit that required only 7 straight lines to produce an image with some resemblance to the letter "E". Figure 8.45(b) shows the result of lowering the maximum allowable deviation value from 60 pixels to 40 pixels, requiring a closer fit between the straight lines and the actual edge. There were a total of 9 straight lines used to approximate the letter "E" in this image. Figures 8.45(c) and (d) are the straight line fits for maximum allowable deviations of 30 and 1, respectively. Figure 8.45(c) required a total of 14 lines, which did a reasonable job at matching the original letter "E". The reduction of the maximum allowable deviation to 1 pixel produced the image shown in Figure 8.45(d), which required 25 lines. Notice that the curves have been added to the three strokes of the letter "E".

Reducing the maximum allowable deviation to zero only increases the number of straight lines used to represent this letter from 25 to 26. The description of the letter "E" using 26 straight lines requires a total of 52 x and y coordinate values. The original edge of Figure 8.41(a) contained a total of 1024 pixels, each requiring an x and y coordinate value. The letter "E" given in Figure 8.45(d) results in a compression ratio of approximately 39 to 1. Beyond the fact that this image requires much less storage, the use of 52 descriptors to define the letter as composed to 1024 pixels makes the recognition of this letter much more realistic. Reducing the number of components required to identify an object reduces the complexity of the object recognition system.

Region description can generally be separated into two methods. The first method of region description defines the set descriptors that provide some information about the geometrical shape of an object. The second method describes the surface area of an object in terms of its texture. As stated earlier, the smoothness or roughness of an object plays an important role in region segmentation by providing an easy method of separating the object based upon its surface structure. The most commonly used geometrical region descriptor is that of *area*. Let R be composed of the set of pixels $p_i = x_i, y_i$ defining an object. Then the area A of this object is simply the total number of pixels that are required to define this object:

$$A = \sum_{i \in R} 1 \ . \tag{8.79}$$

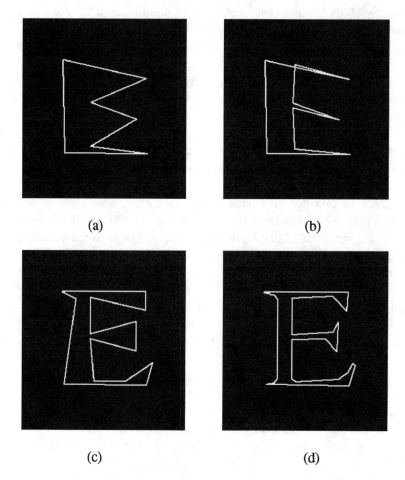

(a)   (b)

(c)   (d)

Figure 8.45: Example of straight edge fitting: the reconstruction of
              Figure 8.41(a) using 7 lines, (b) 9 lines, (c) 14 lines,
              and (d) 25 lines.

Given N pixels (x_0, y_0), (x_1, y_1), $\cdots$ (x_{N-1}, y_{N-1}) defining an object, the
centroid or *geometrical center* of an object is given by

$$x_{centroid} = \frac{1}{N} \sum_{i=0}^{N-1} x_i \tag{8.80}$$

and

$$y_{centroid} = \frac{1}{N} \sum_{i=0}^{N-1} y_i . \tag{8.81}$$

Another geometrical feature of interest is the *length of the perimeter P* of an object's boundary. Let $b_{x,\ i}$, $b_{y,\ i}$ be the coordinates in the x and y directions of the boundary pixels of an object composed of M $(M < N)$ pixels determined using any one of the edge detection methods that are available (Section 8.2). The perimeter of this object can then be written as the total Euclidean length of the boundary

$$P = \sum_{i=0}^{M-2} \sqrt{(b_{x,\ i+1} - b_{x,\ i})^2 + (b_{y,\ i+1} - b_{y,\ i})^2} \ . \tag{8.82}$$

The maximum width and height of an object is determined by scanning an image pixel by pixel. The difference between the first and last rows of the object gives the maximum height. As each row is scanned, the difference between the minimum and maximum x coordinates gives the width of the object. Another method of determining the height and width of an object is to find the smallest rectangle that completely encloses the object. The height and width of this rectangle give the maximum height and width of the object. If the major axis of an object is not oriented along the x or y direction, its direction can be determined by the use of the Hotelling transform given in Chapter 2, which is then used to rotate the object along either the x or y direction prior to finding the smallest rectangle that will enclose the object. Given the maximum height and width of an object, its *aspect ratio* is then defined as

$$\text{ASPECT} = \frac{width_{max}}{height_{max}} \ . \tag{8.83}$$

Consider the enclosing of an irregularly shaped object within a rectangle whose boundary just intersects the maximum width and height of the object. The ratio of the object's area A to the rectangle's area, known as the *rectangularity* of an object, describes how well the object fits into the rectangle:

$$\text{RECTANGULARITY} = \frac{A}{WH} \ , \tag{8.84}$$

where W and H are the width and height of the smallest enclosing rectangle. Related to the rectangularity of an object is the *circularity* of an object, defined as the ratio of an object's perimeter P to its area A:

$$\text{CIRCULARITY} = \frac{P^2}{A} \ . \tag{8.85}$$

Circularity defines the complexity of an object's boundary. As the complexity of an object's boundary increases, so does its circularity value. The smallest value of Equation (8.85) is when an object is completely circular. Also used to describe the size of an object are the *minimum radius* and *minimum diameter* of an enclosing circle. These parameters are similar to the rectangle's minimum width and height parameters in that they describe the radius and diameter of the minimum sized circle that can enclose an object.

The *curvature* of an object's boundary defines the changes in the direction of the boundary. The magnitude of the curvature is defined in terms of the second derivative of its coordinate values. Let x_i and y_i be the *ith* coordinate location describing an object's boundary. Then the curvature of this boundary at the *ith* position is defined as

$$C_i = \sqrt{(2 \cdot x_i - x_{i-1} - x_{i+1})^2 + (2 \cdot y_i - y_{i-1} - y_{i+1})^2} \quad , \tag{8.86}$$

where $i = 1, 2, \cdots M - 2$. Because Equation (8.86) gives the variation of a curve from point to point, the average of this change normalized to the length of the boundary P is defined as *energy of curvature* :

$$\text{ENERGY} = \frac{1}{P} \sum_{i=1}^{M-2} C_i^2 \quad , \tag{8.87}$$

where M is the total number of pixels defining the boundary.

Two other object descriptors that are commonly used are the *average* or *mean distance* of the object pixels from the centroid of an object,

$$\text{MEAN}_{dist} = \frac{1}{N} \sum_{i=0}^{N-1} \sqrt{(x_i - x_{centroid})^2 + (y_i - y_{centroid})^2} , \tag{8.88}$$

and the variance of this distance ,

$$\text{VAR}_{dist} = \frac{1}{N} \sum_{i=0}^{N-1} [\sqrt{(x_i - x_{centroid})^2 + (y_i - y_{centroid})^2} - \text{MEAN}_{dist}]^2 ,$$

$$\tag{8.89}$$

where N is the total number of object pixels and $x_{centroid}$ and $y_{centroid}$ are given by Equations (8.80) and (8.81).

Another descriptor that is sometimes used to describe an object is that of topological features. An object within an image can be composed of more than one connected features or may contain interior holes, as shown in Figures 8.46(b) and (a). Figure 8.46(a) is an example of an object defined using one connected region containing two interior holes, while Figure 8.46(b) is an object described by two connected components with no interior holes. The difference between the number of connected components C and the number of holes H defines the topological descriptor known as the *Euler number E*:

$$E = C - H \ . \tag{8.90}$$

Euler numbers for the objects given in Figures 8.46(a) and (b) reduce to -1 and 2, respectively.

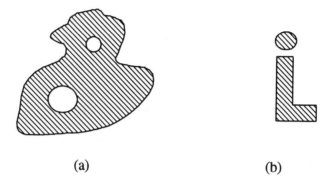

(a)   (b)

Figure 8.46:   Examples of two different types of connected objects:
(a) one connected component with two holes and (b)
two connected components with no holes.

Given an object containing C connected regions and H holes with a contour that can be described with V vertices and D edges between these vertices, then the number of closed surface regions S contained within the object is

$$S = C - H + D - V \ . \tag{8.91}$$

Figure 8.47(a) shows an object containing $V = 11$, $C = 1$, $H = 1$, and $D = 13$. Using Equation (8.91) yields the number of closed surface regions as 2, which agrees with the number of closed contours shown in Figure 8.47.

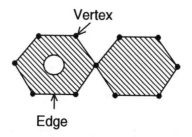

Figure 8.47: An object with 11 vertices, 1 connected region, 1 hole,
13 edges, and 2 closed surface areas.

There exist many other object descriptors which are commonly used, some of which are listed here for reference. An object can also be defined in terms of the *moments* of an image $f(x, y)$. The *ith* and *jth* moment of an $N \times M$ grayscale image is defined as

$$m_{i,j} = \sum_{y=0}^{M-1} \sum_{x=0}^{N-1} x^i y^j f(x, y) \ . \tag{8.92}$$

For an $N \times M$ binary image containing graylevels 0 and 1, Equation (8.92) reduces to

$$m_{i,j} = \sum_{y=0}^{M-1} \sum_{x=0}^{N-1} x^i y^j \ . \tag{8.93}$$

The mean x and y coordinate locations, known as the *center of gravity*, are defined from Equation (8.92) as

$$\bar{x} = \frac{m_{1,0}}{m_{0,0}} \tag{8.94}$$

and

$$\bar{y} = \frac{m_{0,1}}{m_{0,0}} \ . \tag{8.95}$$

The *central moments* of an $N \times M$ image can be written in terms of its mean x and y coordinate locations as

$$\mu_{i,j} = \sum_{y=0}^{M-1} \sum_{x=0}^{N-1} (x - \bar{x})^i (y - \bar{y})^j f(x, y) \ , \tag{8.96}$$

which reduces to

$$\mu_{i,j} = \sum_{y=0}^{M-1} \sum_{x=0}^{N-1} (x - \bar{x})^i (y - \bar{y})^j \tag{8.97}$$

for an $N \times M$ binary image. Given an object located within a binary image, the angle α of the *major axis* of the object relative to the x axis is defined in terms of its central moments as

$$\alpha = 0.5 \tan^{-1}\left(\frac{2\,\mu_{1,\,1}}{\mu_{2,\,0} - \mu_{0,\,2}}\right). \tag{8.98}$$

Also, the size of the object in the x and y directions is given by $\mu_{2,0}$ and $\mu_{0,2}$. A measure of the *size* of an object is then defined as

$$S = \mu_{2,\,0} + \mu_{0,\,2}\,. \tag{8.99}$$

A very important set of descriptors of an object is based upon its surface texture. The simplest method of describing a texture is via the use of the histogram h_i of the graylevels of the pixels within the object:

$$h_i = \frac{n_i}{N_t} \qquad \text{for } 0 \le i < G_{\max}\,, \tag{8.100}$$

where N_t is the total number of pixels used in the histogram calculation and n_i is the number of pixels at the *ith* graylevel. The *nth order moments* of the graylevel distribution of an object are then defined using Equation (8.100) as

$$m^n = \sum_{i\,=\,0}^{G_{\max}} i^n \cdot h_i\,. \tag{8.101}$$

Also computed from the graylevel distribution of an object's pixels are the *centralized moments:*

$$\mu^n = \sum_{i\,=\,0}^{G_{\max}} (i - \bar{m})^n \cdot h_i\,, \tag{8.102}$$

where $\bar{m}$ is the mean or average graylevel of the object, defined from Equation (8.101) as

$$\bar{m} = \sum_{i\,=\,0}^{G_{\max}} i \cdot h_i\,. \tag{8.103}$$

Of particular interest is the variance of the object pixel's graylevels,

$$\sigma^2 = \sum_{i\,=\,0}^{G_{\max}} (i - \bar{m})^2 \cdot h_i\,, \tag{8.104}$$

from which the *smoothness descriptor* can be computed,

$$S = \frac{\sigma^2}{1 + \sigma^2} \ . \tag{8.105}$$

The smoothness descriptor gives the relative roughness of an object's surface. For $\sigma^2 = 0$, the smoothness descriptor is zero, indicating a smooth surface, while for $\sigma^2 >> 1$, the smoothness descriptor approaches 1, indicating a very rough surface. Two other central moment descriptors that are commonly used are the *skewness* and *kurtosis*, which are defined from Equation (8.102) for $n = 3$ and 4, respectively. The skewness describes the symmetry of the graylevel histogram, while the kurtosis describes the shape of the tail(s) of the histogram.

The *entropy* of the graylevel fluctuations can also be used to describe the roughness of an object's surface:

$$\text{ENTROPY} = - \sum_{i=0}^{G_{max}} h_i \cdot \log \ (h_i) \ , \tag{8.106}$$

where the log is the logarithmic function, usually in base e or 2. For an object with a constant uniform graylevel region, the entropy is zero. On the other hand, the entropy is maximum for graylevel variations that are uniformly distributed. As the entropy changes from zero to its maximum value, the texture of an object changes from smooth to rough.

The main limitation of basing the description of an object's texture solely on the graylevel distribution of its pixels is that it provides no spatial information about the texture. Typically a new function is formed that depends on the difference between two points located within an object's surface. Consider a function that depends on the graylevel difference between two pixels at the coordinate locations x, y and $x + d_x, y + d_y$:

$$w(d_x, d_y) = |f(x + d_x, y + d_y) - f(x, y)| \ , \tag{8.107}$$

where x, y are the set of coordinates describing the location of each pixel within an object. For each set of distance d_x, d_y values there exists a graylevel histogram $h_i(d_x, d_y)$ defining the distribution of graylevels as given by $w(d_x, d_y)$.

The mean of $w(d_x, d_y)$ can then be defined as

$$\bar{m}_{d_x, d_y} = \sum_{i=0}^{G_{max}} i \cdot h_i(d_x, d_y) \ . \tag{8.108}$$

The relative value of $\bar{m}_{d_x, d_y}$ describes the size of an object's texture, with large values describing a fine texture. The other descriptors that are commonly used, which are based upon the graylevel difference function given in Equation (8.108), are the variance,

$$\sigma^2_{d_x,d_y} = \sum_{i=0}^{G_{max}} (i - \bar{m}_{d_x,d_y})^2 \cdot h_i(d_x, d_y) \ , \tag{8.109}$$

and the entropy,

$$\text{ENTROPY}_{d_x,d_y} = - \sum_{i=0}^{G_{max}} h_i(d_x, d_y) \cdot \log[h_i(d_x, d_y)] \ . \tag{8.110}$$

The variance describes the amount of graylevel fluctuations for a given separation distance of d_x, d_y between the pixels. As with Equation (8.106), the entropy of graylevel differences is maximum for a uniform distribution for $h_i(d_x, d_y)$.

Another set of texture descriptors that are based upon the distances between pixels uses the joint histogram between two pixels separated by a distance d_x, d_y. Let $C_{i,j}$ define the joint histogram between the two pixels $f(x_i, y_i)$ and $f(x_j, y_j)$ at a separation distance of d_x, d_y. This joint histogram can then be computed using

$$C_{i,j} = \frac{n_{i,j}}{n_t} \quad \text{for } i, j \in \text{Object pixels} \ , \tag{8.111}$$

where $n_{i,j}$ is the total number of occurrences of pixels within the object with graylevels $f(x_i, y_i)$ and $f(x_j, y_j)$ and n_t is the total number of pixels describing the object. If the object occupies the entire region, such as an $N \times M$ image of a given texture, then Equation (8.111) produces a two-dimensional $N \times M$ matrix referred to as the *graylevel occurrence matrix*, with elements given by $C_{i,j}$. There have been several descriptors defined that are used to describe the surface texture of an object. The first descriptor is known as the *maximum probability,*

$$P_{max} = \max[C_{i,j}] \tag{8.112}$$

which gives the most often occurring graylevels of the pair of pixels $f(x_i, y_i)$ and $f(x_i + d_x, y_i + d_y)$ separated by the distance d_x, d_y.

The *entropy* descriptor defines the randomness of the graylevel occurrence matrix,

$$\text{ENTROPY}_{C_{i,j}} = - \sum_{j=0}^{M-1} \sum_{i=0}^{N-1} C_{i,j} \cdot \log(C_{i,j}) \ , \tag{8.113}$$

and is maximum for a uniform distribution of $C_{i,j}$'s. The *element difference moment* descriptor of order n is

$$\text{EDM}_n = \sum_{j=0}^{M-1} \sum_{i=0}^{N-1} \frac{C_{i,j}}{(i-j)^n} \quad i \neq j, \tag{8.114}$$

where this descriptor takes on small values for course textures. The last measure that is based upon the graylevel occurrence matrix is the *uniformity* descriptor

$$\text{UNIFORMITY} = \sum_{j=0}^{M-1} \sum_{i=0}^{N-1} [C_{i,j}]^2 . \tag{8.115}$$

The final two typically used texture descriptors are based upon Fourier spectral analysis and are defined via the *autocorrelation* of the texture image

$$R_f(x, y) = \frac{1}{(N-x, M-y)} \sum_{j=0}^{M-1-y} \sum_{i=0}^{N-1-x} f(i, j) \cdot f(i+x, j+y) , \tag{8.116}$$

and its *power spectral density*, which is defined by the *Wiener-Khintchine* theorem as

$$P_f(n, m) = \frac{1}{NM} \sum_{y=0}^{M-1} \sum_{x=0}^{N-1} R_f(x, y) \cdot e^{-j2\pi(nx/N + my/M)} , \tag{8.117}$$

which is simply the two-dimensional discrete Fourier transform of the autocorrelation function given in Equation (8.116).

Several important features about a texture can be described by its autocorrelation function or its power spectrum. If the autocorrelation function is periodic with a fundamental spacing of d_x, d_y, then the texture is periodic with this spacing. The width of the autocorrelation function gives the average size of the texture, with a smaller width corresponding to a finer spatial structure. An autocorrelation function that can be represented by an impulse corresponds to a power spectral density, which is a constant for all frequencies (often referred to as being white). This autocorrelation function is typical of spatially uncorrelated textures. The autocorrelation function $R_f(0, 0)$ evaluated at zero corresponds to the second graylevel moment as defined by Equation (8.101) with $n = 2$. The variance of the graylevel fluctuations is then given by $R_f(0, 0) - \bar{m}^2$.

CHAPTER 9

Image Compression

This chapter presents the fundamental concepts behind image compression. Images contain an enormous amount of data that must be stored and manipulated. Image compression is based upon the removal of any redundant data that may be present within an image. This chapter is divided into three sections. The first section presents the fundamentals of image compression, giving the background required to understand the next two sections. The second section discusses error-free image compression and is based upon compressing an image so that when it is uncompressed, the original image is obtained. The final section discusses lossy compression methods, which use information about the visual limitations of the human eye to remove information that will not be noticeable from an image. Uncompression of a lossy compressed image does not yield the original image but a slightly degraded version of it.

9.1 Compression Fundamentals

The amount of data required to store an image can become quite enormous for some image processing applications. For example, consider the storage of NTSC color video for a two hour movie in which the three RGB color components have been digitized. If each color component is represented by 8 bits per pixel with an image size equal to 640×480, a single frame of video will require $3 \times 640 \times 480 = 921,600$ bytes of data. NTSC video produces 30 frames per second, which yields a total storage requirement for a two hour movie of 2×10^{11} bytes of data. This is an enormous amount of data to store. Without image compression, the digital storage of NTSC color video would not be possible.

Several applications of electronic image processing have driven the implementation of image compression. The storage and analysis of medical X-ray images require the use of error free compression. For diagnostic reasons, these images cannot be degraded by the compression process. Recently, millions of feet of newsreel and movie film that were taken in the early part of this century have been converted to digital format to prevent the loss of the documentation of important historical events.   To reduce the storage requirements, this digitized film was compressed prior to storage. Similar to the storage of these newsreels and movie film is the digital conversion of rare old books for preservation. A single page of a text (6" × 9") scanned using 1 byte per dot and at a resolution of 300 dots per inch requires 4,860,000 bytes of storage. If an average book is about 300 pages, a single book can require on the order of 1.5 gigabytes of storage. Image compression methods can easily reduce the areas within the book containing only text by a factor of 100, making the digital storage of books possible.

The goal of image compression is to reduce the amount of data required to represent the information present within an image. Consider the following two sentences, which need to convey the information that a box has fallen from a table to the floor:

> "The big rectangular box, which was located at the edge of the
> pretty round table, fell swiftly to the floor and rolled about the
> floor for several seconds."

> "The rectangular box on the round table fell to the floor."

Both sentences relay the information about the box falling to the floor, but the first sentence includes additional words. Many of the words in the first sentence can be eliminated without changing the information about the box falling that is relayed to the reader. The concept of data is different from the concept of information. Data is the representation of information, while information is the concepts and ideas that are relayed to an observer. The more compact the data representation, the better the information can be transferred and understood.

The above two sentence example shows the use of two different data representations to convey the information about a box falling to the floor. The words in each of the sentences are the data that are used to relay this information. The first sentence requires 28 words, while the second sentence requires only 11 words to relay the information. Given two data representations D_1 and D_2 of the same information, requiring d_1 and d_2 units of data, the compression ratio is then defined as

$$C = \frac{d_1}{d_2} \ . \tag{9.1}$$

If $d_1 > d_2$, then more data is required to represent the information using the D_1 data representation as compared to the D_2 data representation. For the two sentence example given above, $d_1 = 28$ and $d_2 = 11$, the compression ratio gained by using the second sentence to relay the information that a box has fallen to the floor is 2.54.

Compression of data to represent information is based upon the removal of redundant data, which is defined as data that is not needed to represent a given information. For example, in the first sentence, the words "big" and "pretty" provide no information about the box falling to the floor and can be removed from the sentence without changing that information. Given two data representations D_1 and D_2 of the same information, requiring d_1 and d_2 units of data and with $d_1 > d_2$, the relative data redundancy between these two representations in percentage is

$$R = \frac{d_1 - d_2}{d_1} \times 100\% \ , \qquad (9.2)$$

or in terms of the compression ratio,

$$R = \frac{C - 1}{C} \times 100\% \ . \qquad (9.3)$$

For the two sentence example, the relative data redundancy reduces to 60.7%. In other words, 60.7% of the words (17 words) in the first sentence are redundant and can be eliminated without sacrificing the information to be relayed to the reader. Eleven words can be used to relay the same information, as illustrated by the second sentence.

In electronic image processing, the unit of data that is used to represent an image is typically a bit and the information is the features that are required to describe an image. The goal of image compression is then to reduce the number of bytes necessary to represent the features present within an image. There are three types of data redundancies that are common and that can be easily reduced. The first is *coding redundancy*. Typically, an image is represented using a constant number of bits for each graylevel within an image. Table 9.1 gives the graylevel distribution for a 128×128 by eight graylevel image, with graylevel 0 representing black and graylevel 7 representing white. The first column gives the eight graylevels, with the adjacent column giving the number of pixels within the image at that graylevel. This image is of low brightness and contrast, since 80% of the pixels have graylevels less than half the full graylevel scale. The next column gives the histogram values for each graylevel. As defined in Chapter 3, the histogram of an $N \times M$ image is

$$h_i = \frac{n_i}{NM} \ , \qquad \text{for } 0 \le i < G_{\max} \ , \qquad (9.4)$$

where n_i is the number of pixels at graylevel i, NM is the total number of pixels within the image, and G_{max} is the maximum graylevel contained within the image. The next column shows the coding of each of the eight graylevels using a constant binary code containing three bits, while the last column gives the coding of each pixel using a variable length code. The concept behind variable length coding is to reduce the number of bits that are required to represent the graylevels within an image. The graylevels that occur the most within the image are given the smallest length codes, while the lowest probability graylevels are given the longest length codes. Variable length coding of an image reduces the total number of bits required to represent an image. As shown in Table 9.1, for the two most probable graylevels of 1 and 2, only two bits are used to represent these graylevels as compared to three bits using the standard binary code. The lowest probable graylevels in Table 9.1 are assigned the longest codes of length 4.

Table 9.1: An example of coding redundancy

Graylevels	n_i	h_i	Binary coding	Variable Length
0	1116	0.0681	000	00
1	4513	0.2754	001	01
2	5420	0.3308	010	10
3	2149	0.1312	011	1100
4	1389	0.0848	100	1101
5	917	0.0560	101	1110
6	654	0.0399	110	111100
7	226	0.0138	111	111101
n_t	16384			

A measurement of the coding redundancy of an image is the average number of bits required to represent an image, defined in terms of the image's histogram as

$$B_{avg} = \sum_{i=0}^{G_{max}} B_i \cdot h_i , \qquad (9.5)$$

where B_i is the number of bits used to represent the *ith* graylevel. For the 3-bit binary and the variable length codes given in Table 9.1, the average number of bits required to code this eight graylevel image is 3 bits for binary coding and 2.76 bits for variable length coding. Hence, the number of bits required to represent this 128×128 image using the variable length code given in Table 9.1 is 45,200 bits, while the number of bits required to represent this image using the standard 3-bit binary code is 49,152 bits. The reduction in the number of bits required to represent this image using a variable length code as compared to a fixed binary code has resulted in a compression ratio of 1.087 to 1. The relative coding redundancy of using the 3-bit binary code over the variable length code of Table 9.1 is 8.04%.

The number of bits necessary to encode an image depends on the information content present within the image. For example, if an image contains only two graylevels then this image can be represented using 1 bit. On the other hand, if an image contains graylevels that are uniformly distributed, the best coding that can be obtained is to use the standard m-bit binary code. Based upon the information theory developed by Shannon in the 1940s, the lower bound in the average number of bits required to represent an image is given by the information present within an image and can be measured in terms of the image's graylevel entropy:

$$\text{ENTROPY} = - \sum_{i=0}^{G_{max}} h_i \cdot \log_2(h_i) \quad , \tag{9.6}$$

where $\log_2()$ is the base 2 logarithm function and the entropy is given in the number of bits. For the 128×128 image given in Table 9.1, the lowest average number of bits required to represent this image is 2.49. The best compression ratio that can be obtained by reducing the coding redundancy present within the 8 graylevel image of Table 9.1, using a variable length code as compared to the fixed length m-bit binary code, is 1.20. The use of the fixed length m-bit binary code produces a coding redundancy of 16.8%. The optimum variable length coding method is that of Huffman coding, which will be presented in the next section in the discussion of error-free compression techniques.

The next type of redundancy found within images is that of *spatial redundancy*. In most images there exists a high degree of graylevel similarity between neighboring pixels. Figure 9.1(a) shows an image of a teacup and saucer, containing low spatial details and low spatial frequencies. Figure 9.1(b), on the other hand, is an image of a farm plow placed in front of a stone wall, containing a large amount of spatial detail. There is a high degree of correlation between the adjacent pixels of Figure 9.1(a) as compared to correlation between the adjacent pixels of Figure 9.1(b). The autocorrelation function describes the average graylevel similarity between neighboring pixels. The autocorrelation of an image is defined as

$$R_f(x, y) = \frac{1}{(N - x, M - y)} \sum_{j=0}^{M-1-y} \sum_{i=0}^{N-1-x} f(i, j) \cdot f(i + x, j + y) \tag{9.7}$$

and is related to its power spectral density through the Wiener-Khintchine theorem as

(a)

(b)

Figure 9.1: Two images with different spatial detail: (a) a
            low spatial detail image and (b) a high spatial
            detail image.

$$P_f(n, m) = \frac{1}{NM} \sum_{y=0}^{M-1} \sum_{x=0}^{N-1} R_f(x, y) \cdot e^{-j2\pi(nx/N + my/M)} , \qquad (9.8)$$

which is simply the two-dimensional discrete Fourier transform of the autocorrelation function given in Equation (9.7). If the autocorrelation function is periodic with a fundamental spacing of d_x, d_y, then the image contains objects that are periodically spaced. The width of the autocorrelation function describes the graylevel similarity of neighboring pixels. Wide autocorrelation functions imply that there are large regions within the image that contain very similar graylevels. On the other hand, narrow autocorrelation functions describe images containing small regions with similar graylevels. Hence, the autocorrelation function for Figure 9.1(a) is much wider than that for Figure 9.1(b). In particular, an image with an impulse autocorrelation function describes an image with no spatial correlation, which is a simple random graylevel image.

The goal of image compression based upon spatial redundancy is to reduce the number of pixels required to represent an image by coding the image in such a way as to eliminate neighboring pixels that are very similar in graylevel. The amount of spatial redundancy present in an image depends on the spatial details contained within an image. An image with a narrow autocorrelation function, which implies an image with high spatial frequencies, has fewer spatial redundancies than an image with a wide autocorrelation function. One method of image compressing a low spatial detail image is to describe the image using a coding scheme that gives the graylevel of a pixel followed by the number of pixels with the same graylevel value. Figure 9.2(a) is a 128×128 binary image of a 64×64 white square placed in the center of a black background. Using 1 bit per pixel yields a total of 16,384 bits required to represent this image.

Figure 9.2(b) is a plot of the graylevel distribution in the horizontal direction for $y = 64$. Consider coding this line by dividing the line into sequences of constant graylevels. From Figure 9.2(b) there are a total of three constant graylevel sequences. The first sequence is at graylevel G_b, the second sequence is at graylevel G_w, and the third sequence is at graylevel G_b. Next, each of these sequences is coded using the graylevel and the length of the sequence. A total of 6 values are required to represent this one horizontal line: $(G_b, 32)$, $(G_w, 64)$, and $(G_b, 32)$. Using 1 bit to represent the graylevel value and 7 bits to represent the length of the sequence yields a total of 24 bits required to represent this one horizontal line. Performing this coding scheme for each of the 128 horizontal lines produces the following two coding sequences. The first 32 and the last 32 horizontal lines are easily coded using one graylevel sequence of $(G_b, 128)$, while the center 64 lines are coded using the three graylevel sequences of $(G_b, 32)$, $(G_w, 64)$, and $(G_b, 32)$. The total number of bits required to represent this image is 2048 bits. This corresponds to an 8 to 1 compression ratio over simply using 1 bit per pixel. This corresponds to a spatial redundancy of 87.5%. In other words, 87.5% of the bits required to represent this image using 1 bit per pixel are redundant. The process of dividing an image into a sequence of constant graylevels is known as *run length coding* and will be further discussed in the next section, covering error-free compression.

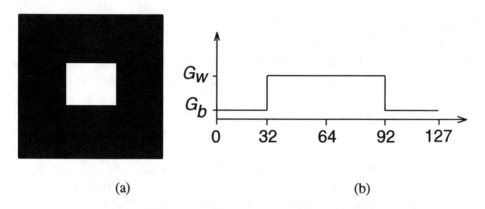

(a)   (b)

Figure 9.2: An example of image compression for removing spatial
             redundancy: (a) the binary image of a white square and
             (b) a graylevel profile obtained in the horizontal direction
             for $y = 64$.

Visual redundancy is based upon the limitations of the human visual system, for which all data is not required to represent an image. As described in Chapter 1, the human eye is poor at determining the absolute brightness of an object. The brightness of an object depends not only on its illumination but also on its size and intensity relative to adjacent objects within the image. The simultaneous contrast example given in Figure 1.13 shows how the interpretation of an object's brightness depends on the surrounding background. Also described in Chapter 1 is the poor response the eye has in determining the intensities in the vicinity of an edge. The Mach band example given in Figure 1.12 shows how the human visual system distorts the intensities in the neighborhood of an edge. The distortion of an edge by the human visual system is an important feature that allows the removal of data from an image without any observed degradation within the image. Unlike coding and spatial redundancies, which allow error-free compression of an image, the process of compressing an image using visual redundancies produces a new image that is a degraded version of the original uncompressed image. Compression that reduces visual redundancies is nonreversible in that visual information is lost during the compression process.

The elimination of visual redundancies to produce a compressed image can be accomplished using several different methods. The first two are based upon nonuniform sampling and quantization of an image. A fine spatial resolution is needed near edges to properly position the edge in the sampled image. In smooth graylevel regions, the spatial resolution of the image is not as critical and can be sampled using a lower spatial resolution. The nonuniform sampling of an image is accomplished by changing the spatial sampling interval within an image depending on whether a sharp graylevel transition is detected. Near edges a fine spatial sampling is used and in smooth graylevel regions a course spatial sampling is used. Another method of removing visual redundancies, which is similar to nonuniform sampling, is that of nonuniform quantization. Since the human visual system is insensitive to graylevels near edges, nonuniform quantization of an image reduces the number of graylevels used to represent an edge. For example, near edges only 3 bits might be used as compared to 8 bits in smooth graylevel regions.

Consider a 2 to 1 compression of the 128×128 by 256 graylevel image of Figure 9.1(a) by reducing the number of bits representing the graylevels within this image from 8 bits to 4 bits. Figure 9.3(a) shows the results of using only 16 different graylevels. Notice the false contouring that is observable in this image as a result of using only 16 graylevels. The compression method used to generate Figure 9.3(a) eliminated the lower 4 bits of the 8 bits of graylevels that represented each pixel, while preserving the upper 4 bits. These upper 4 bits became the values used to represent the compressed image. Uncompressing this 4-bit graylevel image involved using these 4 bits to form the upper 4 bits of the 8-bits of graylevels, while setting the lower 4 bits to zero. Since the human visual system is very sensitive to edges, the edges observed in Figure 9.3(a) as a result of the false contours are very objectionable. Since the eye is not as sensitive to random noise, one way of removing these false contours is to add random noise to the original graylevel image prior to compression. Noise that was uniformly distributed between 0 and 15 was added to the original graylevel image of Figure 9.1(a). This image was then compressed by eliminating the lower 4 bits of the 8 bits used to represent the graylevels for each pixel. Figure 9.3(b) shows the results of adding random noise prior to compressing the image. Notice the lack of edges due to the false contours; these have been eliminated at the slight cost of adding a grainy appearance to the image. This grainy appearance is far less objectionable than the edges that were present.

The fidelity of an image after compression is an important aspect of image compression. Compression methods that are lossless produce images that are an exact replica of the original uncompressed image. Lossy compression methods that remove visual redundancies produce uncompressed images that have lost visual information during the compression process. One measure of a compressed image's fidelity is that of the mean square error between the compressed image $f_C(x, y)$ and the original image $f(x, y)$:

$$\text{MSE} = \frac{1}{NM} \sum_{y=0}^{M-1} \sum_{x=0}^{N-1} (f(x, y) - f_C(x, y))^2, \quad (9.9)$$

(a)

(b)

Figure 9.3: Examples of removing visual redundancy: (a) the
compression of the image given in Figure 9.1(a) using
only 4 bits and (b) the adding of random noise prior to
compressing Figure 9.1(a) to 4 bits.

where both the compressed and the original images are of size $N \times M$. The smaller the MSE is, the closer the compressed image is to the original image. Equation (9.9) can also be normalized with respect to the mean square error of the original image:

$$
\text{NMSE} = \frac{\displaystyle\sum_{y=0}^{M-1} \sum_{x=0}^{N-1} (f(x, y) - f_c(x, y))^2}{\displaystyle\sum_{y=0}^{M-1} \sum_{x=0}^{N-1} f(x, y)^2} . \tag{9.10}
$$

The signal-to-noise ratio (SNR) of the compressed image is then given by

$$
\text{SNR} = \frac{\displaystyle\sum_{y=0}^{M-1} \sum_{x=0}^{N-1} f_c(x, y)^2}{\displaystyle\sum_{y=0}^{M-1} \sum_{x=0}^{N-1} (f(x, y) - f_c(x, y))^2} . \tag{9.11}
$$

The closer the compressed image is to the original image, the higher the signal to noise ratio will be. The main difficulty with using the MSE or the NMSE as a measure of image quality is that in many instances these values do not match the quality that is perceived by the human visual system. A better set of fidelity criteria is based upon the quality criteria that are commonly used to evaluate the performance of several types of nonlinear filters. The first of these fidelity criteria is the *peak-to-peak signal-to-noise ratio*, defined as

$$
\text{PSNR} = \frac{NM \, (f_{max} - f_{min})}{\displaystyle\sum_{y=0}^{M-1} \sum_{x=0}^{N-1} (f(x, y) - f_c(x, y))^2} , \tag{9.12}
$$

where f_{max} and f_{min} are the maximum and minimum graylevels present within the original image, respectively. Equation (9.12) describes the SNR of the compressed image relative to the maximum peak-to-peak variations in the original image.

Another fidelity criteria that can be used is the peak-to-peak SNR defined within the homogeneous regions PSNR_h of an image. Only pixels that are defined within homogeneous (slowly varying graylevel) regions are included within the calculation:

$$\text{PSNR}_h = \frac{N_h \, (f_{\max} - f_{\min})}{\sum\limits_{y} \sum\limits_{x} (f(x, y) - f_c(x, y))^2} \qquad x, y \in \text{homogeneous regions}, \quad (9.13)$$

where N_h is the number of homogeneous pixels used within the calculation of Equation (9.13). Equation (9.13) describes the performance of the compressed image in comparison to the original image within smooth graylevel regions.

Similar to Equation (9.13) is the *peak-to-peak signal-to-noise ratio* PSNR_e within edge regions of an uncompressed image and is defined as

$$\text{PSNR}_e = \frac{N_e \, (f_{\max} - f_{\min})}{\sum\limits_{y} \sum\limits_{x} (f(x, y) - f_c(x, y))^2} \qquad x, y \in \text{edge regions}, \quad (9.14)$$

where N_e is the number of edge pixels used within the calculation of Equation (9.14). Equation (9.14) describes how closely a compressed image follows the edges present within the original image. The quality of an uncompressed image as measured by Equation (9.14) relates directly to the quality perceived by the human visual system, which is very sensitive to edge degradation. Maximizing Equation (9.14) usually produces the best perceived compressed image. The normalized MSEs for the compressed images given in Figure 9.3(a) and (b) as compared to the original graylevel image given in Figure 9.1(a) are 2.94% and 3.708%, respectively. Even though the NMSE for Figure 9.3(b) is larger than that for Figure 9.3(a), the perceived image quality is considerably better for Figure 9.3(b) due to the elimination of the false contours present in Figure 9.3(a). The NMSE for these two examples illustrates the difficulty of using this parameter for measuring the perceived image quality.

The process of image compression can be explained via the block diagram given in Figure 9.4. In general, image compression can be separated into two processes: the encoder and the decoder. The encoder takes an input image and compresses it, while the decoder uncompresses a compressed image. The goal of the encoder is to eliminate any coding, spatial, or visual redundancies that might be present within the original graylevel image. The encoder can be separated into three functional processes, as shown in Figure 9.4(a). The purpose of the mapping function is to transform the original image into a new domain that provides an easy method of eliminating spatial redundancies that may be present within the image. The quantizer process reduces the number of values that are generated from the mapping function by digitizing these values to a smaller subset. It is in this stage that the visual redundancies of the original image are reduced. The final stage, data coding, reduces data redundancies by assigning variable length codes to the output values from the quantizer. Of these three stages, both the mapping function and the data encoder are completely reversible. Hence, these two stages do not remove any information from the image. The quantizer stage, on the other hand, does remove information from the image, resulting in the uncompressed image being a degraded version of the original graylevel image. Error-free compression requires that the quantizer be removed from the encoder and that only the mapping function and the data encoder be used.

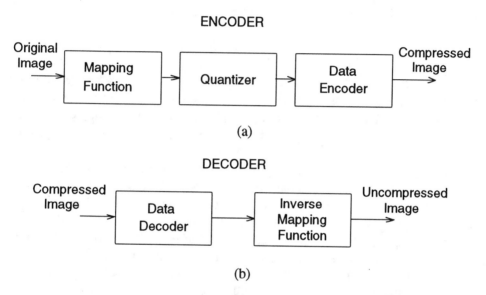

ENCODER

(a)

DECODER

(b)

Figure 9.4: A block diagram illustrating image compression:
(a) the encoder model and (b) the decoder model.

The decoder stage shown in Figure 9.3(b) takes a compressed image and uncompresses it to produce a new graylevel image that represents the original image. For error-free compression, this output image is an exact replica of the original image. For an image compressed using lossy compression, this output image is a degraded version of the original image. The purpose of the data decoder is to perform the reverse operation of the data encoder, producing values that are equal to the output of the quantizer or to the output of the mapping function, depending on whether error-free or lossy compression was used. The final stage of the decoder applies the inverse to the mapping function given in the encoder to obtain the uncompressed image.

9.2 Error-Free Compression Methods

For error-free compression, the encoder block diagram contains only the two functional blocks of the mapping function and the data encoder. Most error-free compression methods are based upon removing the coding redundancies that are present within an image by using a variable length code. The most common code used to represent digital images is that of the standard binary code. For an $N \times M$ image containing G different graylevels, the number of bits B required per pixel is

$$B = \log_2(G) \ . \tag{9.15}$$

Hence, the total number of bits required for this image is

$$B_{total} = N \cdot M \cdot B \ . \tag{9.16}$$

For a 256 graylevel image, Equation (9.15) yields a total of 8 bits required to represent each pixel. For a typical sized 256 graylevel image of 640×480, the total number of bits equals 2,457,600. Independent of the image's graylevel histogram, the average number of bits per pixel is fixed and is given by Equation (9.15).

Coding an image using the fixed binary code produces coding redundancies. A better method of coding an image is to use a variable length code that assigns the shortest code to the most frequently occurring graylevels within an image and the longest codes to the least frequently occurring graylevels. There are many types of variable length codes, but the *Huffman* code is by far the best. This code yields an average number of bits that is slightly greater than predicted by the entropy contained within the image. In particular, the average number of bits B_{avg} required to represent a pixel will lie between

$$ENTROPY < B_{avg} < ENTROPY + 1 \ , \tag{9.17}$$

where the entropy of the image is defined by Equation (9.6). Determination of the Huffman code requires knowledge about the graylevel distribution within an image.

The first step in computing the Huffman code is to compute the graylevel histogram of the image using Equation (9.4). Next, a table is formed with the graylevels arranged in a column by descending order of probability of occurrence. Figure 9.5(a) gives the distribution of graylevels within an 8 graylevel image and its corresponding histogram. The first two columns of Figure 9.5(b) give the graylevels and the probability of each graylevel within this 8 graylevel image rearranged in descending order of probability. The next step in generating the Huffman code is to reduce the number of probabilities by combining the lowest two probabilities together to form one new probability. A new column is generated using the other unmodified probabilities and the newly formed probability. This column is again produced in descending order of probabilities. The final entry in the third column of Figure 9.5(b) was generated by combining the two probabilities 0.03 and 0.01 to form the new probability 0.04. Further reduction in the probability is continued by generating additional

Graylevels	n_i	h_i
0	3441	0.21
1	4423	0.27
2	3932	0.24
3	1802	0.11
4	1311	0.08
5	819	0.05
6	492	0.03
7	164	0.01
n_t	16384	

(a)

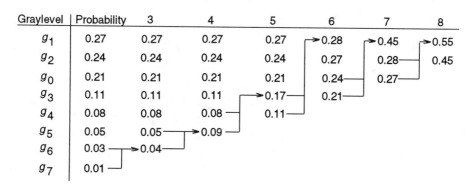

Graylevel	Probability	3	4	5	6	7	8
g_1	0.27	0.27	0.27	0.27	0.28	0.45	0.55
g_2	0.24	0.24	0.24	0.24	0.27	0.28	0.45
g_0	0.21	0.21	0.21	0.21	0.24	0.27	
g_3	0.11	0.11	0.11	0.17	0.21		
g_4	0.08	0.08	0.08	0.11			
g_5	0.05	0.05	0.09				
g_6	0.03	0.04					
g_7	0.01						

(b)

Graylevel	Probability	3	4	5	6	7	8
g_1	0.27 01	0.27 01	0.27 01	0.27 01	0.28 00	0.45 1	0.55 0
g_2	0.24 10	0.24 10	0.24 10	0.24 10	0.27 01	0.28 00	0.45 1
g_0	0.21 11	0.21 11	0.21 11	0.21 11	0.24 10	0.27 01	
g_3	0.11 001	0.11 001	0.11 001	0.17 000	0.21 11		
g_4	0.08 0000	0.08 0000	0.08 0000	0.11 001			
g_5	0.05 00010	0.05 00010	0.09 0001				
g_6	0.03 000110	0.04 00011					
g_7	0.01 000111						

(c)

Figure 9.5: An example of Huffman coding: (a) distribution of graylevels in an 8 graylevel image and its corresponding histogram, (b) phase one of Huffman coding reducing the number of probabilities, and (c) phase two of Huffman coding, assigning the codes to the graylevels.

columns that are formed by combining the two lowest probabilities and then rearranging the new set of probabilities in descending order of their values. The process of reducing the number of probabilities continues until there are only two probabilities remaining.

The fourth column of Figure 9.5(b) was generated by combining the probabilities of 0.04 and 0.05 to produce the new probability 0.09. This new probability and the other 5 unaltered probabilities of column three are rearranged in descending order to produce the values shown in column four. The process of reducing the number of probabilities and rearranging these probabilities in descending order of value to form new columns continues until column eight is generated, containing only two probabilities. At this point the probability reduction process stops. The lines in Figure 9.5(b) show the tracing of the reduction of probabilities from the second column to the last column. These become important during the second phase of the Huffman coding in the assigning of code values to each graylevel.

The second phase of Huffman coding starts at the last column, containing the two probabilities, and working towards the first column assigns a set of variable length codes to each graylevel. During the process of moving from the last column to the second column, each time a probability is encountered that was formed by merging two probabilities together, an additional code symbol is assigned to the two probabilities that form this merged probability. Figure 9.5(c) shows the arbitrary assigning of 0 and 1 to the two probabilities given in the eighth column. Since the bottom two probabilities 0.28 and 0.27 given in column seven formed the 0.55 probability in column eight, with a code value of 0, these two probabilities are assigned an additional code value. Hence, the probabilities of 0.028 and 0.027 are given the two symbol codes of 00 and 01, respectively. Moving from column seven to column six adds one additional code value to the probabilities 0.24 and 0.21 since these probabilities formed the 0.45 probability in column seven. Using the first coding value of 1 assigned to this 0.45 probability and adding the second coding values of 0 and 1 to the 0.24 and 0.21 probabilities yields codes of 10 and 11, respectively. This process of assigning codes continues until the first probability column is reached and all the graylevels have been assigned a variable length code.

Figure 9.5(c) shows the Huffman coding of the 8 graylevel image given in Figure 9.5(a), generated by assigning the shortest codes to the highest probability graylevels and the longest codes to the lowest probability graylevels. Using Equation (9.5) yields an average number of bits per pixel of 2.59. The entropy computed from Equation (9.6) yields a value of 2.55. The Huffman coding of this image is only slightly greater than the best that can be achieved as given by the entropy contained within the image. The Huffman coding of the 8 graylevel image given in Figure 9.5(a) as compared to coding this image using a standard 3-bit binary code has resulted in a compression ratio of 1.16 to 1. In other words, a 14% reduction in coding redundancies has been achieved by using the Huffman code. The coding of the various graylevels within a graylevel image by the Huffman coding algorithm assigns a unique code value to each graylevel. For example, consider the sequence of 0's and 1's generated using the codes given in Figure 9.5(c):

$$1,1,0,0,1,1,0,0,0,0,1,0,0,0,0,1,1,1 \quad . \tag{9.18}$$

Decoding of this image is accomplished using a lookup table that translates the Huffman code to the corresponding graylevel. In searching the binary string given in Equation (9.18), the code value of 11 for graylevel g_0 is the only code that contains only two 1's. The next graylevel value is found by comparing the eight variable length codes given in Figure 9.5(c) with the 16 remaining 0's and 1's in Equation (9.8). The only code that will fit the next set of 0's and 1's is the code of 001, which corresponds to g_3. Scanning this list sequentially yields the set of graylevels g_0, g_3, g_2, g_5, and g_7.

The main limitation of the Huffman code is the complexity of the algorithm in determining the assignment of the variable length codes to the graylevels within an image. This makes the real-time compression of images using Huffman coding difficult. One method of reducing the complexity of the Huffman code is to modify the codes so that graylevels with small probabilities are assigned a fixed code length. The most probable graylevels are assigned a variable length code using the Huffman algorithm, while the least probable graylevels are assigned a fixed length code composed of a flag code and a binary code. This coding method is known as the *modified Huffman* code. The advantage of this code is that it reduces the complexity of assigning variable length codes to graylevels that occur infrequently in an image. For example, consider assigning a set of modified Huffman codes to the graylevels given in Figure 9.5(a). Graylevels with a probability less than 0.10 will be assigned a fixed code length. The first four graylevels are assigned the same codes as Figure 9.5(c) using the standard Huffman coding algorithm, as illustrated in Figure 9.6. The last four graylevels are assigned a fixed binary code with a length of 5. The first three bits of this fixed length code are used to indicate the presence of the fixed length code. The next two code values are used to assign a unique 5 length code to each of the four graylevels, as shown in Figure 9.6. The average number of bits per pixel required to represent this 8 graylevel image using the modified Huffman code is 2.62, which is only slightly higher than the value of 2.58 obtained using the standard Huffman code. The probability reduction process for the modified Huffman code used in this example reduces to two iterations as compared to the six iterations required for the standard Huffman codes given in Figure 9.5(c). The use of the modified Huffman code has drastically reduced the complexity of assigning the Huffman code values, with only a slight increase in the average number of bits per pixel.

Other variable length codes have been developed that assign the shortest length codes to the most probable graylevels and the longest codes to the least probable graylevels. These codes produce an average number of bits per pixel that is slightly larger than the optimum Huffman code, but with much less algorithm complexity. One such code is the *binary shift* code, as shown in Figure 9.7(a) for 16 graylevels. Also shown in Figure 9.7 is the standard 4-bit binary code for comparison with the binary shift code. The binary shift code is generated using only $2^n - 1$ codes that are available from n bits. This leaves one code symbol out of the 2^n possible codes to be used as a flag to indicate the presence of a shifted version of the $2^n - 1$ codes.

Graylevel	Probability
g_1	0.27 01
g_2	0.24 10
g_0	0.21 11
g_3	0.11 001
g_4	0.08 000 00
g_5	0.05 000 10
g_6	0.03 000 01
g_7	0.01 000 11

Figure 9.6: An example of coding an image using the modified Huffman code.

Figure 9.7(a) shows the 2-bit binary shift code. The first three graylevels are generated using the standard 2-bit binary code. The fourth binary code of 11 is used in this example as the shift flag. The 2-bit shift codes for the next three graylevels g_3, g_4, and g_5 are generated using the two bits representing the flag and the 3 binary codes of 00, 01, and 10. Graylevels g_6, g_7, and g_8 are generated in the same manner as the previous 3 shift codes, but this time two sets of the flag indicator code are used. The process of generating shift codes continues until the graylevels to be coded have been assigned a code word. The 2-bit shift code results in short length codes initially but grows in length as the number of graylevels to be encoded increases. The binary shift code can be implemented using any number of bits. Figure 9.7(b) shows a table of 16 graylevels coded using a 3-bit binary shift code. The difference between the 3-bit binary shift code and its 2-bit counterpart is that the 3-bit binary shift code starts with one extra bit but doesn't grow as fast as the 2-bit code.

The n-bit binary shift code starts with n bits and increases in size by n bits after $n - 1$ graylevels have been coded. The larger n is, the larger the length the code starts at, but the slower it grows in length. Consider the binary shift coding of a 256 graylevel image. Clearly if $n = 8$, then the maximum length of the binary code is 8 bits. For $n = 7$, the coding of the most probable graylevels start at 7 bits and grows to 21 bits for the lowest probable graylevels. Which size n-bit binary shift code yields the best reduction in coding redundancies depends on the distribution of graylevels within an image. Images that have graylevel distributions that are narrowly distributed about a small set of graylevels are best coded using a smaller value of n, such as 2 or 3. The highly probable graylevels will be assigned very short length codes, while the least probable graylevels will be assigned longer codes. On the other hand, images that have graylevel distributions that are uniformly distributed about a large number of graylevels are better coded using the n-bit binary shift code with the parameter n set to a large value. This increases the size of the most probable graylevels, but reduces the rate at which the code lengths grow.

Using the 2-bit binary shift code to code the 8 graylevels given in Figure 9.5(a) produces the coding assignment such that the three most probable graylevels are assigned the three 2-bit codes, the next three most probable graylevels are assigned the three 4-bit codes, and the least probable graylevels are assigned two of the three 6-bit codes. This coding assignment yields an average number of bits per pixel of 2.64. This is only slightly higher than that of 2.59 yielded by Huffman coding, yet using a much simpler coding algorithm. The n-bit binary shift code is easily implemented using a lookup table to translate between the standard fixed length binary code and the n-bit binary shift code.

Graylevel	4-bit Binary	Binary Shift	Graylevel	4-Bit Binary	Binary Shift
g_1	0000	00	g_8	1000	11 11 10
g_2	0001	01	g_9	1001	11 11 11 00
g_0	0010	10	g_{10}	1010	11 11 11 01
g_3	0011	11 00	g_{11}	1011	11 11 11 10
g_4	0100	11 01	g_{12}	1100	11 11 11 11 00
g_5	0101	11 10	g_{13}	1101	11 11 11 11 01
g_6	0110	11 11 00	g_{14}	1110	11 11 11 11 10
g_7	0111	11 11 01	g_{15}	1111	11 11 11 11 11 00

(a)

Graylevel	4-bit Binary	Binary Shift	Graylevel	4-Bit Binary	Binary Shift
g_1	0000	000	g_8	1000	111 001
g_2	0001	001	g_9	1001	111 010
g_0	0010	010	g_{10}	1010	111 011
g_3	0011	011	g_{11}	1011	111 100
g_4	0100	100	g_{12}	1100	111 101
g_5	0101	101	g_{13}	1101	111 110
g_6	0110	110	g_{14}	1110	111 111 000
g_7	0111	111 000	g_{15}	1111	111 111 001

(b)

Figure 9.7: Examples of the binary shift code: (a) the 2-bit binary shift code and (b) the 3-bit binary shift code for 16 graylevels

Another near optimum variable length code is the *n-bit B* code. Like the *n*-bit binary shift code, this code can also be implemented using a variable number of bits. Figures 9.8(a) and (b) give the *n*-bit *B* codes for $n = 1$ and $n = 2$, respectively. The smaller the value of *n*, the smaller the code length will be initially, but the faster the code length will increase as more graylevels are coded. The *B* coefficient given in Figures 9.8(a) and (b) takes on the value of 0 or 1 and is used to separate the coded words representing the graylevels. The value of *B* remains constant over a given code word but alternates in value between 0 and 1 for each adjacent code word. For example, coding the sequence of graylevels $\{g_1, g_4, g_9, g_{14}\}$ yields the code sequence of {B0, B1B0, B0B1B1, B0B0B0B0}. Since the value of B must alternate between words, and starting with 0, the final coding of these four graylevels must be {00, 1110, 000101, 10101010}.

Graylevel	4-bit Binary	1-bit B Code	Graylevel	4-Bit Binary	1-bit B Code
g_1	0000	B0	g_8	1000	B0B1B0
g_2	0001	B1	g_9	1001	B0B1B1
g_0	0010	B0B0	g_{10}	1010	B1B0B0
g_3	0011	B0B1	g_{11}	1011	B1B0B1
g_4	0100	B1B0	g_{12}	1100	B1B1B0
g_5	0101	B1B1	g_{13}	1101	B1B1B1
g_6	0110	B0B0B0	g_{14}	1110	B0B0B0B0
g_7	0111	B0B0B1	g_{15}	1111	B0B0B0B1

(a)

Graylevel	4-bit Binary	2-bit B Code	Graylevel	4-Bit Binary	2-bit B Code
g_1	0000	B00	g_8	1000	B01B00
g_2	0001	B01	g_9	1001	B01B01
g_0	0010	B10	g_{10}	1010	B01B10
g_3	0011	B11	g_{11}	1011	B01B11
g_4	0100	B00B00	g_{12}	1100	B10B00
g_5	0101	B00B01	g_{13}	1101	B10B01
g_6	0110	B00B10	g_{14}	1110	B10B10
g_7	0111	B00B11	g_{15}	1111	B10B11

(b)

Figure 9.8: Examples of the *n*-bit *B* code: (a) the 1-bit *B* code
and (b) the 2-bit *B* code for 16 graylevels

Using the 1-bit B code on the graylevels given in Figure 9.5(a) by assigning the shortest code to the most probable graylevel, and then assigning the other codes in such a way that the lengths of the codes increase as the probability of the graylevels decrease results in an average number of bits per pixel of 3.06. This example shows a case where one of the variable length codes can yield an average number of bits per pixel that exceeds that obtained simply using a fixed binary code. In this example, the fixed binary code of 3 bits outperforms the 1-bit B code. If, on the other hand, the graylevel distribution of the 8 graylevels were slightly different, such that g_1 had a probability of 0.37 and g_0 had a probability of 0.11, the average number of bits per pixel reduces to 2.86, which is now an improvement over the 3-bit binary code. The coding efficiency of these variable length codes depends on the distribution of graylevels within an image. Except for the Huffman code, the other nonoptimal variable length codes can yield an average number of bits per pixel that exceeds that of the comparable fixed length binary code. Care must be taken when using a variable length code, which is nonoptimal to reduce coding redundancies.

Another coding method that is commonly used, known as *arithmetic* or *Elias* coding, assigns a single code to a sequence of graylevels. Unlike the previously mentioned variable length codes, which assign a unique code value to each graylevel (known as block coding), the arithmetic code assigns a single code word to a sequence of graylevels. Because there is no one-to-one unique mapping of a code word to a single graylevel, arithmetic coding is commonly referred to as a nonblock code. Implementation of the arithmetic code requires computing the probability distribution of the graylevels of the image to be coded. Consider the scanning of an $N \times M$ image row by row. Each row is then subdivided into an integer number of sequences that form small groups of adjacent pixels. For example, an image with rows containing 400 pixels can be subdivided into sequences of graylevels containing 5 pixels. Consider the sequence of 5 graylevels as

$$g_{10}, \ g_{15}, \ g_{18}, \ g_{14}, \ g_{12} \ , \tag{9.19}$$

with probabilities of 0.2, 0.05, 0.05, 0.1, and 0.1, respectively. Assuming that the occurrence of each graylevel is independent of the other graylevels, the probability of the five graylevel sequences occurring within the image is the product of these 5 graylevels and is equal to 5×10^{-6}. The sum of these five probabilities gives the percentage of pixels within the image that contain at least one of these graylevels, which is equal to 0.5.

The arithmetic coding of an n graylevel sequence starts with segmenting a straight line in the interval of 0 to 1 into n different regions, with the length of the first region equal to the probability of the first graylevel in the sequence divided by the sum of the n probabilities. The next region length is then equal to the probability of the second graylevel of the sequence divided by the sum of the

n probabilities. The length of each region is determined in this same manner until the size of each region has been defined. For example, since there are 5 graylevels given in the sequence of Equation (9.19), this straight line is divided into 5 regions. Since the first graylevel g_{10} has a probability of 0.2 and the sum of the probabilities is 0.5, the first region of this straight line is assigned the range of 0 to 0.4. Both graylevels g_{15} and g_{18} have probabilities of 0.05 and hence produce the two regions of 0.4 to 0.5 and 0.5 to 0.6. The last two graylevels, which also have equal probabilities but of 0.1, produce the two regions from 0.6. to 0.8 and from 0.8 to 1.0.

The first line of Figure 9.9 shows the subdivision of this line from 0 to 1 into five regions based upon the probabilities of the 5 graylevel sequence given in Equation (9.19). The next step in encoding a sequence of graylevels using arithmetic coding is to generate a new line with its scale values varying from 0 to the value separating the first and second regions of the first line. This line is then again subdivided into n regions, with the length of the first region proportional to the probability of the first graylevel of the sequence divided by the sum of the n probabilities. The next region's length is then proportional to the probability of the second graylevel of the sequence divided by the sum of the n probabilities. The length of each region is determined in this same manner until the size of each region has been defined.

The second line given in Figure 9.9 ranges from 0 to 0.4, which corresponds to the probability of the first graylevel g_{10} given in Equation (9.19) divided by the sum of the n probabilities. This line is subdivided into the 5 regions corresponding in length in proportion to the probability of each of the five graylevels divided by the sum of the five probabilities. The range of values for the third line in the arithmetic coding sequence is generated from the range of values used to represent the second symbol in the n graylevel sequence of the second line. This line is then sub-divided using the same approach that was used to sub-divide the two previous lines. The third line in Figure 9.9 was generated from the range of 0.16 to 0.2, which is the range of values used to represent g_{15} in the second line. The process of creating lines that range in value based upon the probability of the *ith* graylevel as defined by the previous line continues for the first $n - 1$ graylevels of the n graylevel sequence. At this point, the range of values defining the final region of this *nth* line (which represents the final graylevel in the n graylevel sequence) gives the arithmetic code for this n graylevel sequence. For the 5 graylevel example given in Equation (9.19) and Figure 9.9, the arithmetic code is given by the range 0.18304 to 0.1832. Any value in this range can be used to represent the 5 graylevel sequence. For example, the value of 0.1831 can be used. The only requirement is that the minimum number of digits used must be large enough to uniquely define this range of graylevels. Hence, a total of 5 digits are needed to define this 5 graylevel sequence.

The approach used to generate the arithmetic code for an arbitrary n graylevel sequence can be expanded to arithmetic coding of an n binary sequence. As with the n graylevel sequence, the first line in the sequence is

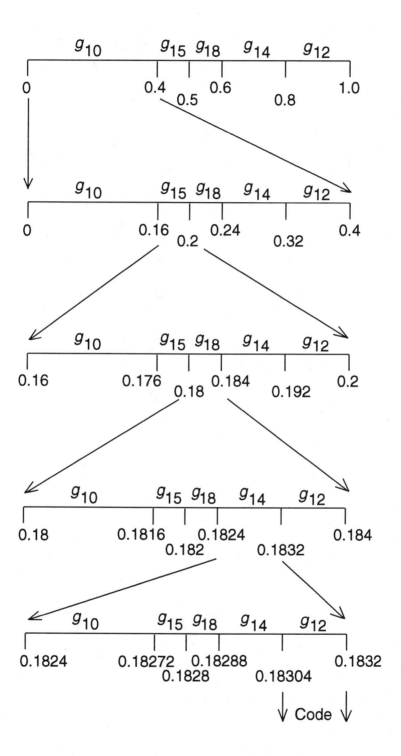

Figure 9.9: An example of arithmetic coding of Equation (9.19).

scaled between 0 and 1. This line is then subdivided into two regions, with the length of the first region equal to the probability p. This divides this line into the two regions of 0 to p and p to 1. The next line in the arithmetic coding sequence is then generated from the first line, with its range defined from 0 to p, if this first binary value is 0 or from p to 1 if the first binary value is 1. Consider the four graylevel binary sequence of 1011, with the probability of a 1 equal to 0.25 and the probability of a 0 equal to 0.75. Figure 9.10 shows the subdivision of this first line into the two regions of 0 to 0.75 and 0.75 to 1. Since the sequence 1011 begins with a 1, the values for the second line given in Figure 9.10 range from 0.75 to 1.

The second line in the arithmetic coding sequence is also subdivided into two regions, with the size of each region given the same proportions as the first line. This subdivides the second line into the two regions of p to $p - p^2$ and $p - p^2$ to 1 if the first binary graylevel is 1. If the first graylevel of the sequence were 0 instead of 1, this second line would be subdivided into the two regions of 0 to p^2 and p^2 to p. The third line in the arithmetic sequence is then generated in the same manner as the second line. The process of generating lines and subdividing these lines into two regions based upon the probability of the 0's and 1's given in the sequence continues until the last graylevel is reached. At this point, the range in the last line, representing the last graylevel, defines the range of values that can be used to represent the n graylevel sequence. Figure 9.10 shows that for the graylevel sequence of 1011 the range of values defining this sequence is 0.92578125 to 0.9375. This range can be represented in binary by expanding the decimal number D in the form of

$$D = B_1 2^{-1} + B_2 2^{-2} + B_3 2^{-3} + B_4 2^{-4} + B_5 2^{-5} \cdots , \qquad (9.20)$$

where the set of B_i's gives the binary bits. The number of bits used to represent D must be large enough to uniquely define the two regions within the last line of the arithmetic coding sequence. The approximate number of bits required to implement an n-bit sequence using arithmetic coding is approximately equal to $-\log_2(p_t)$, where p_t is the probability of the n-bit sequence. For the sequence of 1011, the probability of this sequence is $0.75 \cdot 0.25^3 = 0.0117$. The approximate number of bits required for this sequence is 7 bits. The lower the probability of occurrence of a sequence of bits is, the larger the number of bits that are required to represent this sequence. On the other hand, the larger the probability of occurrence for a given sequence is, the smaller the number of bits that will be required to represent this sequence. For example, the sequence of 0000 has a probability of $0.75^3 = 0.4219$, which requires 2 bits to represent this sequence.

The main limitation of arithmetic coding is the arithmetic accuracy that must be maintained during the coding process. As the number of graylevels used to define a sequence increases, the higher the arithmetic accuracy must be to perform arithmetic coding of this sequence. One method that is typically used to alleviate this accuracy requirement is to rescale each line so that the minimum region has a length of one and then to use a fixed number of bits to represent the boundaries of the other regions.

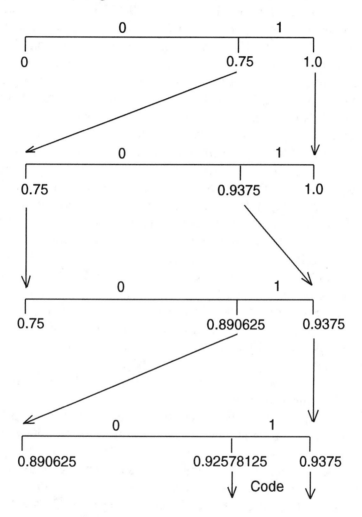

Figure 9.10: An example of arithmetic coding of the binary sequence 1011.

Another method of coding an image, known as *bit-plane* coding, is based upon separating an n-bit graylevel image into n binary images, with each binary image individually representing the spatial distribution of each of the n bits. For example, consider an 8-bit graylevel image containing the bits b_0, b_1, b_2,$\cdots$ b_7. The graylevels in this image are then computed from each of these 8 bits using

$$g = b_0 \cdot 2^0 + b_1 \cdot 2^1 + b_2 \cdot 2^2 + b_3 \cdot 2^3 + b_4 \cdot 2^4 + b_5 \cdot 2^5 + b_6 \cdot 2^6 + b_7 \cdot 2^7 .$$

(9.21)

The goal behind bit-plane coding is to produce a set of binary images that can then be compressed on an individual basis. Figure 9.11(a) shows a 256 graylevel image that is separated into 8 individual binary images, with each bit representing one of the eight bit-planes. Figure 9.11(b) shows the binary image corresponding to the most significant bit b_7, where each of the 256 graylevels contained within Figure 9.11(a) have been coded using an 8-bit fixed binary code. Black in this binary image corresponds to the original image's graylevels of 0 to 127, while white corresponds to the graylevels of 128 to 255 in the original image. Notice the amount of spatial redundancy present within this binary image. This image is easily coded using run length coding or arithmetic coding. Figure 9.11(d) is the bit-plane corresponding to the next lower bit of b_6. This binary image shows less spatial redundancy than the most significant bit-plane image given in Figure 9.11(b). Figures 9.11(f), (h), (j), (l), (n), and (p) give the bit-plane images for the bits b_5, b_4, b_3, b_2, b_1, and b_0.

As the order of the bits used to generate these eight bit-plane images decreases, the spatial redundancies present within the binary images decrease. In fact, the noticeable spatial features of the boy's face are no longer observable in the three least significant bit-plane images of Figures (l), (n), and (p). In most complex images, the lower bit-planes are usually decorrelated, showing very little spatial redundancy. This is why the addition of a small amount of uniformly distributed noise to the 256 graylevel image of Figure 9.1(a) was able to improve the visual quality of this image after quantizing it to 4 bits, producing only sixteen graylevels. This pseudo-randomizing of the lower order 4 bits of the graylevel image was able to remove the false contours from the quantized image, which is more visually objectionable than the slight increase in the grainy appearance of the quantized image.

The spatial correlation in the lower order bit-plane images can be increased by using a different coding approach than the standard n-bit binary code. The difficulty with the binary code is that for small graylevel changes a large number of bits can change. The lower the number of bits that change, transitioning from one graylevel to the next, the higher the spatial redundancies will be in the set of bit-plane images. Consider the fixed binary coding of a 256 graylevel image using 8 bits. Next, consider graylevels 0 and 1. The 8-bit binary codes for these graylevels are 00000000 and 00000001, respectively. This small change in graylevel has affected only the least significant bit-plane leaving the other seven bit-planes unchanged. Now, consider graylevels 127 and 128, producing the 8-bit binary codes of 01111111 and 10000000, respectively. This same small change in graylevel has now resulted in the change of all 8 bits. Coding an image in such a way that a unit change in graylevel produces only a single bit change in the coded representation increases the spatial redundancies of the bit-plane images.

A very commonly used code that provides an easy means of error checking is that of the *binary gray* code. This code is popular because it allows

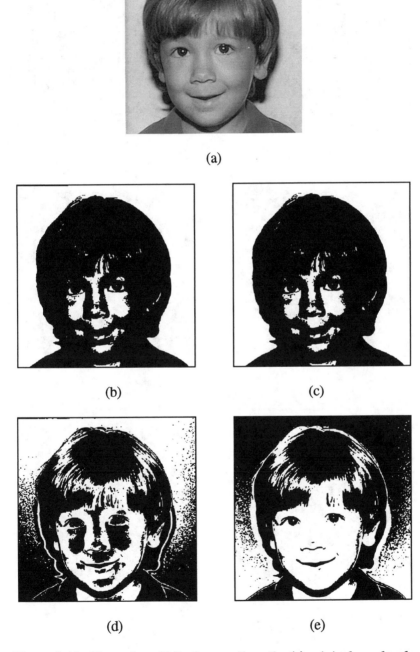

(a)

(b) (c)

(d) (e)

Figure 9.11: Examples of bit-plane coding: the (a) original graylevel, (b) binary code bit 7, (c) gray code bit 7, (d) binary code bit 6, and (e) gray code bit 6 images.

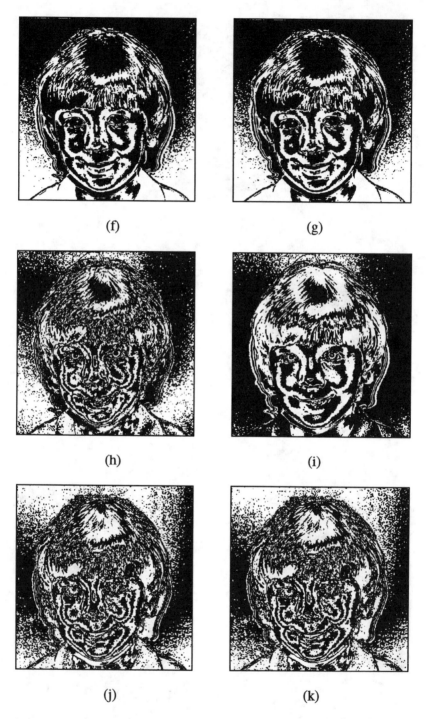

Figure 9.11: Examples of bit-plane coding: the (f) binary code bit 5,
(g) gray code bit 5, (h) binary code bit 4, (i) gray code
bit 4, (j) binary code bit 3, and (k) gray code bit 3 images.

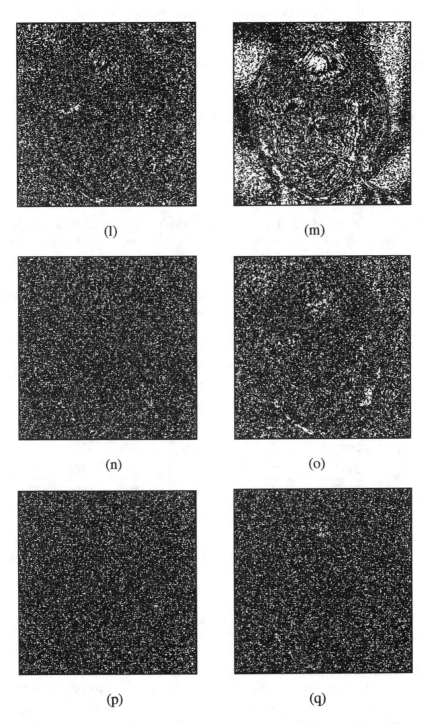

Figure 9.11: Examples of bit-plane coding: the (l) binary code bit 2,
(m) gray code bit 2, (n) binary code bit 1, (o) gray code
bit 1, (*p*) binary code bit 0, and (q) gray code bit 0 images.

only a single bit change between adjacent code words. Because of this feature, this code is commonly used within digital position sensors to represent the digital coded words of position. Since most position sensors produce position values that are linearly increasing or decreasing in position, the use of the gray code guarantees that digital codes for adjacent positions will vary by only 1 bit. If more than 1 bit has changed from one position to the next, an error has occurred and the position value is in error. The *ith* bit of an *n*-bit binary gray code bg_i is easily computed from the fixed *n*-bit binary code using

$$bg_i = b_i \text{ EX-OR } b_{i+1} \qquad \text{for } i = 0, 1, 2 \cdots n - 2$$
$$bg_{n-1} = b_{n-1} , \qquad\qquad\qquad\qquad (9.22)$$

where b_i is the *ith* bit of the graylevel to be coded. Figure 9.12 shows the binary gray coding of 16 graylevels using a total of 4 bits. In the left-hand column is the standard fixed 4-bit binary code, while in the right-hand column is the corresponding 4-bit binary gray code created using Equation (9.22) with $n = 4$. Comparing each adjacent code word for the gray code shows that there is only a 1-bit change in the code words of adjacent graylevels.

Graylevel	4-bit Binary	4-bit Gray	Graylevel	4-Bit Binary	4-bit Gray
g_1	0000	0000	g_8	1000	1100
g_2	0001	0001	g_9	1001	1101
g_0	0010	0011	g_{10}	1010	1111
g_3	0011	0010	g_{11}	1011	1110
g_4	0100	0110	g_{12}	1100	1010
g_5	0101	0111	g_{13}	1101	1011
g_6	0110	0101	g_{14}	1110	1001
g_7	0111	0100	g_{15}	1111	1000

Figure 9.12: The assignment of the 4-bit gray code and the 4-bit binary code to 16 graylevels.

Figure 9.11(c) shows the most significant bit-plane image of Figure 9.11(a) created using the 8-bit binary gray code computed from Equation (9.22). There is very little difference between this image and the most significant bit-plane image given in Figure 9.11(b), created using the 8-bit fixed binary code. The spatial redundancies in both these bit-plane images are about the same. The next lower order bit-plane image created using the binary gray code is shown in Figure 9.11(e). As seen in comparing Figures 9.11(d) and (e), this binary gray coded bit-plane image has a higher spatial redundancy than the same order bit-plane image created using the standard fixed binary code. Figures 9.11(g), (i), (k), (m), (o), and (q) correspond to bg_5, bg_4, bg_3, bg_2, bg_1, and bg_0 bit-plane images coded using the binary gray code. As with the fixed binary code bit-plane images, the lower the order of the bit-plane image, the lower the spatial redundancy present within the image will be. In comparing the fixed binary code bit-plane images to the binary gray code bit-plane images, the use of the gray coding has increased the spatial redundancies present in the lower order bit-plane images. In particular, spatial features of the boy's face are still present in the bg_2 bit-plane image given in Figure 9.11(m). None of the spatial features of the boy's face are observable in the same order bg_2 bit-plane image using the fixed binary code shown in Figure 9.11(l).

The coding of an n-bit graylevel image into n binary images, with each image representing one of the graylevel image's bits, provides a means of compressing an image using run length coding to remove the spatial redundancies present within the binary image. Run length coding of an image cannot be used effectively on graylevel images. Compression of an image using run length coding relies on long sequences of pixels with the same graylevel value. Images that contain large regions of constant graylevel compress well using run length coding. The most significant bit-plane images of Figures 9.11(b) and (c) compress well using run length coding. On the other hand, the least significant bit-plane images given in Figures 9.11(p) and (q) barely compress at all using run length coding due to the lack of spatial redundancies present within both of these images. Clearly, since binary gray coding of a graylevel image produces bit-plane images with higher spatial redundancies, the binary gray code is the preferred coding method prior to applying run length coding to reduce the spatial redundancy. The binary gray coding followed by run length coding of each bit-plane image yields better compression than coding each bit-plane using the standard fixed binary code followed by run length coding. Run length coding is also a standard of the CCITT (Consultative Committee of the International Telephone and Telegraph), which has been included in its Group 3 for use in the transmission of binary images over phone lines, and in particular for use in facsimile, or fax, machines. Hence, the runlength coding of a grayscale image using its n bit-plane images provides a simple method of lossless compression.

The concept of run length coding was briefly discussed in illustrating a compression method that compressed a white square against a black background in the binary image of Figure 9.2(a). Consider the sequence of N pixels g_0, g_1, g_2, $\cdots$ g_{N-1} from a binary image. Run length coding produces a mapping of these pixels into a set of vectors containing two elements:

$$(g_0, g_1, g_2, \cdots g_{N-1}) \rightarrow (gr_0; L_0), (gr_1; L_1), \cdots (gr_q; L_q) , \qquad (9.23)$$

where the graylevels g_0, g_1, g_2, $\cdots$ g_{N-1} can take on only the graylevels of G_a or G_b since these graylevels are from a binary image. Each vector describes a sequence of contiguous pixels with the same graylevel. The first element of this two element vector $(gr_i; L_i)$ gives the graylevel gr_i of the first pixel in the sequence, while L_i gives the number of pixels defined by the sequence.

The one-dimensional run length coding of an $N \times M$ binary image starts with scanning the binary image to be compressed row by row and individually run length coding each row. Starting with the leftmost pixel of the first row and counting the number of connected pixels that have the same graylevel forms the first code vector, given by gr_0 and L_0. The scanning of the row continues, generating the next vector defined by the next set of contiguous pixels with graylevels gr_1 and of length L_1. The generation of these run length vectors continues until all pixels within the first row have been coded. The process of run length coding continues with the second row by dividing this row into a set of sequences with contiguous pixels of the same graylevel. Run length coding of the image is completed when all the rows have been coded. Even though the image was coded in a row-like manner, the image could also have been coded in a column-like manner. The difference is that run length coding an image based on rows reduces spatial redundancy in the horizontal direction, while run length coding an image using columns reduces the spatial redundancy in the vertical direction.

The probability h_i of each one-dimensional run length vector encoding a sequence of pixels at graylevel gr_i can be determined by dividing the number of pixels L_i defining the vector element by the total number of pixels within the image at graylevel gr_i. Consider the collection of all run length vectors that have coded all pixels within the image at graylevel G_a. The entropy E_a associated with these vectors can then be computed using

$$E_a = - \sum_{i=0}^{J-1} ha_i \cdot \log_2(ha_i) \; . \tag{9.24}$$

In Equation (9.24) the parameter J is the number of run length vectors defining the pixels within the image at graylevel G_a and ha_i defines the probability of each run length vector within the image with a length La_i. This probability ha_i can then be determined from

$$ha_i = \frac{NLa_i}{J} \; , \tag{9.25}$$

where NLa_i is the number of run length vectors with length La_i that are used to encode the pixels with graylevel G_a.

Using the same approach in defining the entropy for the run length of the pixels at graylevel G_b yields the entropy for the run length vectors representing the pixels within the image at graylevel G_b as

$$E_b = - \sum_{i=0}^{K-1} hb_i \cdot \log_2(hb_i) \tag{9.26}$$

and with probabilities hb_i of

$$hb_i = \frac{NLb_i}{K} \ , \tag{9.27}$$

where K is the number of run length vectors coding the pixels with graylevel G_b and NLb_i is the total number of run length vectors with a length of Lb_i. The approximate entropy for this binary image using run length coding is

$$E_{rl} = \frac{E_a + E_b}{L_a + L_b} \ , \tag{9.28}$$

where L_a and L_b are the average lengths for the run length vectors representing pixels at graylevels G_a and G_b, given by

$$L_a = \frac{1}{J} \sum_{i=0}^{J} La_i \tag{9.29}$$

and

$$L_b = \frac{1}{K} \sum_{i=0}^{K} Lb_i \ . \tag{9.30}$$

Equation (9.28) defines an estimate of the average number of bits per pixel required to encode the $J + K$ run length vectors using a variable length code for each run length vector.

Figure 9.13 shows an example of a 7×7 binary image, which requires a total of 49 bits to represent an image in uncompressed form. Run length encoding this image row by row starting with the topleft most pixel yields the first run length vector $(0; 7)$ for the first row. The next three rows produce the same run length vector $(1; 7)$, while the two following rows produce the run length vectors $(1; 4)$ and $(0; 3)$. Finally, the last row yields the run length vector $(0; 7)$. The total sequence of the run length vectors for this 7×7 image is

$$(0; 7), (1;7), (1; 7), (1; 7), (1; 4), (0; 3), (1; 4), (0; 3), (0; 7) \ . \tag{9.31}$$

If one bit is used to represent the graylevel and 3 bits are used to define the length element, then a total of 36 bits are required to represent this 7×7 image using run length encoding. Run length encoding of this image has resulted in a compression ratio of 1.36 to 1. If the run length vectors are computed in the vertical direction using columns, the sequence of run length vectors is

$$(0; 1), (1; 5), (0; 1), (0; 1), (1; 5), (0; 1), (0; 1), (1; 5), (0; 1), (0; 1), (1; 5),$$
$$(0; 1), (0; 1), (1; 3), (0; 3), (0; 1), (1; 3), (0; 3), (0; 1), (1; 3), (0; 3) \ . \qquad (9.32)$$

The vertical run length coding of this image has produced 21 run length vectors, requiring a total of 84 bits. Run length coding of this image in the vertical direction has resulted in an increase in the number of bits required to represent the image as a result of less spatial redundancy present in the vertical direction. Images with low spatial redundancies can produce run length codes that are larger than using a fixed n-bit binary code.

0	0	0	0	0	0	0
1	1	1	1	1	1	1
1	1	1	1	1	1	1
1	1	1	1	1	1	1
1	1	1	1	0	0	0
1	1	1	1	0	0	0
0	0	0	0	0	0	0

Figure 9.13: An example of run length encoding of a binary image.

A further improvement in the one-dimensional run length encoding of a binary image can be accomplished by performing zig-zag scanning as shown in Figure 9.14. This scanning method when used with run length encoding further reduces the spatial redundancies present within a binary image. The run length vectors for Figure 9.13 using the zig-zag scan given in Figure 9.14 reduce to

$$(0; 7), (1; 25), (0; 6), (1; 4), (0; 7) \ . \qquad (9.33)$$

The total number of bits required to represent this set of run length vectors using 1 bit for the graylevel element and 5 bits for the length element becomes 30 bits. A close inspection of Equation (9.33) shows that there are 5 unique run length vectors that can be coded using only 3 bits. The 49 bit uncompressed image of Figure 9.13 has now been reduced to 15 bits, resulting in a compression ratio of 3.27 to 1. A further reduction in the compression ratio is obtained by encoding the run length vectors using a variable length code, with the shortest codes assigned to the most probable run length vectors. This produces an average number of bits per pixel of 0.31, which approaches the estimate of 0.12 predicted by Equation (9.28). Another approach of run length encoding is to use a two-dimensional run length code. Most of these codes are beyond the scope of this book and the interested reader is referred to several of the texts given in the bibliography, such as *Digital Image Processing Algorithms* by I. Pitas.

Figure 9.14: Scanning a binary image using a zig-zag scan.

Another method of compressing a binary image by reducing the spatial redundancies present within the image is to edge detect the image, storing the closed contours describing the regions within the image at graylevel G_O. Consider the 256×256 binary image of the letter "E" given in Figure 9.15(a). A total of 65,536 bits are required to represent this uncompressed image. Edge detection of this image, as shown in Figure 9.15(b), produces a total of 1024 edge pixels, which can be described by a set of x, y coordinate values. Using 8 bits to define the x coordinate and 8 bits to describe the y coordinate reduces the number of bits required to code this image to 16,384 bits, yielding a compression ratio of 4 to 1. Further compression can be obtained if the contours of the letter "E" are coded using one of the contour representation methods given in Chapter 8. The contour of the letter "E" can be described using Fourier descriptors. Figure 9.15(c) shows the edge contour of the letter "E" described from the first 201 Fourier descriptors. Using floating point numbers to represent the real and imaginary components of each Fourier descriptor requires 64 bits (4 *bytes* per real number) to describe a Fourier descriptor. Hence, using Fourier descriptors, the letter "E" in Figure 9.15(a) can be coded using 12,864 bits.

Using the 8-chain code description to describe the contour of the letter "E" given in Figure 9.15(b) requires an initial x, y coordinate point and then 3 bits to represent each of the chain code descriptors. Using 8 bits for the initial x coordinate, 8 bits for the initial y coordinate, and then 1023 chain code descriptors of 3 bits completely describes the contour and reduces the number of bits required to code this image to 3,085 bits, a further improvement in the compression ratio to 21.2 to 1. The contour describing the letter "E" can also be described with a set of straight lines. Figure 9.15(d) show the contour image of the letter "E" generated from 25 straight lines. Since each line requires a slope and an intercept, if floating numbers are used, requiring 32 bits per real number, the number of bits required to represent the letter "E" reduces to 1,600 bits. This results in a compression ratio of 41 to 1. These reductions in the number of bits required to represent the letter "E" were not without a price. The contours given

(a)

(b)

Figure 9.15: Examples of compressing an image using the closed
                     contours present within the image: (a) the original
                     binary image and (b) its edge detected image.

(c)

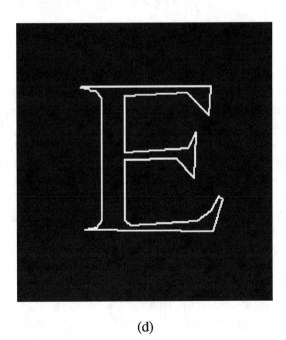

(d)

Figure 9.15: Examples of compressing an image using the closed contours present within the image: the contour images described using (c) 201 Fourier descriptors and (d) 25 straight lines.

by the Fourier descriptors and the straight lines are distorted versions of the original contour, resulting in lossy compression. If error-free compression is desired, the 8-chain code description should be used, since it yields a compressed description of a line with no errors, or all of the Fourier descriptors should be used. Finally, to reconstruct the uncompressed image from the set of closed contours, each closed contour is regenerated within the image from its contour descriptors. Next, the pixels within the image that are enclosed by these closed contours are set to graylevel G_O.

Another method of lossless compression that reduces the spatial redundancies present within a graylevel image is that of *lossless predictive coding*. The concept behind lossless predictive coding is that adjacent pixels that are spatially redundant have very similar graylevels. Instead of coding the graylevels of the pixels directly, the difference in graylevels between adjacent pixels is obtained and then coded. If the adjacent pixels are very similar in graylevel, this difference will be very small and will require fewer bits to code as compared to coding the graylevels directly. Figure 9.16 shows the block diagram for the encoder and the decoder for lossless predictive coding. The goal of the predictor is to provide as close an estimate $f_{est}(x, y)$ as possible to the input image $f(x, y)$. The difference between the predictor output and the input image defines the error $e(x, y)$ in the estimation given by the predictor:

$$e(x, y) = f(x, y) - f_{est}(x, y) \ . \tag{9.34}$$

This error image $e(x, y)$ is then coded using either Huffman coding or one of the variable length codes to produce the compressed image. To uncompress an image that was compressed using lossless predictive coding, the inverse mapping to that given by the encoder (the decoder) is used to regenerate the error image $e(x, y)$, as shown in Figure 9.16(b). Next, the predictor generates an estimate of the original image $f_{est}(x, y)$. The sum of the error image $e(x, y)$ and the estimate image from the predictor $f_{est}(x, y)$ produces the uncompressed image, which is an exact replica of the original graylevel image.

The most difficult part of lossless predictive compression is the selection of the predictor. Typically, the predictor is defined as a function of the previous set of input pixels from the input image. Starting at the top left-hand corner of the input image and proceeding down to the bottom right-hand corner, the previous pixels input to the predictor are defined as the pixels contained in the previous rows and to the left of the present horizontal position. In other words, the output of the predictor in terms of the previous input pixels is defined as

$$f_{est}(x, y) = \text{int}\left[\sum_{k=1}^{K} \sum_{j=1}^{J} c_{j,\,k} \cdot f(x-k, y-j) + 0.5 \right], \tag{9.35}$$

where $c_{j,k}$ are the coefficients of the predictor and the parameters J and K determine the number of previous input pixels that are used in the calculation of the predictor output, which defines the order of the predictor. Typically, it has been found that there is only a slight improvement in predictor quality by using more than three previous input pixels (a third order predictor). Since the input graylevel image contains graylevels that are integers in value, the 0.5 term and the int[] function rounds the output of the predictor to the nearest integer graylevel.

In the compression of NTSC video images using lossless predictive coding, the predictor given by Equation (9.35) is expanded to three dimensions,

$$f_{est}(x, y) = \text{int}\left[\sum_{n=1}^{N} \sum_{k=1}^{K} \sum_{j=1}^{J} c_{j,k,n} \cdot f(x-k, y-j, t-n) + 0.5 \right], \quad (9.36)$$

to include N previous temporal frames of NTSC video. In NTSC video, there is very little change in the spatial content within adjacent frames. Much of this spatial information is redundant and can be easily compressed using lossless predictive compression. Instead of transmitting the entire frame of video, only the difference between video frames is transmitted. Each individual frame of video is then reconstructed (uncompressed) using the decoder given in Figure 9.16(b).

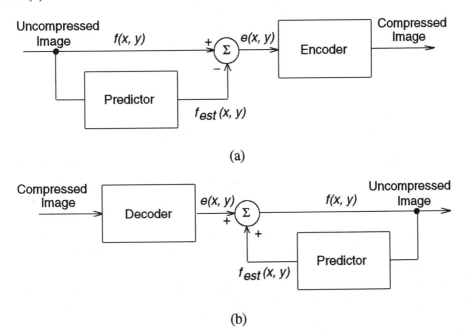

Figure 9.16: The block diagram for lossless predictive coding: (a) the encoder and (b) the decoder.

Consider the compression of the 5 × 5 image given in Figure 9.17 using the first order linear predictor that depends on the previous horizontal pixel's graylevel:

$$f_{est}(x, y) = f(x - 1, y).$$
(9.37)

This image is scanned row by row starting with the leftmost pixel of each row and recording its graylevel value and then recording the difference in graylevels between the following adjacent pixels. The lossless predictive coding of these 5 rows produces the following sequence:

$$
\begin{array}{llllll}
0: & 0, & 2, & 1, & 1 \\
1: & 0, & 1, & 1, & 0 \\
1: & 0, & 2, & 1, & 0 \\
1: & 1, & 1, & 1, & 0 \\
1: & -1, & 2, & 2, & 0 \ , & \qquad (9.38)
\end{array}
$$

where the elements of the first column in Equation (9.38) are the graylevels given in the first column of the 5 × 5 image of Figure 9.17.

0	0	2	3	4
1	1	2	3	3
1	1	3	4	4
1	2	3	4	4
1	0	2	4	4

Figure 9.17: An example of lossless predictive coding.

The coding of this 5 row predictive coded sequence using 2 bits for the difference graylevels and 3 bits to represent the graylevel of the first pixel encountered in each row results in a total of 55 bits required to encode this 5 × 5 image. Using the fixed 3-bit binary code results in a total of 75 bits. Lossless predictive coding of this 5 × 5 image as compared to using the fixed 3-bit binary code has yielded a compression ratio of 1.36 to 1. Compression of an image using Equation (9.37) is commonly referred to as differential coding, in that the difference graylevel is coded and saved as the compressed image. Uncompressing the 5 × 5 predictive coded sequence given in Equation (9.38) starts by using the column of graylevels given in this equation to form the first column of graylevels in the uncompressed image. Then in a row-like manner, scanning from left to right, the other pixels are generated by adding the difference graylevel to the graylevel of the pixel to the left. For example, the first row given in Equation (9.38) is uncompressed using the following additions: 0, (0 + 0), (0 + 2), (2 + 1), (3 + 1).

Further compression of this image is possible if a zig-zag scan as shown in Figure 9.14 is used and the predictor given in Equation (9.37) is modified to

$$f_{est}(x, y) = \begin{cases} f(x - 1, y) & \text{for } y \text{ even} \\ f(x + 1, y) & \text{for } y \text{ odd} \end{cases} . \tag{9.39}$$

Using a zig-zag scan removes the requirement of storing the initial graylevel of each row and further reduces the number of bits required to encode the image. The storage of only the first pixel's graylevel is required to uncompress the predictive coded image. Using a zig-zag scan on the 5×5 image of Figure 9.17 yields the initial graylevel of zero and the following sequence of 24 difference graylevels:

$$0: 0, 2, 1, 1, -1, 0, -1, -1, 0, 0, 0, 2, 1, 0, 0, 0, -1, -1, -1, 0, -1, 2, 2, 0. \tag{9.40}$$

The histogram and the Huffman coding of these 24 difference graylevels are given in Figure 9.18. The average number of bits required to represent these difference graylevels computed from Equation (9.5) is 1.87 bits. Using 3 bits for the initial graylevel of 0 yields the total of 48 bits required to encode this 5×5 image using lossless predictive coding followed by Huffman coding, producing a compression ratio of 1.56 to 1.

Graylevel	Probability	
g_{-1}	0.292	00
g_0	0.416	1
g_1	0.125	011
g_2	0.167	010

Figure 9.18: The Huffman coding of the difference graylevels given in Equation (9.40).

Figure 9.19(a) shows a 480×480 by 256 graylevel image of a young girl containing large regions of spatial redundancies, which is particularly noticeable in the background region. Figure 9.19(b) is its corresponding graylevel histogram, showing the distribution of graylevels present within the image. The standard deviation of these graylevels and the average graylevel of this image are computed from this histogram as 92.8 and 55.38, respectively. Using Equation (9.6) the entropy is computed as 7.31 bits. The best that optimum coding of this image can yield is an average bit per pixel of 7.31 bits. Figure 9.19(c) shows the error image $e(x, y)$ obtained by lossless predictive coding of this 480×480

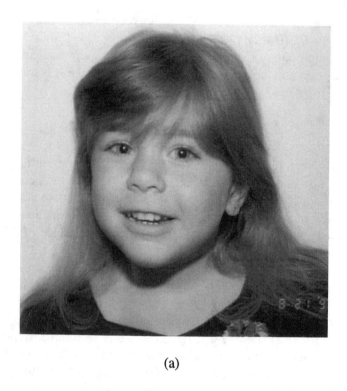

(a)

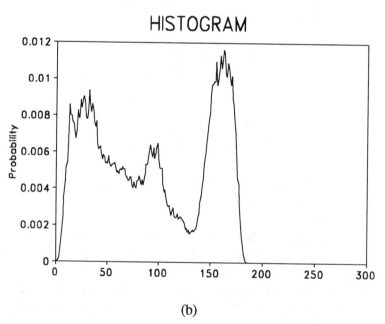

(b)

Figure 9.19: Example of lossless predictive coding: (a) the original
image and (b) its corresponding histogram.

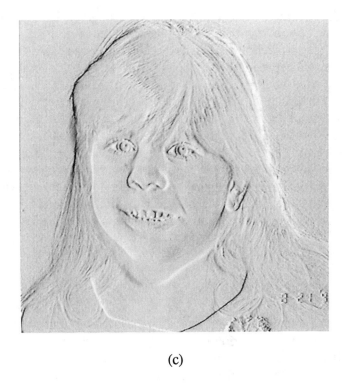

(c)

HISTOGRAM

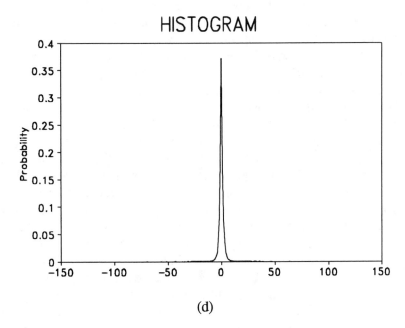

(d)

Figure 9.19: Example of lossless predictive coding: (c) the error
graylevel image and (d) its corresponding histogram.

by 256 graylevel image using the linear predictor given in Equation (9.37). Encoding of the image was then accomplished using a row scan from left to right, storing the graylevel of the first pixel of each row and then computing the graylevel difference between adjacent pixels. This is the same lossless predicted coding method used to encode the 5×5 image of Figure 9.17, producing the sequence given in Equation (9.38). Figure 9.19(c) shows the error image $e(x, y)$, which was generated by scaling the range of graylevel differences from -255 through 255 to 0 through 255 so that they could be displayed as an image. The contrast of this error image was then scaled by 17 to enhance the spatial features present. Notice how well the predictor performed within homogeneous regions.

Figure 9.19(d) is the histogram of the graylevel differences shown in Figure 9.19(c) prior to any scaling. Note the small distribution of graylevels present within the error image $e(x, y)$. Because of the sharpness of this histogram about a graylevel difference of 0, the histogram for $e(x, y)$ is quite often modeled using the discrete version of the *Laplacian* PDF,

$$h_i(e_k) = \frac{1}{\sigma \sqrt{2}} e^{-2^{1/2}|e_k|/\sigma} \, , \qquad\qquad (9.41)$$

where σ is the standard deviation of the graylevel differences and e_k is the kth graylevel difference value. Computation of the mean and standard deviation from the graylevel difference histogram given in Figure 9.19(d) yields values of 0 and 2.46, respectively. Computation of the entropy for these graylevel differences yields 2.92 bits. This decrease in the entropy for the graylevel differences as compared to the entropy of the original graylevel image shows the reduction in the spatial redundancies. Ignoring the overhead bits required to store the initial graylevel of each row, which is small compared to the total number of bits required to encode the error image $e(x, y)$, the approximate compression ratio gained over using an optimum variable length code directly on the original graylevel image is approximately $7.31 / 2.92 = 2.5$.

The reduction in the number of bits required to encode the error image $e(x, y)$ depends on how good the predictor $f_{est}(x, y)$ is at estimating the uncompressed image's graylevels $f(x, y)$. In an ideal sense, if the estimator is perfect, the error $e(x, y)$ would always be zero and the number of bits required to encode this image would also be zero. In essence, no bits are needed to encode this image because the perfect estimator gives the exact uncompressed image and hence can predict the exact replica of the original graylevel image during the uncompression process. In real world situations, the perfect predictor does not exist and the best that can be obtained is to use an estimator that minimizes the output error $e(x, y)$ or graylevel difference during the lossless predictive encoding. In other words, the goal is to find the set of coefficients used by the predictor as given in Equation (9.35) that minimizes the mean square error (MSE) of the graylevel differences $e(x, y)$.

The solution for the optimum predictor as presented here assumes the use of floating point arithmetic, which increases the computational complexity of the predictor slightly over using integer arithmetic. Hence, the definition of the predictor given in Equation (9.35) will be rewritten by dropping the int[] function and 0.5 term, which is used to round the predictor to the nearest integer,

$$f_{est}(x, y) = \sum_{k=1}^{K} \sum_{j=1}^{J} c_{j, k} \cdot f(x - k, y - j) . \tag{9.42}$$

Consider the formation of two vectors of length JK from the coefficients $c_{j, k}$ and the input image $f(x, y)$. The first one-dimensional vector $\mathbf{F}$, a column matrix representing the input image, is formed by arranging the elements of $f(x - k, y - j)$ in the following manner

$$\mathbf{F_n}^{\mathbf{T}} = [\, f(x - 1, y - 1), f(x - 2, y - 1) \cdots f(x - K, y - 1), f(x - 1, y - 2),$$
$$f(x - 2, y - 2), \cdots f(x - K, y - 2), \cdots f(x - 1, y - J), \cdots f(x - K, y - J) \,] , \tag{9.43}$$

where $\mathbf{T}$ means matrix transpose and the *nth* position in $\mathbf{F_n}$ is defined in terms of j, k as

$$n = k + j \cdot K . \tag{9.44}$$

In addition, j and k can be written in terms of n as

$$j = \text{int}\left[\frac{n}{K}\right] \tag{9.45}$$

and

$$k = n - \text{int}\left[\frac{n}{K}\right] , \tag{9.46}$$

where $k = 1, 2, 3, \cdots K$ and $j = 1, 2, 3, \cdots J$. The second vector $\mathbf{C_n}$, also a column matrix, is formed from the coefficients $c_{j, k}$ by arranging these elements in $\mathbf{C_n}$ as follows

$$\mathbf{C_n}^{\mathbf{T}} = [\, c_{x - 1, y - 1}, c_{x - 2, y - 1} \cdots c_{x - K, y - 1}, c_{x - 1, y - 2},$$
$$c_{x - 2, y - 2}, \cdots c_{x - K, y - 2}, \cdots c_{x - 1, y - J}, \cdots c_{x - K, y - J} \,] , \tag{9.47}$$

where the *nth* position in $\mathbf{C_n}$ is also given by Equation (9.44).

Using matrix multiplication and the definitions of $\mathbf{C_n}$ and $\mathbf{F_n}^{\mathbf{T}}$ given in Equations (9.43) and (9.47), the output of the predictor is then defined as

$$f_{est}(x, y) = \mathbf{C_n}^{\mathbf{T}} \mathbf{F_n} = \mathbf{F_n}^{\mathbf{T}} \mathbf{C_n} . \tag{9.48}$$

The error image $e(x, y)$ between the predictor output and the input image at the coordinate x, y is

$$e(x, y) = f(x, y) - f_{est}(x, y) \ . \tag{9.49}$$

Substituting Equation (9.48) into (9.49) yields

$$e(x, y) = f(x, y) - \mathbf{C_n}^\mathbf{T} \mathbf{F_n} \tag{9.50}$$

or

$$e(x, y) = f(x, y) - \mathbf{F_n}^\mathbf{T} \mathbf{C_n} \ \cdot \tag{9.51}$$

The approach in deriving the optimum predictor requires squaring either Equations (9.50) or (9.51) and then taking the expected value to obtain the mean square error. Equations (9.50) and (9.51) are nothing more than the standard equations used within the digital signal processing field to describe an adaptive filter with weights given by $\mathbf{C_n}$ and inputs given by $\mathbf{F_n}$. The approach presented here to find the optimum set of coefficients $\mathbf{C_n}$ is derived from the text *Adaptive Signal Processing* by Widrow and Stearns given in the bibliography. Squaring Equation (9.51) yields

$$e(x, y)^2 = f(x, y)^2 + \mathbf{C_n}^\mathbf{T} \mathbf{F_n} \mathbf{F_n}^\mathbf{T} \mathbf{C_n} - 2 f(x, y) \mathbf{F_n}^\mathbf{T} \mathbf{C_n} \ . \tag{9.52}$$

Taking the expected value of both sides of Equation (9.52) gives the **MSE** of the predictor in terms of the input image $f(x, y)$:

$$\text{MSE} = E[e(x, y)^2] = E[f(x, y)^2 + \mathbf{C_n}^\mathbf{T} \mathbf{F_n} \mathbf{F_n}^\mathbf{T} \mathbf{C_n} - 2 f(x, y) \mathbf{F_n}^\mathbf{T} \mathbf{C_n}] \ , \tag{9.53}$$

or

$$\text{MSE} = E[f(x, y)^2] + \mathbf{C_n}^\mathbf{T} E[\mathbf{F_n} \mathbf{F_n}^\mathbf{T}] \mathbf{C_n} - 2 E[f(x, y) \mathbf{F_n}^\mathbf{T}] \mathbf{C_n} \ . \tag{9.54}$$

But $E[\mathbf{F_n}\mathbf{F_n}^\mathbf{T}]$ is the autocorrelation of the previous K, J input image pixels included in the predictor computation, defined by

$$R = E[\mathbf{F_n} \mathbf{F_n}^T] = E \begin{bmatrix} f_1^2 & f_2 f_1 & f_3 f_1 & \cdots & f_K f_1 \\ f_1 f_2 & f_2^2 & f_3 f_2 & \cdots & f_K f_2 \\ \vdots & \vdots & \vdots & & \vdots \\ f_1 f_J & f_2 f_J & f_3 f_J & \cdots & f_K f_J \end{bmatrix} \ , \tag{9.55}$$

where f_n is defined as

$$f_n = f(x - n - \text{int}[n/K], y - \text{int}[n/K]) . \tag{9.56}$$

For example, if $n = 5$, $K = 3$, and $J = 3$, $f_5 = f(x - 2, y - 1)$. In a similar manner, $E[f(x, y) \, \mathbf{F}_n^\mathbf{T}]$ is the correlation of the input image's pixels at the coordinate x, y with the previous pixels included in the predictor computation. Let the column vector $\mathbf{G}$ of length JK define

$$\mathbf{G} = E[f(x, y) \, \boldsymbol{F_n}] = E \begin{bmatrix} f(x, y) f_1 \\ f(x, y) f_2 \\ f(x, y) f_3 \\ \vdots \\ f(x, y) f_{KJ} \end{bmatrix}, \tag{9.57}$$

where f_n is defined by Equation (9.56). Substituting Equations (9.55) and (9.57) into Equation (9.54) yields the definition of the MSE as

$$\text{MSE} = E[f(x, y)^2] + \mathbf{C}_n^\mathbf{T} \, \mathbf{R} \, \mathbf{C}_n - 2 \, \mathbf{G}^\mathbf{T} \, \mathbf{C}_n . \tag{9.58}$$

The goal is to find the set of coefficients $\mathbf{C}_n$ that yields the minimum **MSE**. This is obtained by finding the gradient of Equation (9.58) with respect to the coefficients $\mathbf{C}_n$. Let

$$\boldsymbol{\nabla} = \begin{bmatrix} \dfrac{\partial \, \text{MSE}}{\partial \, C_{1,\,1}} \\[2mm] \dfrac{\partial \, \text{MSE}}{\partial \, C_{2,\,1}} \\[2mm] \vdots \\[2mm] \dfrac{\partial \, \text{MSE}}{\partial \, C_{K,\,J}} \end{bmatrix} \tag{9.59}$$

be the gradient vector, in which the elements are composed of the partial derivatives of the MSE in respect to the predictor coefficients $\mathbf{C}_n$. Taking the gradient of Equation (9.58) yields

$$\boldsymbol{\nabla} = 2 \, \mathbf{R} \, \mathbf{C}_n - 2 \, \mathbf{G} . \tag{9.60}$$

The optimum estimation of the input image from the predictor occurs when the gradient is equal to zero ($\boldsymbol{\nabla} = \mathbf{0}$), which occurs when the predictor coefficients are equal to

$$\mathbf{C_n} = \mathbf{R}^{-1}\, \mathbf{G} \ . \tag{9.61}$$

The optimum set of coefficients $\mathbf{C_n}$ requires computation of the inverse of the autocorrelation matrix of the previous K, J pixels from the input image.

Since the average value of an image cannot be zero, this tends to bias the graylevel differences $e(x, y)$. This is easily seen by taking the expected value of Equation (9.50):

$$E[\, e(x, y)\,] = E[\,f(x, y)\,] - \mathbf{C_n}^{\mathbf{T}}\, E[\,\mathbf{F_n}\,] \ . \tag{9.62}$$

But, the expected values of $f(x, y)$ and $\mathbf{F_n}$ are equal to the mean graylevel $\bar{m}$ of the image. Hence, Equation (9.62) reduces to

$$E[\, e(x, y)\,] = \bar{m} - \mathbf{C_n}^{\mathbf{T}}\, \bar{\mathbf{m}}_{\mathbf{n}} \ , \tag{9.63}$$

where $\bar{\mathbf{m}}_{\mathbf{n}}$ is a column matrix of length JK in which all of the elements are equal to $\bar{m}$:

$$\bar{\mathbf{m}}_{\mathbf{n}}^{\mathbf{T}} = [\, \bar{m},\ \bar{m},\ \bar{m}, \cdots m\,] \ . \tag{9.64}$$

Performing the matrix multiplication in Equation (9.63) yields

$$E[\, e(x, y)\,] = \bar{m} \cdot (1 - \sum_{k=1}^{K} \sum_{j=1}^{J} c_{j,\,k}) \ . \tag{9.65}$$

Under one-dimensional signal processing conditions, the mean $\bar{m}$ is typically zero and the error $e(x, y)$ becomes zero. For electronic image processing applications, the mean $\bar{m}$ is nonzero and the error $e(x, y)$ becomes biased. A way around this problem is to set the right-hand term in Equation (9.65) to zero by setting the sum of the weights equal to 1:

$$\sum_{k=1}^{K} \sum_{j=1}^{J} c_{j,\,k} = 1 \ . \tag{9.66}$$

Care must be used in setting the weights equal to one. Lossless predictive coding is very sensitive to noise. One graylevel difference that is in error can affect the graylevel of every other pixel following this error. To guarantee that if an error does occur, it will propagate to as few pixels as possible, typically the sum of the weights add to slightly less than one. For example, the sum of the weights may add to 0.97 instead of one. It also should be mentioned that the method of lossless predictive coding on an individual row basis has the added advantage of limiting an error to the row in which the error has occurred, as compared to coding the image using a zig-zag scan pattern. Once an error has occurred in a zig-zag pattern, it can propagate to several different rows.

The solution of Equation (9.61) is a formidable task to accomplish in real-time image compression. It requires the computation of the autocorrelation function of the previous K, J input pixels, from which the inverse of this autocorrelation matrix must be obtained. In addition, the autocorrelation of the pixel at the coordinates x, y with the previous input pixels must also be computed. For most complex images, the correlation between pixels exists for small distances, which defines small regions of similar graylevels within the image. The autocorrelation of an input image with this type of correlation can be modeled as being separable in the vertical and horizontal directions:

$$E[f(x, y)\,f(x - k, y - j)] = \bar{m}^2 + \sigma^2\,\rho h(k)\cdot\rho v(j)\,, \qquad (9.67)$$

where σ^2 is the variance of the input image and $\rho h(k)$ and $\rho v(j)$ are the correlation coefficients in the horizontal and vertical directions, respectively. The solution of the optimum predictor as defined by Equation (9.61) using Equation (9.67) gives the fourth order predictor of

$$\begin{aligned} f_{est}(x, y) = \ &\rho h(1)\cdot f(x - 1, y) + \rho v(1)\cdot f(x, y - 1) - \\ &\rho h(1)\cdot \rho v(1)\cdot f(x - 1, y - 1) + 0\cdot f(x + 1, y - 1)\,. \quad (9.68) \end{aligned}$$

There are many possible predictors that can be implemented using Equation (9.68). Some of these are presented in Equations (9.69) through (9.73):

$$f_{est}(x, y) = 0.97\,f(x - 1, y) \qquad\qquad (first\ order) \qquad (9.69)$$

$$f_{est}(x, y) = 0.97\,f(x, y - 1) \qquad\qquad (first\ order) \qquad (9.70)$$

$$f_{est}(x, y) = 0.5\,f(x - 1, y) + 0.5\,f(x, y - 1) \qquad (second\ order) \quad (9.71)$$

$$\begin{aligned} f_{est}(x, y) = \ &0.75\,f(x - 1, y) - 0.5\,f(x - 1, y - 1) \\ &+ 0.75\,f(x, y - 1) \qquad\qquad (third\ order) \qquad (9.72) \end{aligned}$$

$$f_{est}(x, y) = f(x - 1, y) - f(x - 1, y - 1) + f(x, y - 1) \quad (third\ order)\,. \quad (9.73)$$

Figure 9.20 shows four example error images generated using the optimum predictors given by Equation (9.70) through (9.73) to predict the image given in Figure 9.19(a). Each figure was generated by scaling the range of graylevel differences from −255 through 255 to 0 through 255 so that these errors could be displayed as an image. The contrast of this image was then scaled by 17 to enhance the spatial features present within these images and also to provide a fair

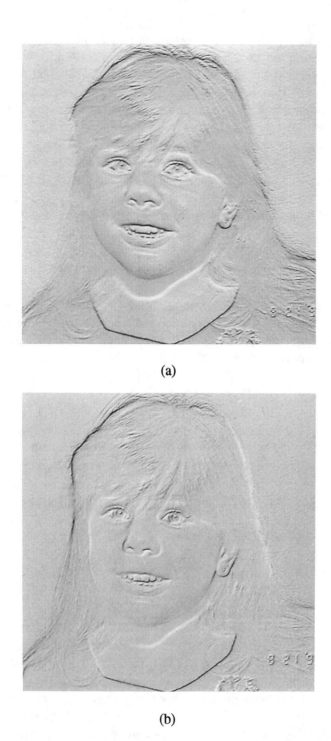

(a)

(b)

Figure 9.20: Examples of using different predictors on Figure 9.19(a): the errors associated with the predictors of (a) Equation (9.70) and (b) Equation (9.71).

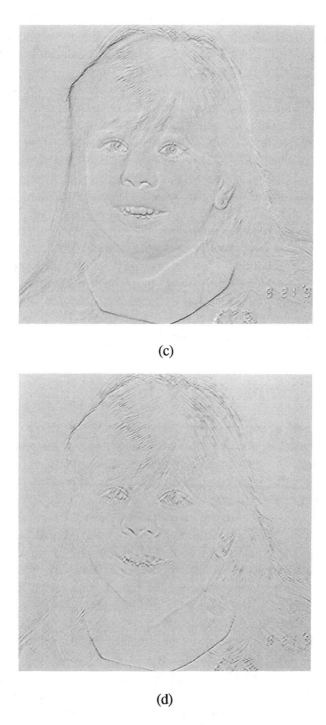

(c)

(d)

Figure 9.20: Examples of using different predictors on Figure 9.19(a): the errors associated with the predictors of (c) Equation (9.72) and (d) Equation (9.73).

comparison with the error image given in Figure 9.19(c). Notice how well these predictors performed within homogeneous regions. Figure 9.20(a) was generated using Equation (9.70), which removed spatial redundancies in the vertical direction. The standard deviation and entropy of this error image are 2.55 and 3.0. Figure 9.20(b) was generated using the second order predictor given in Equation (9.71). This predictor removed spatial redundancies in both the vertical and horizontal directions, producing a lower standard deviation and entropy of 1.85 and 2.42, respectively. Figure 9.20(c) was generated from the third order predictor given in Equation (9.72), yielding a standard deviation and entropy of 1.71 and 2.37. Finally, Figure 9.20(d) was generated from the third order predictor given in Equation (9.73), which yielded a standard deviation and entropy of 2.21 and 2.89. Since the third order predictor of Equation (9.72) produced the lowest standard deviation and entropy for the image of Figure 9.19(a), this predictor yields the best possible compression. The only difficulty of using this predictor is that it requires floating point arithmetic. For real image applications, a better predictor to use is that given in Equation (9.73). Even though this predictor yielded a slightly larger standard deviation and entropy than the two predictors given in Equations (9.71) and (9.72), its advantage is that is does not require floating point arithmetic.

9.3 Lossy Compression Methods

Unlike error-free compression, lossy compression removes information during the compression process. The uncompressing of an image from error-free compression methods yields an exact replica of the original image. The best that can be obtained with lossy compression is an uncompressed image that is a degraded version of the original uncompressed image. The approach taken by lossy compression is to reduce the visual redundancies present within an image. Lossy compression methods use information about the peculiarities of the human visual system to remove as much information as possible from an image without changing the perceived visual quality of the image. There are predominantly two methods of lossy image compression. The first is based upon lossy predictive coding and the second is based upon transform coding. Of the various types of transforms that exist, the discrete cosine transform is by far the most popular transform used to compress an image because of its information packing capability and its computational simplicity.

Figure 9.21(a) gives a block diagram for compressing an image using the lossy predictive compression method. Missing from the lossless prediction block diagram of Figure 9.16(a), but present here, is the quantizer. It is the process of quantization that removes information from the image, making this compression method lossy. In fact, if the quantizer is removed from the block diagram, this block diagram reduces to the lossless prediction method given in

Figure 9.16(a). With the quantizer removed, $q(x, y) = e(x, y)$ and $f_e(x, y)$ becomes

$$f_e(x, y) = f_{est}(x, y) + e(x, y) . \qquad (9.74)$$

But, the sum of the error and the output from the predictor given in the right-hand side of Equation (9.74) is nothing more than the original uncompressed image $f(x, y)$. The input to the predictor is simply the original uncompressed image, which is exactly the same input to the predictor given in Figure 9.16(a). Hence, the removal of the quantizer has reduced the lossy prediction block diagram of Figure 9.21(a) to that of the lossless prediction block diagram given in Figure 9.16(a). The decoder shown in Figure 9.21(b) takes the lossy compressed image and uncompresses it using the same predictor that was used to compress it. A comparison of this decoder with the one used to uncompress an image using lossless compression shows that the two decoders are identical. To summarize, the only function in Figure 9.21 that makes this predictive compression method lossy is the insertion of the quantizer in the encoder.

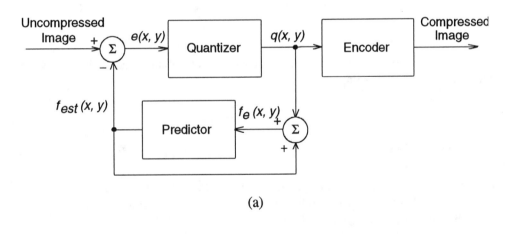

(a)

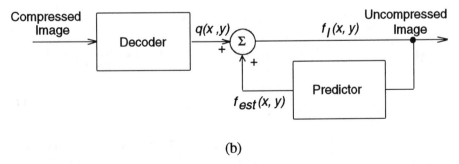

(b)

Figure 9.21: The block diagram for lossy predictive coding: (a) the encoder and (b) the decoder.

With the quantizer placed between the error output and the input to the predictor, the input to the predictor becomes a noisy version of the original input image,

$$f_e(x, y) = q(x, y) + f_{est}(x, y),\qquad(9.75)$$

where the error output is defined as the difference between the input image and the output of the predictor,

$$e(x, y) = f(x, y) - f_{est}(x, y).\qquad(9.76)$$

The function of the quantizer is to reduce the number of code words required to represent the compressed image by reducing the visual redundancies present within the original input image. The quantizer maps a set of M error values $e(x, y)$ to a new set of N values $q(x, y)$ such that $N < M$.

Consider the simplest of quantizers, that of the *1-bit binary delta quantizer*:

$$q(x, y) = \begin{cases} A & \text{for } e(x, y) > 0 \\ -A & \text{for } e(x, y) \leq 0 \end{cases}\qquad(9.77)$$

This quantizer generates two possible words of values A and $-A$, depending on whether $e(x, y)$ is less than or equal to zero or greater than zero. If the predictor is of first order, for example,

$$f_{est}(x, y) = f(x - 1, y),\qquad(9.78)$$

then the lossy prediction method given in Figure 9.21(a) reduces to the well known *delta modulator*.

The effect that a single-bit quantizer has on the compression of an image can be illustrated by tracing through the steps required to compress the following one-dimensional sequence $f(x)$ of numbers:

$$1, 1, 1, 2, 6, 6, 5, 4, 4, 4.\qquad(9.79)$$

The 1-bit quantizer given in Equation (9.77) can be modeled in one dimension as

$$q(x) = \begin{cases} A & \text{for } e(x) > 0 \\ -A & \text{for } e(x) \leq 0 \end{cases}\qquad(9.80)$$

and the first order predictor given in Equation (9.78) can also be written in one dimension as $f_{est}(x) = f(x - 1)$. Setting $A = 1$ and assuming an initial value for the predictor in both the encoder and decoder of 1 (i.e. $f_e(0) = 1$) produces the values shown in Table 9.2 for the various functions (in one dimension) listed in

Figures 9.21(a) and (b). The first value, 1, in the sequence given in Equation (9.79) is not compressed and becomes the initial condition for both predictors. The difference between the second value, 1, in this sequence of numbers and the output of the predictor, which is also 1, is formed to produce an error $e(x)$ of 0. An error of 0 results in a quantizer output of -1, which when summed with the present predictor output of 1 generates $f_e(x) = 0$. This value of 0 becomes the next output from the predictor. The difference between the next number, 1, within the sequence and the new output of the predictor, which is 0, is formed to generate the second error value of 1. This produces a new quantizer value of 1, which when added to $f_{est}(x)$ produces the next input to the predictor of 1. This cycle continues until all the numbers within the sequence have been encoded, producing a set of quantizer outputs that represents the compressed version of the input sequence of numbers. Table 9.2 shows the results of compressing the other values given in the sequence of Equation (9.79).

Table 9.2: An example of compressing and uncompressing a one-dimensional sequence of numbers using delta modulation.

Encoder					Decoder	
$f(x)$	$f_{est}(x)$	$e(x)$	$f_e(x)$	$q(x)$	$f_{est}(x)$	$f_i(x)$
1	–	–	1	–	–	1
1	1	0	0	-1	1	0
1	0	1	1	1	0	1
2	1	1	2	1	1	2
6	2	4	3	1	2	3
6	3	3	4	1	3	4
5	4	1	5	1	4	5
4	5	-1	4	-1	5	4
4	4	0	3	-1	4	3
4	3	1	4	1	3	4

Uncompressing the input sequence of numbers given in Equation (9.79) is accomplished by initializing the decoder predictor with the first sequence value of 1. The uncompressed sequence is then generated by adding the output of the predictor to the quantizer values. For example, the second value for the uncompressed sequence is found by adding the output of the predictor, which is 1, to the quantizer output, which is -1, yielding an uncompressed value of 0. Figure 9.22 shows the uncompressing of an input step that was compressed using delta modulation. Within the flat regions of the input to the left and right of the step, the reconstructed signal varies above and below the actual input values, producing *granular noise*. Within the region of the input step, the quantizer was unable to follow the slope of the step, resulting in *step overload*. Both the granular noise and step overload are a function of the quantizer value A.

Increasing this parameter reduces slope overload, enabling the uncompressed signal to track the input step. But increasing *A* also increases the granular noise that is present in the uncompressed signal. In terms of images, slope overload results in a blurred image, while granular noise adds a grainy appearance to an uncompressed image.

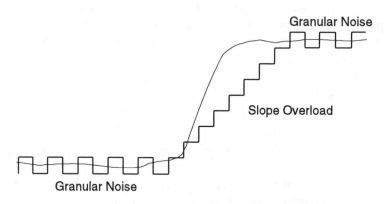

Figure 9.22: An example of delta modulation.

Figure 9.23 shows examples of using delta modulation with various values of *A* to compress a 480×480 by 256 graylevel image. The original image is given in Figure 9.23(a). Each of the 480 rows within the image was individually compressed by storing the graylevel of the first pixel within each row followed by the sequence of quantizer values. A total of $480 \cdot 8$ bits are required to store the graylevel of the first pixel within each row and a total of $479 \cdot 480$ bits are required to store the 480 quantizer sequences (one sequence for each row). Since eight bits per pixel were used to represent the original image, the use of delta modulation to compress this image has resulted in a compression ratio of 7.88 to 1. This compression ratio is much higher than the compression ratios achieved using lossless compression. High compression ratios are possible with lossy compression, but at the cost of degrading the reconstructed uncompressed image. As a rule of thumb, the higher the compression ratio, the more degradation that is noticeable in the uncompressed image.

Figure 9.23(b) shows the result of using delta modulation to compress the image given in Figure 9.23(a) with the quantizer value $A = 1$. Notice the blurring present within this image due to the inability of the encoder to track the sharp edges (slope overload) present within the image. Figure 9.23(c) is the uncompressed image generated from the delta modulated compressed image with $A = 10$. The blurring has been reduced in this image, at the cost of adding a slightly grainy appearance to the image, which is noticeable in the cheek regions of the young girl's face. The full sharpness of the original image has not been completely restored in this image. The glint in the girl's eyes present in the original image is barely observable in this image. Figure 9.23(d) is the uncompressed image generated from the compressed image using $A = 20$.

(a)

(b)

Figure 9.23: Examples of compressing an image using delta modulation: (a) the original image and (b) the compressed image with $A = 1$.

(c)

(d)

Figure 9.23: Examples of compressing an image using delta modulation:
the compressed image with (c) $A = 10$ and (d) $A = 20$.

Increasing the quantizer value to 20 did improve the sharpness of this image, but at the cost of adding a grainy appearance. The grainy appearance of this image is clearly visible and has now become objectionable.

The quality of an uncompressed image depends on the quantizer used to digitize the error function $e(x, y)$. Clearly, digitizing this error function using more bits improves the quality of the uncompressed image. But, adding more bits of resolution to the quantizer reduces the amount of compression that is possible. Figure 9.19(d) is a typical histogram of an error function, showing the distribution of graylevel differences, where the most probable graylevel differences lie close to 0. It was previously stated that the distribution of graylevel differences can be modeled using the Laplacian PDF as defined in Equation (9.41). Quantization of these graylevel differences into a set of values must include all possible graylevel differences in the quantization process. For example, for an 8-bit image, which can take on 256 graylevels, the quantizer must digitize the range of graylevel differences in the range of −255 to 255. An N-bit uniform quantizer digitizes the range of graylevel differences into 2^N equally sized regions. For example, if an 8-bit quantizer is used to represent the graylevel differences in the range of −255 to 255, this range of graylevel differences will be divided into 256 equally sized regions, which are all equal in size with a value of 2.

Clearly, dividing the range of graylevel differences into N equally sized regions is not the optimum quantizer solution. Decreasing the size of each of these regions to improve the resolution of the quantizer reduces the quantization error introduced into the compression process but also reduces the overall compression ratio. A better quantizer can be obtained by using a nonuniform quantizer that divides the range of graylevels into unequally sized regions, depending on the distribution of graylevel differences. Since the distribution of graylevel differences can be modeled using the Laplacian PDF, which gives a high probability of occurrence around zero, an optimum quantizer would use small digitization ranges within the vicinity of 0 and increase the size of the ranges for graylevel differences far removed from 0. This quantization approach uses a fine resolution in the regions where most of the graylevel differences occur and a course resolution in the regions where the graylevel differences have a low probability of occurrence. Nonuniform quantizers of this type reduce the error associated with the quantization process.

The most common of the nonuniform quantizers is the *Max-Lloyd quantizer*, which is based upon the distribution of the input values to the quantizer. Figure 9.24 shows the definition of a typical quantizer. There are two steps in quantizing a continuous function. The first step is to use a set of $N + 1$ decision values d_i to quantize a continuous input $e(x, y)$ into a set of N code words. These code words can simply be a set of n integer numbers,

$$n = \text{int}\left[\frac{e(x, y) - d_0}{d_N - d_0}(N - 1)\right], \tag{9.81}$$

where $d_N - d_0$ defines the range of inputs to the quantizer. These decision values are equally spaced for a uniform quantizer and unequally spaced for the Max-Lloyd quantizer. The second step converts these N code words into a set of discrete values representing the input to quantizer, where these discrete output values are defined as

$$q_i = \frac{d_i + d_{i+1}}{2} \quad \text{for } i = 0, 1, 2, \dots N - 1 \ . \tag{9.82}$$

The output $q(x, y)$ of the quantization process, illustrated by Figure 9.24, is just one of the N different possible values for q_i described by Equation (9.82).

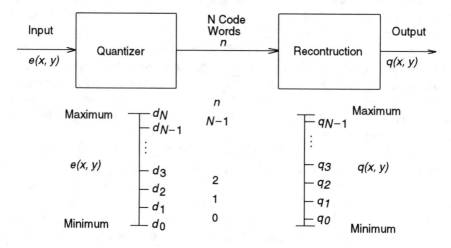

Figure 9.24: A block diagram explaining the process of quantization.

For a given continuous input e in the range of d_i to d_{i+1}, the output from the quantizer will be q_i. The mean square error associated with this input is given by

$$E_i = \int_{d_i}^{d_{i+1}} (e - q_i)^2 \, p_e(e) \, de \ , \tag{9.83}$$

where $p_e(e)$ is the PDF of the continuous input. The total error E_t for all possible input values reduces to the sum of Equation (9.83) over the N decision ranges:

$$E_t = \sum_{i=0}^{N-1} \int_{d_i}^{d_{i+1}} (e - q_i)^2 \, p_e(e) \, de \ . \tag{9.84}$$

The goal is to reduce this error E_t to the smallest possible value. Differentiating Equation (9.84) using Leibnitz's rule produces

$$\frac{\partial E_t}{\partial d_i} = (d_i - q_i - 1)^2 \, p_e(d_i) - (d_i - q_i)^2 \, p_e(d_i) \; , \tag{9.85}$$

and then setting this result to zero yields the first equation that must be satisfied to minimize the total quantizer error:

$$d_i = \frac{q_i + q_i - 1}{2} \; . \tag{9.86}$$

In a similar manner, differentiating Equation (9.84) with respect to q_i yields

$$\frac{\partial E_t}{\partial q_i} = -2 \int\limits_{d_i}^{d_i + 1} (e - q_i) \, p_e(e) \, de \; . \tag{9.87}$$

Setting Equation (9.87) to zero and solving for q_i gives the second equation that must be satisfied to minimize the error associated with the quantization process:

$$q_i = \frac{\displaystyle\int\limits_{d_i}^{d_i + 1} e \, p_e(e) \, de_i}{\displaystyle\int\limits_{d_i}^{d_i + 1} p_e(e) \, de} \; . \tag{9.88}$$

Except for a few probability density functions, there are no closed form solutions that satisfy Equations (9.86) and (9.88). Typically, these two equations are solved iteratively using a computer. An initial set of $N + 1$ decision values is chosen that subdivides the range of the input into N regions. This initial set of decision values can simply divide the input into N equally sized regions. Next, Equation (9.88) is used to compute the reconstruction values q_i. These values are then substituted into Equation (9.86) to produce a new set of decision values. This new set of decision values is then substituted back into Equation (9.88). This iterative process continues until the change in the decision values is less than a given threshold level. Typically, this iterative process converges within a small number of iterations. Tables 9.3(a) and (b) give the output decision values and the reconstruction values for the Max-Lloyd quantizer generated using a unit variance Laplacian probability density function. Table 9.3(a) was generated using 4 digitization levels (2 bits), while Table 9.3(b) was generated using 8 digitization levels (3 bits). Notice how the size of the decision regions decreases, improving the resolution of the quantizer for input values near 0. These two tables are easily scaled for a Laplacian PDF with any variance by multiplying the decision and the reconstruction values given in the table by the standard deviation of the Laplacian PDF.

Table 9.3: The Max-Lloyd quantizer for a unit variance: (a) $N = 4$ and (b) $N = 8$.

d_i	d_{i+1}	n	q_i
$-\infty$	-1.13	0	-1.83
-1.13	0.00	1	-0.42
0.00	1.13	2	0.42
1.13	∞	3	1.83

(a)

d_i	d_{i+1}	n	q_i
$-\infty$	-2.38	0	-3.09
-2.38	-1.25	1	-1.67
-1.25	-0.53	2	-0.83
-0.53	0.0	3	-0.23
0.0	0.53	4	0.23
0.53	1.25	5	0.83
1.25	2.38	6	1.67
2.38	∞	7	3.09

(b)

Figure 9.25 gives examples of the application of a lossy predictor with a Max-Lloyd quantizer to compress the image of Figure 9.19(a). Figure 9.25(a) is the uncompressed image generated with the 4 quantization level Max-Lloyd predictor given in Table 9.3(a) and the third order optimum predictor given in Equation (9.72). The actual decision and range values used for the 4 decision and reconstruction values were obtained by multiplying the values given in Table 9.3(a) by the standard deviation of the error, 1.71, which was obtained from the error values given in Figure 9.20(c). Figure 9.25(b) is the corresponding error image generated by subtracting the differences between the original image given in Figure 9.19(a) and the uncompressed image of Figure 9.25(a) and then scaling these differences between 0 and 255 so that they can be displayed as an image.

(a)

(b)

Figure 9.25: Examples of compressing Figure 9.19(a) using lossy prediction: (a) the uncompressed image for a Max-Lloyd quantizer of $N = 4$ and (b) its corresponding error.

(c)

(d)

Figure 9.25: Examples of compressing Figure 9.19(a) using lossy
                prediction: (c) the uncompressed image for a Max-Lloyd
                quantizer of $N = 8$ and (d) its corresponding error.

In comparing the original image given in Figure 9.19(a) to the lossy uncompressed image, the uncompressed image has been slightly blurred. The fine spatial details of the image have been lost. In fact, the high spatial frequencies present within the original graylevel image have been removed from the uncompressed image. The presence of high error values within the error image of Figure 9.25(b) near the edges also shows that the lossy predictor was unable to follow the sharp edges present within the image.

Figure 9.25(c) is the uncompressed image of Figure 9.19(a) obtained using the optimum predictor given in Equation (9.72) and the 8 quantization level Max-Lloyd quantizer given in Table 9.3(b). The actual decision and reconstruction values were obtained in the same manner as for the 4 quantization level Max-Lloyd quantizer, by scaling the values given in Table 9.3(b) by 1.71. Figure 9.25(d) is the corresponding error image for this lossy predictor. In comparing the uncompressed image given in Figure 9.25(c) to the uncompressed image of Figure 9.25(a), the uncompressed image using the 8 quantization level Max-Lloyd predictor has resulted in a sharper image and is of better quality. In comparing the error image given in Figure 9.25(d) to the error image given in Figure 9.25(b), the error image for the lossy predictor generated with the 8 quantization Max-Lloyd quantizer shows less error between the original image and the uncompressed image within the edge regions. As more quantization levels are added to the quantizer the better the lossy predictor can track edges as a result of reducing slope overload.

Another common method of lossy compression is that of *transform coding*. Here an image is transformed using one of the many image transforms given in Chapter 2. The goal is to transform the image to produce a new set of coefficients that are decorrelated. This results in a set of coefficients, in which most of the information present in the image has been packed into a small number of coefficients. The coefficients that provide little image information can then be discarded, resulting in a lossy compressed version of the original image. Figure 9.26(a) shows the steps involved in compressing an image using transform coding. The first step of the compressor is to transform the image to obtain the decorrelated coefficients. The next step is the part of the compression process that can make transform coding lossy. It is in this stage that coefficients containing low image information content are removed and the remaining coefficients quantized to a given number of code words. The final step is to encode these remaining quantized coefficients using one of the optimum variable length codes such as the Huffman code.

Figure 9.26(b) shows the decompressor used to obtain the uncompressed image. The decoder extracts the code words, which are then applied to the dequantizer to reconstruct the set of quantized transform coefficients. The set of transform coefficients that represents the image is then produced from the quantized coefficients generated from the dequantizer step and by setting the truncated coefficients to zero. This set of coefficients is then inverse transformed to yield a lossy version of the original image. Transform coding of an image does not always have to produce a lossy version of the original image. Removing the quantization and truncation step given in Figure 9.26(a) produces a compressor that is no longer lossy. Image compression using transform coding is still possible without the step of quantization and truncation, but with much lower compression ratios.

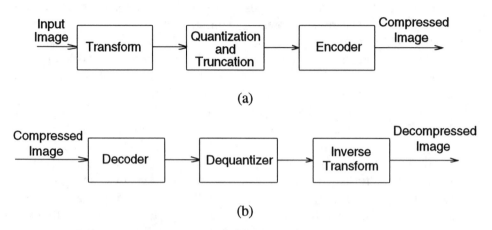

Figure 9.26: The block diagram showing the steps involved in transform coding an image: (a) the compressor and (b) the decompressor.

Of the image transforms that exist, the Karhunen-Loève is by far the best at packing the most image information into the fewest number of coefficients. The computation of the Karhunen-Loève requires the computation of the autocorrelation matrix **R** of the input image, from which the eigenvalues λ_i's and the eigenvector matrix **Q** of the autocorrelation matrix are then computed. Defining an image as a two-dimensional matrix **F**, the transformation that yields the Karhunen-Loève transform is

$$\mathbf{G} = \mathbf{Q}\,\mathbf{F}\,\mathbf{Q}^{\mathbf{T}} \quad , \tag{9.89}$$

where **T** means transpose. The major limitation of the Karhunen-Loève transform is the computation complexity required by computing the autocorrelation of the input image, from which the eigenvalues and eigenvector matrix are computed. Unlike the other transforms presented in Chapter 2, there is no unique kernel given for the Karhunen-Loève transform. The kernel used to perform the transformation is based upon the input image. Due to its computational complexity, the Karhunen-Loève transform is rarely used to compress an image. Other transforms that provide nearly the same image information packing capability are used.

The three most common image transforms that are used to transform code an image are the Hadamard-Walsh transform, the discrete Fourier transform, and the discrete cosine transform (each of which was previously described in detail in Chapter 2). The Walsh transform in two dimensions for an $N \times N$ image is defined as

$$F(n, m) = \frac{1}{N} \sum_{y=0}^{N-1} \sum_{x=0}^{N-1} f(x, y)$$

$$\cdot \prod_{i=0}^{q-1} (-1)^{b_i(x) \, b_{q-1-i}(n) + b_i(y) \, b_{q-1-i}(m)}, \qquad (9.90)$$

and its inverse as

$$f(x, y) = \frac{1}{N} \sum_{m=0}^{N-1} \sum_{n=0}^{N-1} F(n, m)$$

$$\cdot \prod_{i=0}^{q-1} (-1)^{b_i(x) \, b_{q-1-i}(n) + b_i(y) \, b_{q-1-i}(m)}, \qquad (9.91)$$

where $b_i(x)$ is a binary bit representation. In a similar manner, the Hadamard transform defined in two dimensions for an $N \times N$ image is

$$F(n, m) = \frac{1}{N} \sum_{y=0}^{N-1} \sum_{x=0}^{N-1} f(x, y) \cdot (-1)^{\sum_{i=0}^{q-1} b_i(x) \, a_i(n) + b_i(y) \, a_i(m)} \qquad (9.92)$$

and its inverse as

$$f(x, y) = \frac{1}{N} \sum_{m=0}^{N-1} \sum_{n=0}^{N-1} F(n, m) \cdot (-1)^{\sum_{i=0}^{q-1} b_i(x) \, a_i(n) + b_i(y) \, a_i(m)}, \qquad (9.93)$$

where $b_i(x)$ is the binary bit representation and the term $a_i(n)$ is given by

$$a_0(n) = b_{q-1}(n),$$
$$a_1(n) = b_{q-1}(n) + b_{q-2}(n)$$
$$\vdots$$
$$a_{q-1}(n) = b_1(n) + b_0(n) \, . \qquad (9.94)$$

Equations (9.90) through (9.94) provide the least information packing capability of the three transforms mentioned, yet these transforms are popular for image compression due to their computational simplicity. The forward and inverse kernels for both the Hadamard and the Walsh transforms can be computed using integer addition. The computational requirements of these two transforms is simple enough that real-time compression using them is possible.

Both the discrete Fourier and the discrete cosine transforms provide high image information packing capability. In fact, the packing capability of the DCT is very near that of the Karhunen-Loève transform, but without its computational complexity. As defined in Chapter 2, the DFT is defined as

$$F(n, m) \;=\; \frac{1}{NM} \sum_{y=0}^{M-1} \sum_{x=0}^{N-1} f(x, y) \cdot e^{-j2\pi(nx/N + my/M)} \tag{9.95}$$

and its inverse as

$$f(x, y) \;=\; \sum_{m=0}^{M-1} \sum_{n=0}^{N-1} F(n, m) \cdot e^{j2\pi(nx/N + my/M)} \;. \tag{9.96}$$

The DFT has been used to transform code an image because of its familiarity and the ease with which it can be computed using the fast Fourier transform FFT algorithm. Unfortunately, one of the disadvantages of the discrete Fourier transform as compared to the Hadamard-Walsh and discrete cosine transforms is that the kernel is complex. This produces DFT coefficients that are also complex, which doubles the amount of elements that must be stored and manipulated. This results in adding computational complexity and a reduction in the overall compression efficiency associated with the DFT. Another disadvantage of the DFT over the DCT is that of the periodicity of the transform. With a DFT of length N, it is assumed that the input is also periodic of length N. In other words, for an $N \times N$ image $f(n, m)$

$$F(n, m) = F(n + N, m) = F(n, m + N) = F(n + N, m + N) \;. \tag{9.97}$$

Typically, transform coding of an $N \times N$ image involves subdividing the image into $n \times n$ nonoverlapping regions.   Then transform coding is applied independently to each of these $n \times n$ sub-images. Because of the windowing artifacts associated with the DFT as a result of boundary discontinuities, taking the DFT of an $n \times n$ sub-image can result in high spatial frequencies that are not present within the $n \times n$ sub-image. This results in additional nonzero Fourier coefficients and hence reduces the overall compression capability of the DFT.

The most commonly used transform in image transform coding is the DCT. Because this transform is so popular, it has become the standard transform coding method used in the compression of images using the JPEG standard developed by a joint committee of the International Standardization Organization (ISO) and the CCITT. The DCT transform provides nearly the same information packing capability of the Karhunen-Loève transform and offers the mirror-symmetry required in performing block transform coding of an image. As defined in Chapter 2, the DCT of a square $N \times N$ image is

$$C(n, m) = k_1(n)\, k_2(m) \sum_{y=0}^{N-1} \sum_{x=0}^{N-1} f(x, y) \cdot \cos\left(\pi n\, \frac{x + 1/2}{N}\right) \cdot \cos\left(\pi m\, \frac{y + 1/2}{N}\right),$$

$$(9.98)$$

where n and m are defined from 0 to $N - 1$ and

$$k_1(n) = \begin{cases} \sqrt{\dfrac{1}{N}} & \text{for } n = 0 \\[2mm] \sqrt{\dfrac{2}{N}} & \text{otherwise} \end{cases} \qquad k_2(m) = \begin{cases} \sqrt{\dfrac{1}{N}} & \text{for } m = 0 \\[2mm] \sqrt{\dfrac{2}{N}} & \text{otherwise} \end{cases} \qquad (9.99)$$

The inverse DCT, which uses the same kernel as the forward DCT, is defined as

$$f(x, y) = \sum_{m=0}^{N-1} \sum_{n=0}^{N-1} C(n, m) \cdot \cos\left(\pi x\, \frac{n + 1/2}{N}\right) \cdot \cos\left(\pi y\, \frac{m + 1/2}{N}\right). \qquad (9.100)$$

The DCT offers several advantages over the DFT. Since its kernel is real, the DCT coefficients are also real, reducing the number of coefficients required to encode an image. As compared to the DFT, which requires both real and imaginary coefficients, the computational complexity required to compress an image by truncating only the real set of DCT coefficients is also reduced. As with the DFT, the DCT also assumes that the input image is periodic. The main difference is that for a square $N \times N$ image $f(n, m)$, the DFT assumes a periodicity of N as described by Equation (9.97), while the DCT assumes a periodicity of $2N$:

$$F(n, m) = F(n + 2N, m) = F(n, m + 2N) = F(n + 2N, m + 2N) . \qquad (9.101)$$

The edge effects of the DFT play an important part in the transform coding of an image using smaller sized sub-images. The edge artifacts of the DFT add objectionable windowing artifacts to the uncompressed image, which is equal to the block sub-image size.

Figure 9.27 illustrates the mirror-symmetry of the DCT as compared to the DFT for a one-dimensional signal of length N. Illustrated in Figure 9.27(a) is the periodicity assumed by the DFT, showing that this signal repeats at the rate of N. As a result of this periodicity, boundary discontinuities have occurred at the edge of the signal's sampling window due to the initial sampled value not being equal to the last sampled value. This adds artificial Fourier coefficients to actual Fourier coefficients of the signal. In Chapter 2, Figure 2.2 shows the effect of truncating the Fourier coefficients of a periodic signal that contains a sharp discontinuity. The removal of the high frequency Fourier coefficients results in "ringing" near the discontinuity, known as the Gibbs phenomenon. Several windowing functions have been developed, such as the Hamming and Blackman windows, that force the beginning and end sampled values to the same value, eliminating the discontinuity. Figure 9.27(b) shows the periodicity assumed by the DCT. The periodicity of this signal occurs at the rate of $2N$, and every other period is a mirror-symmetry of the previous and next adjacent periods. This guarantees that the ending and starting sampled values will be equal, eliminating the edge discontinuity that can be present within the DFT. The elimination of these edge discontinuities by the DCT results in more DCT coefficients, which are negligible, and hence provides a higher compression capability.

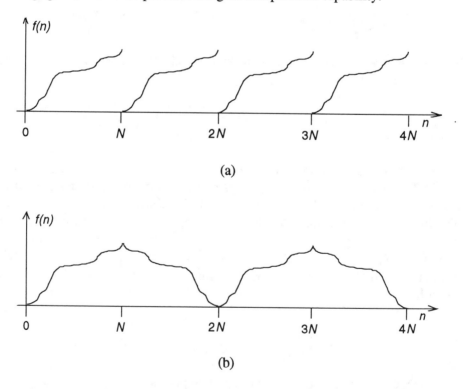

Figure 9.27: The assumed periodicity of (a) the DFT and (b) the DCT.

The typical method in which an image is compressed using the DCT is to divide the image into equally sized nonoverlapping sub-images. The DCT coefficients of each sub-image are individually computed. Next, the negligible DCT coefficient of each sub-image are removed and the remaining coefficients for each sub-image are stored and used to reconstruct the uncompressed image. There are two approaches used to eliminate negligible DCT coefficients. The first method is based on a threshold decision value. For a given threshold value, any DCT coefficients that exceeds this value is saved, since it is large enough in value that it probably contains important information about the original image. The second method of truncating the DCT coefficients is based upon preserving coefficients within a certain zonal region of the DCT. Typically, the low spatial frequency components within an image will dominate the high spatial frequency components. A simple zonal method is to preserve the DCT coefficients within a small region about the DC coefficient and to truncate the other DCT coefficients. Another zonal method is based upon the variance for each DCT coefficient. For an $n \times n$ sub-image, there will be a total of n^2 DCT coefficients. From sub-image to sub-image each of these n^2 DCT coefficients will vary in value. A set of n^2 variances can be computed that represents the change in value of the n^2 DCT coefficients from sub-image to sub-image. Coefficients with large variances must represent DCT coefficients that carry the highest image information and therefore should be retained.

After the desired set of coefficients is selected using either zonal or threshold coding, these coefficients must be quantized and then coded. As previously discussed, the quantizer can be either a uniform quantizer subdividing the range of coefficient values equally or a nonuniform quantizer, such as the optimum Max-Lloyd quantizer, generating a set of variable sized quantization ranges. Once the DCT coefficients have been quantized into M different values, a variable length code such as the Huffman code is used to encode these quantization levels.

The compression of a 256 graylevel image using the JPEG standard uses the DCT to transform code an image. Compressing an image using the JPEG standard starts with subdividing the image into a set of 8×8 sub-images. Next, the DCT coefficients of each sub-image are individually computed. Each of the DCT coefficients within each sub-image are then normalized by an array of 8-bit integers and then rounded to the nearest integer. This normalization array is then stored and used to unnormalize the DCT coefficients during the uncompression process. The first DCT coefficient of each sub-image (known as the DC coefficient) is then encoded using a lossless predictor. The remaining coefficients (known as the AC coefficients) of each sub-image are then ordered into a set of one-dimensional arrays from the second DCT coefficient to the last DCT coefficient of each sub-image using a zig-zag scanning method. Each nonzero coefficient is then encoded using 8 bits. The lower 4 bits give the variable sized quantization range that the coefficient is located in, while the upper 4 bits give the position of the DCT coefficient relative to the previous nonzero DCT coefficient in the one-dimensional array. This is effectively a modified run length coding of the DCT coefficients. If the distance exceeds 15, then this coefficient is represented by several run-length code words. If the upper 4 bits are 0, then this indicates the end of the nonzero DCT coefficients. The rest of the coefficients in the one-dimensional array are equal to zero. The code words generated in the quantization and run-length step are then coded using Huffman coding via a default Huffman code table.

It is during the normalization and the quantization and run-length coding steps that the JPEG compression process becomes lossy. Changing the range of values used within the normalization array changes the number of coefficients that are nonzero, which changes the amount of compression that is possible. Generating an uncompressed JPEG image requires that the set of quantized DCT coefficients be generated from the Huffman codes. Next, these quantized DCT coefficients are unnormalized using the normalization array of 8-bit integer numbers. The inverse DCT of the reconstructed DCT coefficients for each sub-image is computed to yield the uncompressing image components for each sub-image.

Figure 9.28(a) is the JPEG compressed image of Figure 9.19(a) obtained by selecting a compression ratio of 20 to 1. Figure 9.28(b) is its corresponding error image, computed as the difference of the original graylevel image given in Figure 9.19(a) and the JPEG compressed image given in Figure 9.28(a). This error was scaled between 0 and 255 to create an error image, which was then contrast enhanced by a factor of 17 to highlight any errors that might be present. Inspection of this error image shows that for the most part there is no discernible difference between the original image and the uncompressed image. Figure 9.28(c) is the uncompressed image of Figure 9.19(a) obtained with a compression ratio of 30 to 1, and Figure 9.28(d) is its corresponding error image, obtained in the same manner in which Figure 9.28(b) was created. A close inspection of this error image shows that some of the features of the young girl are barely present. However, the difference between the original image and the uncompressed image is not noticeable to a human observer. Figure 9.28(e) is the uncompressed image obtained with a compression ratio of 40 to 1, while Figure 9.28(f) is its corresponding error image. In fact, errors between the original image and the uncompressed image of Figure 9.28(e) are now observable. False contours are present in the background region to the right of the young girl and some block artifacts are present in her left cheek. Overall, the uncompressed image appears to be of lower perceived quality than   the original image of Figure 9.19(a).

The final lossy compression method of interest is that of compressing a color image. In the simplest form, each of the red, green, and blue color components can be compressed separately. This is the approach taken by the JPEG standard in compressing a color image. The main difficulty with this approach is that it does not take into account the correlation that exists between the individual color components, and it ignores the color limitations of the human visual system.

(a)

(b)

Figure 9.28: Examples of applying the JPEG compression
standard to Figure 9.19(a): (a) the uncompressed
image obtained with a compression ratio of 20 to 1
and (b) its corresponding error image.

(c)

(d)

Figure 9.28: Examples of applying the JPEG compression
standard to Figure 9.19(a): (c) the uncompressed
image obtained with a compression ratio of 30 to 1
and (d) its corresponding error image.

(e)

(f)

Figure 9.28: Examples of applying the JPEG compression
standard to Figure 9.19(a): (e) the uncompressed
image obtained with a compression ratio of 40 to 1
and (f) its corresponding error image.

One method of compressing a color image using the correlation that exists between each color component is to treat these components as a three-dimensional vector. A rotation of the color coordinate system is performed using the Karhunen-Loève transform to obtain a new set of three color components that are decorrelated. Since there is no correlation between each of the color components, each can now be compressed separately. As stated earlier, the main limitation of using this compression method is the computational complexity associated with performing the Karhunen-Loève transform.

Another color compression method uses information about the limitations of the human visual system to compress a color image. This compression is based upon the technique specified in the NTSC standard for displaying color images on a color television. During the 1950s the NTSC color television standard was created to ensure compatibility with existing black and white televisions. This required that the color information associated with a color image lie in the same 6 MHz bandwidth that was used for standard black and white television. This compression method required that the RGB color components be transformed into the YIQ color space using

$$\begin{bmatrix} Y \\ I \\ Q \end{bmatrix} = \begin{bmatrix} 0.299 & 0.587 & 0.114 \\ 0.596 & -0.275 & -0.321 \\ 0.212 & -0.523 & 0.311 \end{bmatrix} \begin{bmatrix} R \\ G \\ B \end{bmatrix}. \qquad (9.102)$$

Next, it was realized that because of the spatial limitation of the eye to the I and Q color components and since most of an image's information is contained within the luminance Y component, the I and Q color components could be easily compressed without producing any noticeable artifacts in the color images presented on a color television. Therefore, the NTSC color television standard selected to compress the I and Q color components by spectrally lowpass filtering the I component to 2.5 MHz and the Q component to 0.5 MHz so that their spectral frequencies fit within the 6 MHz frequency space allocated to black and white television. Temporal filtering of these two color components is equivalent to spatially filtering the I and Q color components. The temporal filtered I and Q components along with the Y component are then converted back to the three RGB color components, which modulate the red, blue, and green electron guns of a color CRT, producing a color image:

$$\begin{bmatrix} R \\ G \\ B \end{bmatrix} = \begin{bmatrix} 1.0 & 0.956 & 0.62 \\ 1.0 & -0.272 & -0.647 \\ 1.0 & -1.108 & 1.705 \end{bmatrix} \begin{bmatrix} Y \\ I \\ Q \end{bmatrix}. \qquad (9.103)$$

In electronic image processing, one method of compressing a color image is to use the same approach as devised by the NTSC to compress a color television image into the same information bandwidth as that of black and white television. A digitized color image is transformed into its *YIQ* color components. Next, the *I* and *Q* color components are lowpass filtered to increase the spatial redundancies present within each of these color components. Finally, a lossy compression method is applied to each of *I* and *Q* color components to produce a compressed version of these two components. Finally, the *Y* color component is compressed independently of the *I* and *Q* components. To reconstruct the uncompressed RGB color image, each of the compressed YIQ color components are uncompressed and then substituted into Equation (9.103) to generate the uncompressed RGB color components.

The compression methods presented in this chapter define the basis for the compression standards that exist today. The area of image compression is a growing field that is continuing to evolve. As the demands increase for higher resolution images, so will the demands for improved color image processing methods. For further information on this subject, the interested reader is referred to the text *Digital Compression Techniques* by M. Rabbani and P. W. Jones given in the bibliography, which provides an excellent treatment of other image compression methods not mentioned here.

Bibliography

CHAPTER 1

Abramson, Albert, *The History of Television, 1880 to 1941*, McFarland & Company, Inc., Jefferson, NC, 1987.

Baxes, Gregory A., *Digital Image Processing: A Practical Primer*, Prentice-Hall, Inc., Englewood Cliffs, NJ, 1984.

Green, William B., *Digital Image Processing: A Systems Approach*, Van Nostrand Reinhold Company, New York, 1983.

Gregory, Richard L., *Eye and Brain: The Psychology of Seeing*, Princeton University Press, Princeton, NJ, 1990.

Goldman, Martin, *The Demon in the Aether*, Paul Harris Publishing, Edinburgh, Scotland, 1983.

Gonzalez, Rafael C., and Richard E. Woods, *Digital Image Processing*, Addison-Wesley, Reading, MA, 1992.

McFarlane, Maynard D., "Digital Pictures Fifty Years Ago," *Proc. of the IEEE*, Vol. 60, No. 7, pgs. 768-770, July 1972.

Mast, Gerald, *A Short History of the Movies*, Macmillan Publishing Company, New York, 1986.

Moses, Robert A., M.D., *Adler's Physiology of the Eye*, The C. V. Mosby Company, St. Louis, 1970.

Myler, Harley R., and Arthur R. Weeks, *Computer Imaging Recipes in C*, PTR Prentice Hall, Englewood Cliffs, NJ, 1993.

North, Joseph H., *The Early Development of the Motion Picture (1987-1909)*, ARNO Press, New York, 1973.

Resnikoff, Howard L., *The Illusion of Reality*, Springer-Verlag, New York, 1989.

Tasman, William, M.D., ed., *Duane's Foundations of Clinical Ophthalmology*, Volume 2, J. B. Lippincott Company, Philadelphia, 1994.

Thorne, J. O., ed., *Chambers's Biographical Dictionary*, St. Martin's Press, New York, 1966.

Tissandier, Gaston, *A History and Handbook of Photography*, ARNO Press, New York, 1973.

CHAPTER 2

Antoniou, Andreas, *Digital Filters: Analysis, Design, and Applications*, McGraw-Hill, New York, 1993.

Castleman, Kenneth R., *Digital Image Processing*, Prentice-Hall, Englewood Cliffs, NJ, 1979.

Dougherty, Edward R., and Phillip A. Laplante, *Introduction to Real-Time Imaging*, SPIE Press, Bellingham, WA, 1995.

Gaskill, Jack D., *Linear Systems, Fourier Transforms, and Optics,* John Wiley and Sons, NY, 1978.

Gonzalez, Rafael C., and Richard E. Woods, *Digital Image Processing*, Addison-Wesley, Reading, MA, 1992.

Lim, Jae S., *Two-Dimensional Signal and Image Processing*, PTR Prentice-Hall, Englewood Cliffs, NJ, 1990.

Myler, Harley R., and Arthur R. Weeks, *Computer Imaging Recipes in C*, PTR Prentice Hall, Englewood Cliffs, NJ, 1993.

Myler, Harley R., and Arthur R. Weeks, *The Pocket Handbook of Image Processing Algorithms in C*, PTR Prentice Hall, Englewood Cliffs, NJ, 1993.

Schalkoff, Robert J., *Digital Image Processing and Computer Vision: An Introduction to Theory and Implementations*, John Wiley & Sons, New York, 1989.

Sneddon, Ian N., *The Use of Integral Transforms*, McGraw-Hill, New York, 1972.

Stremler, Ferrel G., *Introduction to Communication Systems*, Addison-Wesley, Reading, MA, 1979.

Tolstov, Georgi P., *Fourier Series*, Prentice-Hall, Englewood Cliffs, NJ, 1962.

Ziemer, R. E., and W. H. Tranter, *Principles of Communications: Systems, Modulation, and Noise*, Houghton Mifflin, Boston, 1976.

CHAPTER 3

Castleman, Kenneth R., *Digital Image Processing*, Prentice-Hall, Englewood Cliffs, NJ, 1979.

Gonzalez, Rafael C., and Richard E. Woods, *Digital Image Processing*, Addison-Wesley, Reading, MA, 1992.

Hall, Ernest L., *Computer Image Processing and Recognition*, Academic Press, New York, 1979.

Lim, Jae S., *Two-Dimensional Signal and Image Processing*, PTR Prentice-Hall, Englewood Cliffs, NJ, 1990.

Myler, Harley R., and Arthur R. Weeks, *Computer Imaging Recipes in C*, PTR Prentice Hall, Englewood Cliffs, NJ, 1993.

Myler, Harley R., and Arthur R. Weeks, *The Pocket Handbook of Image Processing Algorithms in C*, PTR Prentice Hall, Englewood Cliffs, NJ, 1993.

Pitas, Ioannis, *Digital Image Processing Algorithms*, Prentice Hall, New York, 1993.

Schalkoff, Robert J., *Digital Image Processing and Computer Vision: An Introduction to Theory and Implementations*, John Wiley & Sons, New York, 1989.

CHAPTER 4

Gonzalez, Rafael C., and Richard E. Woods, *Digital Image Processing*, Addison-Wesley, Reading, MA, 1992.

Goodman, Joseph W., *Statistical Optics*, John Wiley & Sons, New York, 1985

Hall, Ernest L., *Computer Image Processing and Recognition*, Academic Press, New York, 1979.

Kennedy, E. J., *Operational Amplifier Circuits: Theory and Applications*, Holt, Rinehart and Winston, New York, 1988.

Lim, Jae S., *Two-Dimensional Signal and Image Processing*, PTR Prentice-Hall, Englewood Cliffs, NJ, 1990.

Myler, Harley R., and Arthur R. Weeks, *Computer Imaging Recipes in C*, PTR Prentice Hall, Englewood Cliffs, NJ, 1993.

Myler, Harley R., and Arthur R. Weeks, *The Pocket Handbook of Image Processing Algorithms in C*, PTR Prentice Hall, Englewood Cliffs, NJ, 1993.

Papoulis, Anthanasios, *Probability, Random Variables, and Stochastic Processes*, McGraw-Hill, New York, 1965.

Pitas, Ioannis, *Digital Image Processing Algorithms*, Prentice Hall, New York, 1993.

Proakis, John G., and Dimitris G. Manolakis, *Introduction to Digital Signal Processing*, Macmillan Publishing Company, 1988.

Schalkoff, Robert J., *Digital Image Processing and Computer Vision: An Introduction to Theory and Implementations*, John Wiley & Sons, New York, 1989.

Sid-Ahmed, Maher A., *Image Processing: Theory, Algorithms, and Architectures*, McGraw-Hill, New York, 1995.

Ziemer, R. E., and W. H. Tranter, *Principles of Communications: Systems, Modulation, and Noise*, Houghton Mifflin, Boston, 1976.

CHAPTER 5

Dougherty, Edward R., *An Introduction to Morphological Image Processing*, SPIE Press, Bellingham, WA, 1992.

Gonzalez, Rafael C., and Richard E. Woods, *Digital Image Processing*, Addison-Wesley, Reading, MA, 1992.

Haralick, Robert M., and Linda G. Shapiro, *Computer and Robot Vision: Volume I*, Addison-Wesley, Reading, MA, 1992.

Lim, Jae S., *Two-Dimensional Signal and Image Processing*, PTR Prentice-Hall, Englewood Cliffs, NJ, 1990.

Myler, Harley R., and Arthur R. Weeks, *Computer Imaging Recipes in C*, PTR Prentice Hall, Englewood Cliffs, NJ, 1993.

Myler, Harley R., and Arthur R. Weeks, *The Pocket Handbook of Image Processing Algorithms in C*, PTR Prentice Hall, Englewood Cliffs, NJ, 1993.

Pitas, I., and A. N. Venetsanopoulos, *Nonlinear Digital Filters: Principles and Applications*, Kluwer Academic Publishers, Boston, 1990.

Schalkoff, Robert J., *Digital Image Processing and Computer Vision: An Introduction to Theory and Implementations*, John Wiley & Sons, New York, 1989.

CHAPTER 6

Carterette, Edward C., and Morton P. Friedman, eds., *Handbook of Perception*, Volume 2, Academic Press, New York, 1975.

Dereniak, Eustace L., and Devon G. Crowe, *Optical Radiation Detectors*, John Wiley & Sons, New York, 1994.

Foley, James D., Andries van Dam, Steven K. Feiner, and John F. Hughes, *Computer Graphics: Principles and Practice*, Addison-Wesley, Reading, MA, 1996.

Gonzalez, Rafael C., and Richard E. Woods, *Digital Image Processing*, Addison-Wesley, Reading, MA, 1992.

Hall, Ernest L., *Computer Image Processing and Recognition*, Academic Press, New York, 1979.

Kingston, R. H., *Detection of Optical and Infrared Radiation*, Springer-Verlag, Berlin, 1979.

Martin, A. V. J., *Technical Television*, Prentice-Hall, Englewood Cliffs, NJ, 1962.

Myler, Harley R., and Arthur R. Weeks, *The Pocket Handbook of Image Processing Algorithms in C*, PTR Prentice Hall, Englewood Cliffs, NJ, 1993.

Pitas, Ioannis, *Digital Image Processing Algorithms*, Prentice Hall, New York, 1993.

Russ, John C., *The Image Processing Handbook*, CRC Press, Boca Raton, FL, 1995.

Sid-Ahmed, Maher A., *Image Processing: Theory, Algorithms, and Architectures*, McGraw-Hill, New York, 1995.

Weeks, A. R., C. E. Felix, and H. R. Myler, "Edge Detection of Color Images Using the HSL Color Space", *SPIE Proceedings: Nonlinear Image Processing VI*, Vol. 2424, San Jose, pgs. 291-301, March, 1995.

Weeks, A. R., and T. Kasparis, "Fundamental of Color Image Processing", *DSP Applications*, Vol. 2, No. 10, pgs. 47-60, October, 1993.

Weeks, Arthur R., G. Eric Hague, and Harley R. Myler, "Histogram equalization of 24-bit color images in the color difference (C-Y) color space", *Journal of Electronic Imaging*, Vol. 4, No. 1, pgs. 15-22, January, 1995.

CHAPTER 7

Bracewell, Ronald N., *Two-Dimensional Imaging,* Prentice-Hall, Englewood Cliffs, NJ, 1995.

Burt, Peter J. and Edward H. Adelson, "The Laplacian Pyramid as a Compact Image Code", *IEEE Transactions on Comm.*, Vol. COM-31, No. 4, pgs. 532-540, April, 1983.

Dougherty, Edward R., *An Introduction to Morphological Image Processing*, SPIE Press, Bellingham, WA, 1992.

Dougherty, Edward R., and Phillip A. Laplante, *Introduction to Real-Time Imaging*, SPIE Press, Bellingham, WA, 1995.

Gonzalez, Rafael C., and Richard E. Woods, *Digital Image Processing*, Addison-Wesley, Reading, MA, 1992.

Hou, Hsieh S. and Harry C. Andrews, "Cubic Splines for Image Interpolation and Digital Filtering", *IEEE Transactions on ASSP*, Vol. ASSP-26, No. 6, pgs. 508-517, December, 1978.

Haralick, Robert M., and Linda G. Shapiro, *Computer and Robot Vision: Volume I*, Addison-Wesley, Reading, MA, 1992.

Lee, C. K. and C. H. Li, "Adaptive thresholding via Gaussian Pyramid", *IEEE International Conference on Circuits and Systems*, Shenzhen, China, pgs. 313-316, June, 1993.

Myler, Harley R., and Arthur R. Weeks, *Computer Imaging Recipes in C*, PTR Prentice Hall, Englewood Cliffs, NJ, 1993.

Myler, Harley R., and Arthur R. Weeks, *The Pocket Handbook of Image Processing Algorithms in C*, PTR Prentice Hall, Englewood Cliffs, NJ, 1993.

Pitas, I., and A. N. Venetsanopoulos, *Nonlinear Digital Filters: Principles and Applications*, Kluwer Academic Publishers, Boston, 1990.

Pitas, Ioannis, *Digital Image Processing Algorithms*, Prentice Hall, New York, 1993.

Russ, John C., *The Image Processing Handbook*, CRC Press, Boca Raton, FL, 1995.

Schalkoff, Robert J., *Digital Image Processing and Computer Vision: An Introduction to Theory and Implementations*, John Wiley & Sons, New York, 1989.

CHAPTER 8

Castleman, Kenneth R., *Digital Image Processing*, Prentice-Hall, Englewood Cliffs, NJ, 1979.

Dougherty, Edward R., and Phillip A. Laplante, *Introduction to Real-Time Imaging*, SPIE Press, Bellingham, WA, 1995.

Gonzalez, Rafael C., and Richard E. Woods, *Digital Image Processing*, Addison-Wesley, Reading, MA, 1992.

Haralick, Robert M., and Linda G. Shapiro, *Computer and Robot Vision: Volume I*, Addison-Wesley, Reading, MA, 1992.

Myler, Harley R., and Arthur R. Weeks, *Computer Imaging Recipes in C*, PTR Prentice Hall, Englewood Cliffs, NJ, 1993.

Myler, Harley R., and Arthur R. Weeks, *The Pocket Handbook of Image Processing Algorithms in C*, PTR Prentice Hall, Englewood Cliffs, NJ, 1993.

Pitas, Ioannis, *Digital Image Processing Algorithms*, Prentice Hall, New York, 1993.

Resnikoff, Howard L., *The Illusion of Reality*, Springer-Verlag, New York, 1989.

Russ, John C., *The Image Processing Handbook*, CRC Press, Boca Raton, FL, 1995.

Schalkoff, Robert J., *Digital Image Processing and Computer Vision: An Introduction to Theory and Implementations*, John Wiley & Sons, New York, 1989.

Tou, J. T., and R. C. Gonzalez, *Pattern Recognition Principles*, Addison-Wesley Publishing Company, Reading, 1974.

Weeks, A. R., H. R. Myler, and Michele D. Lewis, "Adaptive thresholding algorithm that maximizes edge features within an image", *Journal of Electronic Imaging*, Vol. 2, No. 4, pgs. 304-313, Nov. 1993.

Ziemer, R. E., and W. H. Tranter, *Principles of Communications: Systems, Modulation, and Noise*, Houghton Mifflin, Boston, 1976.

CHAPTER 9

Dougherty, Edward R., and Phillip A. Laplante, *Introduction to Real-Time Imaging*, SPIE Press, Bellingham, WA, 1995.

Gonzalez, Rafael C., and Richard E. Woods, *Digital Image Processing*, Addison-Wesley, Reading, MA, 1992.

Hall, Ernest L., *Computer Image Processing and Recognition*, Academic Press, New York, 1979.

Lim, Jae S., *Two-Dimensional Signal and Image Processing*, PTR Prentice-Hall, Englewood Cliffs, NJ, 1990.

Martin, A. V. J., *Technical Television*, Prentice-Hall, Englewood Cliffs, NJ, 1962.

Pitas, Ioannis, *Digital Image Processing Algorithms*, Prentice Hall, New York, 1993.

Rabbani, M., and P. W. Jones, *Digital Image Compression Techniques,* SPIE Press, Bellingham, WA, 1991.

Russ, John C., *The Image Processing Handbook*, CRC Press, Boca Raton, FL, 1995.

Schalkoff, Robert J., *Digital Image Processing and Computer Vision: An Introduction to Theory and Implementations*, John Wiley & Sons, New York, 1989.

Widrow, B., and S. D. Stearns, *Adaptive Signal Processing*, Prentice-Hall, Englewood Cliffs, NJ, 1985.

Index

1-bit binary delta quantizer, 524
4-chain code, 452, 453
4-connectivity, 35, 36
4-distance, 36, 37, 443
4-neighbors, 34, 295, 296
8-chain code, 452, 453, 505, 508
    example, 453
8-connectivity, 35, 36
8-distance, 36, 37, 443
8-neighbors, 34, 425, 430, 436, 442,
    443

24-bit color, 240, 257, 276, 285
24-bit color display system, 290, 292,
    293
24-bit color image, 30, 239
24-bit color video acquisition system,
    38, 39
35 mm film, 12
absolute temperature, 230
absorption, 25
adaptive filters, 208
    alpha-trimmed mean, 215, 216
    AWED, 220
    DW-MTM, 213, 214, 215
    MMSE, 208, 210, 211, 212, 213,
        215, 218
    SAM, 218
    two-component filter, 216
adaptive thresholding, 396, 403, 404,
    405, 407, 411, 414
    example, 405, 406, 407, 408, 411,
        412, 413
addition, 95, 96, 102
additive noise, 126, 208, 389, 391, 399
additive zero mean noise, 102
adjacent neighbors, 442
aerospace industry, 11
airplane silhouette, 344, 346
Airy disc, 63
Airy function, 53, 63
aliasing, 51
alpha-trimmed mean filter, 185, 188,
    215, 216, 217
    example, 186, 187, 188, 216, 217
AM, 6
Amstutz, 7

analog-to-digital converter, 39
AND, 99, 327, 328
angular information, 409
Apollo program, 9, 93
area, 359, 360, 365, 367, 387, 403,
    461, 463
area measurement parameter, 360
arithmetic code, 491, 496
    example, 491, 492, 493, 494, 495
arithmetic mean filter, 129
Armstrong, 6
aspect ratio, 463
associative property, 324, 376
Austrian church, 284
autocorrelation function, 163, 470, 475,
    477, 519
autocorrelation matrix, 518, 536
autoscaling, 97
average, 91, 104, 109, 151, 296, 467
average distance, 464
average filter, 129, 130, 131, 136
    example, 133, 134, 135
average number of bits, 474, 484, 487,
    491, 503
AWED filter, 220
B code, 490
background, 388, 389, 391, 394, 396,
    399, 400, 402
band-reject filter, 151, 152
bandpass filter, 151, 152
barrel distortion, 310
Bartlane system, 7, 8
Baudot 5-bit telegraph, 7
Bessel filter, 152
Bessel function, 53
between-class variance, 405, 406
bilinear equations, 311
bilinear interpolation, 314
binary code, 474, 475, 484, 486, 487,
    488, 496, 501, 504
    example, 497, 498, 499, 500
binary gray code, 496, 500, 501
    example, 497, 498, 499, 500
binary image, 30, 388, 391, 396, 407,
    448, 466, 503, 505
binary morphological filters, 294, 316,
    367

closing, 333, 334
dilation, 317
double sided contour edge
 detector, 342
edge detection, 333
end point detector, 357
erosion, 317, 320
geometrical decomposition, 363
hit-or-miss, 348
inside contour edge detector, 340
opening, 333
outside contour edge detector, 337
pattern spectrum, 361
pruning, 355
skeletonization, 333, 342
structuring element, 319
thickening, 354
thinning, 351
binary shift code, 488, 489
 example, 489
binary signal, 396
bit-plane code, 495, 496, 500
 example, 497, 498, 499
blackbody radiator, 230
Blackman window, 58, 540
blobness, 407
blue screening, 279
Blum, 342
Boolean operations
 AND, 99
 EX-OR, 99
 NOT, 99
 OR, 99
border pixels, 131
bowling ball, 414
breakdown value, 139, 177, 182, 188,
 191
brightness, 91, 92, 93, 96, 97, 102, 110,
 112, 123, 137, 198, 205, 221,
 223, 224, 234, 235, 264, 281,
 379, 394, 407, 478
brightness adaptation, 19
Butterworth filter, 152, 153, 158
 circularly symmetric, 154, 155
 highpass, 153, 154, 434
 lowpass, 153
C-Y color model, 265, 280, 293
C-Y color space, 251, 252, 266, 267,
 268, 270, 276, 282
C-Y to RGB transformation, 266
camera obscura, 2
Canaletto, 2

cardinal spline interpolation, 298
caricatures, 407, 414
Cartesian coordinates, 77
cartoon drawing, 414
cat's face, 414
cathode ray tube, 6
cavemen, 1
CCD camera, 5, 12, 147
CCITT, 501, 539
CD-ROM, 12
center of gravity, 466
centering property, 62
central limit theorem, 123
central moments, 466
central pixel, 416
central vision, 17
centralized moments, 467
centroid, 85, 301, 403, 404, 455, 462
chain codes, 387, 452, 454
character recognition, 407, 414
Charles, 2
Chebyshev filter, 152
chemical photography, 3
chroma keying, 278
chromatic filters, 280
chromaticity, 235, 236, 265
CIE, 233, 236, 237, 272
CIE chromaticity diagram, 236, 237,
 238, 243, 257, 272
CIE XYZ color space, 273
CIE XYZ coordinate system, 272
CIE-RGB, 237, 238
CIE: white point, 274, 275
circ function, 52, 53
circular object, 452
circular structuring element, 319, 330,
 333, 337, 359, 363, 367
circularity, 463
circularly symmetric filter, 150, 151,
 224, 434
circularly symmetric image, 61
circumference, 359
city block distance, 36
classification problem, 403
clear objects, 25
close-open filter, 336, 379
 example, 380, 383
closed contour, 436, 440, 455, 465,
 505, 506
closing, 333, 334, 335, 378, 379, 458
 example, 334, 335, 337, 339, 380,
 382

closing properties
    extensive, 336
    idempotent, 336, 378
    increasing, 336, 378
    translation invariant, 335, 378
clusters, 403
CMY color model, 272
CMY to CMYK transformation, 272
CMYK color model, 272
coding
    arithmetic, 491
    binary, 474
    binary gray, 496
    binary shift, 488
    bit-plane, 495
    Elias, 491
    lossless predictive, 508
    lossy predictive, 522
    m-bit, 475
    modified Huffman, 487
    n-bit B, 490
    run length, 478, 501
    transform, 522, 535
    variable length, 474
coding efficiency, 491
coding redundancy, 473, 474, 475, 478,
    482, 484, 486
color, 407
color cube, 238, 260, 263, 267
color edge detector, 284
    example, 285, 286, 287
color histogram equalization, 254, 281,
    282, 283
    example, 252, 253, 283
color image compression, 546
color image processing, 39
color lookup table, 239
color models, 237
    C-Y, 265
    CIE-RGB, 237
    CMY, 272
    CMYK, 272
    HLS, 271
    HSB, 271
    HSI, 237
    HSV, 271
    L*a*b*, 274
    L*u*v*, 274
    NTSC-RGB, 237
    RGB, 238
    U*V*W*, 274
    USC, 274
    UVW, 273
    YIQ, 270
color phosphors, 237
color photographic film, 232
color photography, 4, 280
color printer, 228, 232, 237
color scanner, 11, 228, 276
color spatial filtering
    example, 279
color television, 7, 266, 267, 276
color video, 11
color white balance, 228, 280
    example, 250, 281
colorizing, 288
comb function, 52, 54, 55, 60
Commission Internationale de
    L'Eclairage, 233
commutative property, 324, 374
complementary hue, 236
complementary image, 324
complementary-alpha-trimmed mean
    filter, 188, 189
complex conjugate property, 61
complex surface, 368, 374
compression ratio, 458, 472, 473, 475,
    477, 486, 504, 505, 511, 514,
    526, 542, 543, 544, 545
computer, 228, 233, 240, 407
computer graphics, 297
computer monitor, 232
conditional dilation, 357
    example, 357, 358, 359
cone receptors, 233
connected component labeling, 448
    example, 450, 451
connected components, 465
connectivity, 35, 387, 442, 443, 448,
    450, 453, 465
contour description, 387, 452
contour entropy, 410, 411, 414
contour fitting, 387
contour information, 410
contour smoothing, 319, 323, 333
contra-harmonic mean filter, 198, 199,
    205
    example, 201, 204
contrast, 91, 93, 94, 96, 97, 110, 112,
    221, 224, 253, 281
contrast stretching, 97
convexity, 407
convolution, 129, 137, 144, 160
convolution identity, 55, 57, 62

Cooley, 41, 71
correlation, 124
correlation matrix, 86
cosine function, 41
covariance matrix, 85, 89
CRT, 39, 232, 237, 290
cubic B-spline interpolation, 297, 298,
        303
    example, 304, 305
curvature, 387, 407, 464
daffodil flower, 279
data, 472, 473
data redundancy, 473
DC value, 42, 61, 63
DCT, 75, 522, 537, 538, 539, 541, 542
    kernel, 76
    mirror symmetry, 76
    mirror-symmetry property, 539,
        540
    periodicity, 540
De Forest, 6
decision function, 403
decoder, 482, 483, 508, 523, 535
defense industry, 11
degradation model, 159, 160
    homogeneous, 159
    linear, 159
    space invariant, 159
delta modulation, 524, 526
    example, 525, 526, 527, 528
DFT, 55, 58, 90, 144, 146, 154, 160,
        161, 162, 222, 223, 536, 538,
        540
    centering property, 62, 156
    complex conjugate property, 61
    constant image, 60
    convolution identity, 62, 63
    cosine function, 60
    Gaussian function, 60
    impulse function, 60
    linearity property, 59
    mean filter, 135
    periodicity, 540
    periodicity property, 60
    properties, 59
    rotational property, 61
    scaling property, 61
    separable transform, 70
    shifting property, 60, 135
    symmetric transform, 70
diagonal descriptors, 452
diagonal edge detection, 422, 423

diagonal line detection, 418, 419, 421
diagonal matrix, 86
diagonal-neighbors, 34
Dickson, 4
digital position sensor, 500
digital video camera, 411
digital-to-analog converter, 290
digitization, 29, 30, 31, 32, 33, 117
dilation, 180, 318, 319, 320, 322, 323,
        324, 325, 327, 330, 333, 335,
        337, 357, 359, 369, 372, 374,
        378, 379, 380
    example, 320, 330, 332, 373, 374,
        376
    implementation, 327
dilation properties
    associative, 324, 376
    commutative, 324, 374
    distributive, 325, 326, 377
    increasing, 325, 377
    scaling, 325
    translation invariant, 324, 376
dilation: implementation, 328, 329
Dirichlet problem, 40
discrete cosine transform, 75
disjointed images, 348
dispersion edge filter, 191
    example, 194, 195
distorted grid, 310, 311
distributive property, 325, 326, 377
division, 95, 96, 102
do not cares, 351, 352, 356
double sided contour edge detector,
        342
    example, 340, 341, 342
double windowed modified trimmed
    mean filter, 190
Dougherty, 386
DW-MTM filter, 213
    example, 214, 215
Eastman, 3
eccentricity, 359
edge degradation, 482
edge detection, 144, 191, 205, 284,
        285, 333, 342, 380, 387, 410,
        414, 415, 422, 434, 435, 440,
        452, 463, 505, 506
    example, 206, 285, 286, 287, 434
    Frei and Chen, 427
    frequency domain, 434
    gradient, 422
    Laplacian, 430

Prewitt, 426, 427
Robert's, 426
Sobel, 426
two pass, 435
edge merging, 459
example, 459, 460, 461, 462
edge smoothing, 454
edge splitting, 460
example, 460
edge tracing, 436, 452, 453, 461
example, 436, 437, 438, 439, 440
Edison, 4
Edison Museum, 114, 117, 285
Egyptian civilization, 2
eigenvalues, 84, 86, 89
eigenvector, 86, 536
electromagnetic interference, 122
electromagnetic spectrum, 229, 231,
243
electronic camera, 228, 231
electronic noise, 159
electronic photography, 228
element difference moment, 470
Elias code, 491
elliptical filter, 152
encoder, 482, 483
end pixels, 357
end point detector, 357
example, 357, 358
energy of curvature, 464
entropy, 410, 468, 469, 475, 484, 486,
502, 511, 522
erosion, 180, 320, 321, 322, 323, 324,
325, 326, 327, 333, 335, 337,
343, 347, 365, 367, 369, 371,
372, 374, 378, 379, 380
example, 323, 330, 331, 373, 374,
375
erosion properties
distributive, 326, 377
increasing, 325, 377
scaling, 325
translation invariant, 324, 325, 376
erosion: implementation, 327, 328
error-free compression, 471, 478, 483,
508, 522
Euclidean distance, 36
Euclidean length, 463
Euclidean space, 238
Euler's identity, 44
Euler number, 465
EX-OR, 99

exponential function, 101, 222, 223
extensive property, 336
eye, 13, 16, 17, 19, 20, 21, 233, 234,
264, 279, 471, 478
anterior chamber, 16
aqueous humor, 16
choroid, 16
ciliary muscles, 18
cones, 17, 19, 233
cornea, 13
field-of-view, 17
fovea, 17
lens, 16
muscles/fibers, 16
optic nerve, 16
retina, 16, 19
rods, 17, 19, 233
sclera, 13
spatial resolution, 20
vitreous humor, 16
facsimile machine, 501
false contours, 30, 479
falsecoloring, 288
fax machine, 501
FCC, 6
feature extraction, 347
FFT, 27, 41, 69, 71, 72, 75, 456, 538
fields, 314
filter mask, 130, 141, 143, 144, 174,
175, 176, 177, 182, 188, 189,
191, 197, 198, 201, 215, 307,
415, 416, 418, 422, 423, 425,
426, 431, 433
filter window, 218, 219, 220
fingerprints, 10, 11
first derivative, 295, 422, 424
first order predictor, 510, 524
first order RCL filter, 153
fixed length code, 487
FLIR, 231
fluorescent lighting, 280
FM, 6
Forbes, 4, 229, 234
Forth suspension bridge, 427
forward Fourier transform, 44
forward looking infrared, 231
Fourier, 40
Fourier descriptors, 454, 455, 456, 458,
505, 507, 508
example, 456, 457
Fourier frequency domain, 126, 434
Fourier magnitude components, 42

Fourier series, 41
Fourier transform, 44, 46, 47, 48, 49,
    51, 53, 55, 359, 363, 387, 434,
    454, 455, 456, 470, 477, 536,
    538, 540
fovea, 233
frame buffer, 39
Frei and Chen edge detection, 427
French curve, 297
frequency domain, 90, 121
fundamental frequency, 41
Gaussian filter, 136, 137, 431
    example, 138
Gaussian function, 60, 136
Gaussian noise, 104, 123, 125, 130,
    132, 166, 167, 170, 171, 172,
    182, 185, 186, 188, 189, 190,
    197, 198, 199, 205, 206, 211,
    213, 214, 215, 216, 217, 219,
    249, 279, 389, 391, 394, 399,
    401, 402
    example, 127, 128, 389
Gaussian pyramid, 306, 307
geometric mean filter, 129, 197, 199,
    201
geometrical decomposition, 363, 365,
    367
    example, 365, 366
GEOS, 9
Gibbs phenomenon, 42, 299, 540
Gonzalez, 403
gradient edge detection, 422, 425, 426,
    427, 436
gradient filter, 380, 407
    example, 384
granular noise, 525, 526
granulometric size density function,
    361
granulometric size distribution, 361
granulometric techniques, 359
graph theoretic methods, 440
graylevel definition, 29
graylevel mapping functions, 95
graylevel morphological filters, 294,
    316, 367
    close-open, 379
    closing, 378
    dilation, 372
    erosion, 371
    gradient, 380
    hybrid, 380
    open-close, 379

opening, 378
structuring element, 372
top surface, 368
tophat, 380
umbra, 367
graylevel occurrence matrix, 469, 470
graylevel scaling, 374
graylevel truncation, 374
grayscale definition, 29
grayscale display system, 290, 292
Hadamard kernel, 75, 76
Hadamard transform, 73, 74, 537, 538
Hadamard-Walsh transform, 73, 536,
    538
Hamming window, 58, 540
handwritten characters, 407, 414
harmonic frequencies, 41
harmonic mean filter, 129, 197, 199,
    201
height, 387, 463, 464
hexagonal grid, 26
hieroglyphics, 1
high speed photography, 8
highpass filter, 139, 151, 217, 218,
    219, 223, 224, 227, 305, 434
    example, 143, 145
histogram, 113, 122, 129, 220, 388,
    391, 392, 394, 395, 396, 397,
    399, 400, 402, 403, 404, 411,
    467, 468, 473, 484, 511, 514,
    529
    bimodal, 394, 396, 400, 401, 402
    computation, 109
    desired, 117, 120
    examples, 110, 111, 114
    Gaussian, 124
    negative exponential, 125
    pseudo code, 109
    salt-and-pepper, 125, 126
    uniform, 117, 122
histogram equalization, 13, 111, 112,
    113, 228, 281, 282, 283
    example, 114, 115, 116, 252, 253,
    254, 283
histogram specification, 117
    example, 117, 118, 119, 120
histogram techniques, 90, 109
hit, 347, 348
hit-or-miss, 348, 349, 351, 353, 354,
    357, 359
    example, 349, 350
HLS color space, 271

Holland, 276
homomorphic filter, 173, 221, 222,
    224, 394
    example, 225, 226, 227
horizontal descriptors, 452, 453
horizontal edge detection, 422, 423,
    427, 429, 430, 432
horizontal line detection, 418, 419, 420
horizontal neighbors, 435
hot spot, 174, 320
Hotelling transform, 84, 85, 86, 87, 88,
    89, 295, 463
Hough space, 78, 79, 83
Hough transform, 77, 79, 81, 82, 83
HSB color model, 271
HSI color model, 237, 244, 260, 261,
    263, 264, 265, 266, 267, 275,
    281, 285
HSI to RGB transformation, 262
HSI triangle, 260, 262, 265
HSL color model, 272
HSV color model, 271, 272
Hubble telescope, 10
hue, 235, 237, 257, 259, 261, 262, 263,
    264, 267, 268, 269, 270, 271,
    275, 278, 279, 284, 285, 287
    example, 247, 276, 277
hue noise, 279
Huffman code, 484, 487, 488, 489,
    508, 511, 535, 541, 542
    example, 484, 485, 486
Huffman coding, 475
human civilization, 1
human visual system, 13, 21, 90, 235,
    277, 478, 479, 481, 482, 546
hybrid filter, 380
ideal highpass filter, 152
ideal lowpass filter, 152, 153
idempotent property, 336, 378
illumination function, 221, 224, 227,
    394
image averaging, 104, 105
image compression, 367, 471, 472, 477
image copy, 299
image enhancement, 90, 120, 158
image geometry, 294, 299
    centroid, 301
    magnification, 303
    morphing, 314
    pyramid, 305
    reduction, 303
    rotation, 299, 300

scaling, 301
skewing, 309
translation, 299
truncation, 300
warping, 309
image magnification, 303
    example, 303, 304, 305
image model, 25, 26
image move, 299
image neighbors, 34
image perception, 407
image processing systems, 37
image recognition, 407, 452
image reduction, 303, 306
image representation, 387, 452
image restoration, 13, 158
    example, 159
image rotation, 84
image segmentation, 13, 30, 387, 452
image shading, 173, 221, 225, 226, 394
imaging applications, 10
impulse function, 52, 53, 54, 57, 60
impulse noise, 379
impulse response, 129
incandescent lighting, 280
increasing property, 325, 336, 377, 378
information, 472, 473, 475, 478, 483,
    522, 535, 541
information gain, 407, 409, 410
information theory, 475
infrared, 290
infrared camera, 230, 231
infrared spectrum, 229, 230
infrared wavelength, 231
inside contour edge detector, 340
    example, 340
intensity, 237, 257, 261, 262, 263, 264
internet, 240
intersection, 348, 351, 357, 371
intersection of objects, 316, 319, 322,
    325, 326, 369
intersection of regions, 441
intersection of umbras, 369, 370
inverse filtering, 121, 161, 162, 168,
    172
    example, 162, 163, 164, 165, 166,
    167
in vitro fertilization, 10
ISO, 539
jet fighter silhouette, 344, 346
joint histogram, 469
Jones, 547

JPEG, 75, 240, 539, 541, 542
　　example, 542, 543, 544, 545
JPL, 9
Jupiter, 9
Kaiser window, 58
Karhunen-Loève transform, 75, 84,
　　536, 538, 539, 546
Kennedy Space Center, 91, 141
Kerr effect, 5
kinescope, 4, 6
Kirsch edge detection, 422, 423
KLT, 75
Kodak camera, 3
Kodak Inc., 12, 37
KSC, 91
kurtosis, 468
L*a*b* color model, 274, 275
L*u*v* color model, 274, 275
LANDSAT, 9
Laplace transform, 44
Laplacian edge detection, 430, 431,
　　432
　　example, 432, 433
Laplacian PDF, 514, 529, 531, 532
laser speckle, 125
Leibnitz's rule, 531
lens aberrations, 159
lens equation, 18
lens: barrel distortion, 310
　　example, 312, 313, 314
Leonardo daVinci, 2
light energy, 25, 26
line detection, 414, 415, 418
　　example, 419, 420, 421
line ends, 407
linear filtering, 49, 124, 144
linear geometrical transformation, 312
linear interpolation, 295, 296
linear mapping functions
　　addition, 95
　　autoscaling, 97
　　division, 95
　　multiplication, 95
　　subtraction, 95
linear operations
　　addition, 102
　　division, 102
　　multiplication, 102
　　subtraction, 102
linear plane, 296
linear predictor, 514
linear spatial filtering, 279

linear spatial filters, 129
linear system theory, 62
link list data structure, 448
liquid crystal display, 5
local histogram, 220
local standard deviation, 213
local variance, 210, 211, 213, 215, 217
logarithmic function, 99, 100, 222,
　　223, 475
logical predicate, 441, 442, 443, 446,
　　448
lookup table, 39, 293
lossless predictive code, 508, 509, 510,
　　518
　　example, 510, 511, 512, 513
lossless predictor, 541
lossy compression, 471, 479, 483, 508,
　　522, 523
lossy predictive code, 522
lowpass filter, 45, 48, 130, 140, 151,
　　158, 173, 217, 218, 303, 305,
　　306, 402, 458
luminance, 234, 235, 253, 254, 257,
　　260, 264, 265, 267, 268, 269,
　　270, 275, 280, 281, 282, 283,
　　284, 285, 286
lunar module, 91, 141
m-bit binary code, 475
m-connectivity, 35, 36
Mach, 21
Mach band effect, 21, 22, 478
MAD estimator, 213
magnitude spectrum, 45, 46, 47, 49,
　　61, 63, 146, 148, 150, 151,
　　152, 154
　　Gaussian filter, 136
　　mean filter, 135
major axis, 466
map: United States, 443, 444
mapping function, 482, 483
Mariner spacecraft, 9
Mars, 9
MAT, 342
Matheron, 316, 359
matrix multiplication, 515
Max-Lloyd quantizer, 529, 530, 531,
　　534, 541
　　example, 532, 533, 535
maximal discs, 343
maximum, 174, 177, 180, 185, 188,
　　191, 215, 220, 365, 368, 369,
　　374, 391, 401, 405, 407, 410,

418, 442, 459, 461, 463, 468, 474, 481
maximum filter, 180, 195, 197, 199, 201, 327, 372, 380
    example, 184
maximum probability, 469
maximum structuring element, 365
Maxwell, 4, 229, 231, 234
Maxwell triangle, 260
mean, 122, 189, 467, 514
    Gaussian, 124
    uniform, 123
mean distance, 464
mean filter, 130, 131, 136, 140, 173, 177, 180, 182, 185, 190, 195, 197, 198, 199, 210, 211, 213, 215, 216, 220, 279, 303, 307, 402
    example, 132, 133, 134, 135, 139, 212, 249
medial axis transform, 342
median, 213
median filter, 122, 129, 132, 137, 138, 150, 173, 174, 175, 177, 185, 190, 213, 215, 216, 218, 219, 220, 379
    example, 138, 139, 140, 187, 188, 201, 203
median of the absolute deviations estimator, 213
medical imaging, 10
Mercury, 9
Michelangelo, 2
microsurgery, 10
midpoint filter, 177, 189, 380
    example, 180, 181
minimum, 174, 177, 180, 185, 188, 191, 220, 369, 371, 374, 394, 402, 416, 442, 464, 481
minimum diameter, 464
minimum filter, 180, 182, 195, 197, 199, 201, 327, 372, 380
    example, 183
minimum observable contrast, 20
minimum perceivable contrast, 20, 21
minimum probability of error classifier, 396
minimum radius, 464
minimum sized circle, 464
Minkowski set addition, 316, 317
Minkowski set subtraction, 320
miss, 347, 348

MMSE filter, 208, 210, 211, 213, 215, 218
    example, 211, 212
modem, 240
modified Huffman code, 487
modified trimmed mean filter, 189, 190, 195
    example, 190
moments, 466
morphing, 11, 314
    example, 314, 315
morphological filters, 30, 130, 294, 452
    conditional dilation, 357
morphological image processing, 180, 182
motion pictures, 4, 5, 11
MPE classifier, 396, 402
MSE, 209, 479, 514, 516, 530
multilevel thresholding, 391
    example, 393, 394
multiplication, 95, 96, 102, 106, 108
multiplicative noise, 126, 173, 221, 222
multispectral image processing, 231
n-bit B code, 490
    example, 490
n-bit binary code, 496
n-bit binary shift code, 488, 489
NASA, 8, 9, 10
nearest neighbor interpolation, 295, 296
negative exponential noise, 124, 125
    example, 127
negative operation, 98
negative outlier noise, 197
newsreel film, 5, 11, 472
Newton, 229
NIMBUS, 9
Nipkow, 6
NMSE, 481, 482
noise variance, 210, 211, 217
noisy electronic camera, 102
nonlinear edge detection, 440
nonlinear filters, 260, 279, 440, 481
    adaptive, 173, 208
    adaptive alpha-trimmed mean, 215
    adaptive AWED filter, 220
    adaptive DW-MTM, 213
    adaptive MMSE, 208
    adaptive SAM, 218
    adaptive two-component, 216

alpha-trimmed mean, 185
complementary-alpha-mean, 189
complementary-alpha-trimmed
    mean, 188
contra-harmonic mean, 198
dispersion, 191
double windowed modified
    trimmed mean, 190
geometric mean, 197
harmonic mean, 197
homomorphic, 221
maximum, 182
median, 173, 175, 177
midpoint, 177
minimum, 180, 182
modified trimmed mean, 189, 190
ordered statistic, 191
power mean, 198
quasi-range, 191
range, 191
spatial mean, 197
weighted median, 175, 176, 177
nonlinear geometrical transformation,
    312
nonlinear image processing, 173
nonlinear mapping functions
    exponential, 101
    logarithmic, 99
nonlinear operations, 93
nonlinear spatial filters, 129
    median, 137
nonuniform camera response, 106
nonuniform illumination, 394
nonuniform quantization, 478, 479
nonuniform quantizer, 529
nonuniform sampling, 478, 479
NOT, 99
nth order moments, 467
NTSC, 6, 12, 27, 37, 39, 237, 265, 267,
    276, 471, 509, 546
NTSC RGB color model, 273
NTSC video, 147
NTSC-RGB, 237, 238
NTSC:BW display system, 290, 292
null image, 334
null set, 323, 441
Nyquist sampling theorem, 50, 57, 298
object classes, 403, 404, 405
object features, 396
object identification, 387, 407, 452
object recognition, 367, 387, 414
    example, 415

object: circular, 391, 394
object: homogeneous, 388
object: rectangular, 391, 394
object: square, 389, 391
Oklahoma City, 443
one-dimensional Fourier transform, 44
opaque objects, 25
open-close filter, 336, 379
    example, 380, 383
opening, 333, 334, 355, 359, 360, 361,
    367, 378, 379, 458
    example, 333, 334, 337, 338, 380,
    382
opening properties
    extensive, 336
    idempotent, 336, 378
    increasing, 336, 378
    translation invariant, 335, 378
Opticks, 229
optimum predictor, 514, 515, 516, 517,
    518, 519, 532
    example, 519, 520, 521, 522, 535
optimum quantizer, 529
optimum threshold value, 401, 402,
    403, 405, 407, 410
optimum thresholding, 391, 396, 400,
    401, 403
OR, 99, 327, 328
order statistic filters, 138, 191, 195,
    440
    example, 192
    implementation, 196
order statistics, 173, 174, 180
orientation code, 452
original grid, 310, 311
Otsu, 403, 405
outlier noise, 126, 132, 137, 173, 177,
    185, 188, 191, 195, 202, 203,
    213, 216, 219
outlier pixels, 139, 182, 189
outside contour edge detector, 337
    example, 340
painter's palette, 239
palette color, 239
palette display system, 290, 292, 293
palette table, 239, 240
parametric Wiener Filter, 168
pastel colors, 235
pattern recognition, 403
pattern spectrum, 361, 363
    example, 361, 362, 363, 364
PDF, 112, 122, 297, 361

pel, 26
pepper noise, 180, 182, 197, 198, 201, 204, 336, 379, 380
perceived color, 234
perceptual psychologists, 407
perimeter, 387, 436, 463
periodic noise, 149, 150
periodic square wave, 42, 43
periodic texture, 470
perpendicular line, 459
Perry, 5
Persian Gulf War, 11
persistence of vision, 4
phase preserving filter, 45, 147, 151
phase spectrum, 45, 46, 146
PhotoCD, 12, 37
photoengraving techniques, 8
photography, 3
photopic vision, 17, 233
pinhole camera, 2
Pitas, 221, 333, 386, 440, 504
pixel, 26
pixel replication, 295
Planck, 230
point detection, 414, 415, 416, 417
    example, 416
pointwise operations, 90, 120
    algebraic, 90
    Boolean, 90, 99
    nonlinear, 90, 91
polar coordinates, 77
polygons, 297
polynomial descriptors, 458
poor illumination, 394
    example, 395
Porta, 2
power mean filter, 198, 199, 205
    example, 201, 203
power spectral density, 162
predictor, 508, 509, 514, 515, 517, 524, 525
Prewitt edge detection, 426, 427
primary colors, 238
primary colors of light, 231
primary colors of paint, 232
primary subtractive colors, 232
prism, 229
probability density function, 112
probability of occurrence, 124
pruning, 355, 356
    example, 357, 358, 359
PSD, 162, 168, 297, 470

pseudo-randomizing, 496
pseudocoloring, 39, 255, 256, 288, 289, 293
    example, 289, 290, 291
PSNR, 481
$PSNR_e$, 482
$PSNR_h$, 481
pyramid, 305, 306, 307
    example, 307, 308
quadtrees, 447
quantization, 30, 31, 32, 33
quantizer, 482, 483, 522, 523, 524, 525, 526, 529, 530, 531, 541
quasi-range filter, 191
    example, 194, 195
Rabbani, 547
rainbow, 229
range filter, 191, 380, 440
    example, 193, 205, 207
Ranger spacecraft, 8
RCA, 6
real-time image processing, 39, 90
real-time video, 11
recognition system, 13
rect function, 51, 52, 53, 58, 63, 136
rectangular filter mask, 130
rectangular grid, 26
rectangular object, 452
rectangularity, 463
recursive algorithm, 443, 448
redundant data, 471, 473
reflectance, 25, 221, 394
region description, 452, 461
region growing by pixel aggregation, 442, 444, 445, 448
region segmentation, 387, 440, 441, 452
    example, 443, 444, 445, 447, 449
    growing, 442
    split and merging, 446
Resnikoff, 407
retina, 233
RGB color model, 238, 239, 257, 259, 261, 262, 264, 275, 276, 280, 281, 284, 288, 289, 293, 546
RGB color space, 279
RGB to C-Y transformation, 265
RGB to CIE XYZ transformation, 273
RGB to CMY transformation, 272
RGB to HLS transformation, 271
RGB to HSI transformation, 260, 263
RGB to YIQ transformation, 270, 546

Robert's edge detection, 426
rocket engine, 93, 97
roll-camera, 3
rolling sphere concept, 372
rolling wheel concept, 318, 319, 322,
    324, 333, 337, 372
root image, 379
rotation, 299, 301, 303, 456
    example, 301, 302
rotation transformation, 86, 87
rotational property, 61
Röthenburg, Germany, 227
roundness, 359
RS170, 30, 37, 38, 39
RS170 video acquisition system, 38
run length code, 478, 496, 501, 502,
    503, 542
    example, 503, 504
saccadic motion, 18
salt noise, 182, 198, 201, 336, 379, 380
salt-and-pepper noise, 125, 126, 129,
    132, 139, 173, 174, 175, 177,
    185, 186, 189, 191, 192, 195,
    214, 215, 216, 336, 379, 380,
    381
    example, 127, 128
SAM filter, 218
sampled function, 55, 56, 57, 58
sampling interpolation, 298
Santa Claus, 87
satellite imagery, 314
satellite weather images, 314
saturation, 228, 235, 236, 237, 252,
    253, 257, 261, 262, 263, 264,
    265, 266, 268, 269, 270, 271,
    275, 276, 278, 281, 282, 283,
    284, 285, 286
    example, 246, 276, 277
saturation enhancement algorithm, 277
Saturn, 9
scaling, 301, 303, 455, 514, 519, 535
scaling property, 61, 325
scotopic vision, 17, 20, 233
SEASAT, 9
second derivative, 296, 297, 422, 424,
    430, 431, 432, 464
second graylevel moment, 470
second order interpolation, 297
second order predictor, 522
second order RCL filter, 153
second order statistics, 162
secondary color of light, 231

secondary colors, 238
secondary subtractive colors, 232
seed pixel, 442, 443, 450
segmentation, 387
selective color processing, 277
    example, 248
separable transform, 69
Serra, 316
set difference, 337, 343, 351, 365, 367
Shannon, 475
shape measurement, 359
shift flag, 488
sifting property, 57
signal dependent noise, 213
silver chloride, 2
silver nitrate, 3
simultaneous contrast, 22, 24, 478
sinc function, 52, 53
sinc-function interpolation, 298
sine function, 41
six-point interpolation, 297
    example, 301, 302
size, 467
size measurement, 359
skeletonization, 333, 342, 344, 351,
    355, 452
    example, 344, 345, 346
skewing, 309
    example, 309, 310
skewness, 468
slope overload, 526
smallest enclosing rectangle, 463
smart bombs, 11
smoothness descriptor, 468
Snow White, 11
SNR, 481
Sobel edge detection, 79, 80, 285, 307,
    426, 428, 429
    example, 308, 427, 430
Sombrero function, 63
spatial
    interpolation, 309
spatial convolution, 129, 141, 144, 160,
    307, 327
spatial correlation, 496
spatial decomposition, 441
spatial domain, 90, 121
spatial filtering, 35, 121, 129, 130, 141,
    174, 228, 415, 416
spatial frequency, 49, 50, 61, 130, 135,
    136
spatial frequency filtering, 144, 223

example, 148, 149, 157
spatial interpolation, 294, 295, 299, 300, 303
  cardinal spline, 298
  cubic B-spline, 297
  linear, 295
  nearest neighbor, 295
  pixel replication, 295
  sampling, 298
  second order, 297
  sinc-function, 298
  six-point, 297
  zero order, 295
spatial redundancy, 475, 477, 478, 482, 496, 500, 501, 504, 505, 508, 511, 522
spatial sampling, 26, 28, 29, 59
split and merging, 446, 448, 449
  example, 448
square filter mask, 130
square object, 448
square structuring element, 361, 367
standard deviation, 109, 110, 123, 189, 190, 213, 514, 522, 532
  uniform, 123
Stefan-Boltzmann law, 230
step overload, 525
stick figures, 342
strong edges, 435
structuring element, 319, 320, 322, 324, 325, 326, 327, 329, 333, 334, 344, 348, 349, 350, 354, 359, 363, 365, 372, 373, 374, 376, 380
  decomposition, 326
subtraction, 95, 96, 102, 104, 106, 107
subtractive colors, 232, 272
Sultry image, 411
sunspot activity, 122
surface texture, 387, 388, 452, 461, 467, 470
Surveyor spacecraft, 9
symmetric transform, 69
tail pixels, 355
Talbot, 3
television, 5, 6, 7, 93, 232, 266, 267, 276, 314, 546
template matching, 458
temporal filtering, 546
temporal redundancy, 509
thickening, 354
  example, 354

thinning, 351, 352, 354, 355
  example, 353, 354
third order predictor, 522
three primary colors, 229, 234, 235, 236, 257
three primary colors of light, 232
thresholding, 13, 77, 387, 388, 389, 394, 396, 401, 402, 403, 404, 416, 435, 436, 452, 541
  adaptive, 396
  cardioangiograms, 402
  contour entropy, 407, 414
  example, 389, 390, 392, 393, 396, 397, 398
  optimum, 396
  Otsu, 403, 404, 411, 414
tiepoints, 311, 312, 314
tilt, 407
tint, 228, 279
TIROS, 9
top surface, 369, 371
tophat filter, 380, 385
  example, 384, 385
Tou, 403
transform code, 522, 535, 536
transform kernel, 44, 69, 71, 72, 74
translation, 299, 300, 301, 317, 320, 322, 329, 348, 371, 455
  example, 301, 302
translation invariant property, 324, 325, 335, 378
translation of objects, 319
transmittance, 25
trapezoidal effect, 310
trichromatic coefficients, 235, 236
tristimulus values, 234, 235
true-color images, 11, 239
truncation, 300, 374, 541
  example, 301, 302, 310
Tukey, 41, 71
two pass edge detection, 435
two-component adaptive filter, 216, 217, 219
two-dimensional Fourier transform, 48
U*V*W* color model, 274
U.S. Civil War, 3
ultraviolet spectrum, 229
umbra, 367, 368, 369, 371
uncorrelated noise, 122
uncorrelated texture, 470
uniform illumination, 394

uniform noise, 122, 123, 177, 188, 189, 197, 217, 219
    example, 127
uniform PDF, 112, 468, 469, 489
uniformity, 470
union of maximal discs, 343
union of maximum structuring elements, 365
union of objects, 316, 318, 320, 325, 369
union of regions, 441
union of translations, 317, 320, 323, 330
union of umbras, 369, 371, 372
union of volumes, 368
University of Central Florida, 79
unrealizable colors, 275
unsharp filter, 139, 141
    example, 143, 145
USC model, 274
UVW color model, 273
UVW to XYZ transformation, 273
vacuum tube, 6
valley filter, 385
    example, 385
variable length code, 474, 475, 482, 484, 486, 487, 490, 491, 514, 535
variance, 122, 123, 201, 209, 215, 389, 399, 401, 404, 464, 467, 469, 519, 541
    Gaussian, 124
Venetsanopoulos, 221, 333, 386
Venus, 9
vertical descriptors, 452, 453
vertical edge detection, 422, 423, 428
vertical line detection, 418, 419, 420
video acquisition system, 12
video carrier, 6
video clips, 11, 12
video electronics, 240
Viking spacecraft, 9
visible spectrum, 229, 231, 233, 235, 290
visual information, 407
visual redundancy, 478, 479, 482, 522, 524
    example, 480
volume, 367, 368
Walsh kernel, 75, 76
Walsh transform, 72, 73, 537, 538
    separable, 73

symmetric, 73
Walt Disney Productions, 11
warping, 309, 310, 312, 314
    example, 312, 313
weak edges, 435
Weber ratio, 20
Weeks, 407
weighted median filter, 175, 176, 177, 195
    example, 176, 178, 179
Westinghouse Inc., 6
width, 387, 463, 464, 477
Wiener filter, 13, 121, 162, 168, 172
    example, 169, 170, 171
Wiener-Khintchine theorem, 162, 470
World War II, 8
X-rays, 10, 472
XYZ to L*a*b* transformation, 274
XYZ to U*V*W* transformation, 274
XYZ to UVW transformation, 273
YIQ color model, 245, 270, 280, 546, 547
YIQ to RGB transformation, 270, 546
Young, 229, 234
zero mean Gaussian noise, 104
zero mean uncorrelated Gaussian noise, 132, 172
zero mean uncorrelated noise, 104
zero order interpolation, 295
zero phase filter, 147
zig-zag scan, 504, 505, 511, 519, 541
zonal method, 541

Arthur R. Weeks, Jr. received his M.S.E and Ph.D. degrees in Electrical Engineering from the University of Central Florida in Orlando in 1983 and 1987, respectively. After completion of his Ph.D., he spent approximately one year at the Royal Signals and Radar Establishment in Malvern, England studying enhanced backscattering and laser beam propagation. He joined the Electrical and Computer Engineering Department at the University of Central Florida in 1989, where he is now an Associate professor. He has written numerous papers on image processing and laser beam propagation and has co-authored two additional texts in image processing, *Computer Imaging Recipes in C* and *The Pocket Handbook of Image Processing Algorithms in C*. He has taught numerous short courses for SPIE in image processing and is currently an associate editor of the SPIE/IS&T Journal of Electronic Imaging. His current research interests include color image processing, image enhancement using nonlinear filters, medical imaging, and optical image reconstruction. Dr. Weeks is also a member of the IEEE and SPIE.